Design of
Breakwaters

Design of Breakwaters

Proceedings of the conference *Breakwaters '88* organized by the Institution of Civil Engineers and held in Eastbourne on 4 – 6 May 1988

┓┏ Thomas Telford, London

Conference organized by the Institution of Civil Engineers

Organizing Committee: P. Lacey (Chairman), N. W. H. Allsop,
J. E. Clifford, J. D. Mettam, F. O'Hara, W. A. Price, I. W. Stickland

First published 1988

British Library Cataloguing in Publication Data
Breakwaters '88 *(Conference: Eastbourne, England)*
   Design of breakwaters.
   1. Breakwaters. Design & construction
   I. Title   II. Institution of Civil
   Engineers
   627'.24

ISBN 0 7277 1351 5

Published for the Institution of Civil Engineers by Thomas Telford Ltd, Thomas Telford House,
1 Heron Quay, London E14 9XF.

Printed in Great Britain by Billing and Sons Ltd, Worcester.

D
627 . 24
BRE

Contents

1. The use of physical and mathematical models in the optimisation of breakwater layout

Dr J. V. SMALLMAN and P. J. BERESFORD, Hydraulics Research, Wallingford

SYNOPSIS. The primary function of a harbour breakwater is to provide shelter for vessels at their moorings. Optimising the layout of the breakwaters to ameliorate the effects of severe wave action is one of the most important factors affecting harbour design. Physical and computational models of wave disturbance are commonly used to examine breakwater layout. This paper describes some of the modelling techniques which are presently in use, and gives examples of their application.

INTRODUCTION

1. The viability of any harbour or port installation, whether existing or proposed, depends heavily on attracting and maintaining a high usage rate of facilities. For commercial ports it is necessary to have efficient throughput of cargo and fast turnaround of vessels. One way of achieving this is by ensuring berth conditions are consistent with acceptably small levels of downtime. In harbours intended for use by smaller craft a relatively low level of wave activity needs to be ensured even in severe storms. This is because the vessels using such harbours can only tolerate small movements at their moorings, and in most cases (e.g. in a marina) it is not practicable to move boats to more sheltered positions in adverse weather.

2. Breakwaters are most often used to reduce the effects of severe wave action inside harbours. In designing a harbour the engineer will aim for a breakwater layout which offers good protection at moorings, and satisfies the economic constraints. There are a variety of physical and mathematical models which can be used to optimise breakwater layout in harbour design.

3. As a starting point a brief outline is given of the different types of model which can be used in harbour design studies. This includes a discussion of some of the recent advances in modelling techniques. We then examine the process of selection of the appropriate model(s) for a study at a particular site. To conclude examples are given of the practical application of physical and

mathematical models in the design of breakwater layout in ports and harbours.

MODELLING TECHNIQUES

4. Once the preliminary design for new harbour facilities has been made, and the wave conditions near its entrance have been derived, the next stage is to examine the layout in a model study. To achieve optimum hydraulic conditions at moorings in either a large commercial port, or a smaller harbour, some modifications will almost certainly be required to the length and alignment of the breakwaters. In making such modifications in a model there must be a high degree of certainty that similar effects will ensue when making the equivalent change in the real situation. That is, there must be confidence that the type of model which is being used represents well the important processes which occur in nature. With this in mind we outline the salient features of both physical and various computational models, and assess their value with reference to comparisons which have been made between model test results and field measurements.

Physical models

5. A random wave physical model of wave disturbance is the most comprehensive type of model available for optimising breakwater layout in the design of a port or harbour. A physical model of wave disturbance will typically represent an area of coastline either side of the harbour, the harbour itself and an offshore area extending to a water depth of about 20m to 30m. The scales are normally in the range 1:60 to 1:120 and the modelling is controlled by Froude scaling. Waves are generated using mobile wave machines which can be moved easily to allow a wide range of incident waves to be tested. Random waves have been used in physical models for over ten years, and have been found to give much more realistic results than the previously used regular waves.

6. An important feature of using random waves is that they allow wave grouping effects to be represented. In a random sea grouping of waves occurs as a natural consequence of the assumed Gaussian nature of the water surface elevation. These wave groups are associated with the energy, at long periods, which is known as set down (ref. 1). In a large commercial port set down may cause resonance in the harbour, or be responsible for the significant horizontal movement of vessels moored by conventional methods. Clearly, if set down is correctly represented in the harbour model the breakwater layout can be modified so as to minimise these effects.

7. Much research in recent years has been directed at the correct representation of set down in physical models of wave disturbance. The wave makers most commonly used in physical models at present generate long crested waves.

2

A device to compensate for set down in long crested seas has been developed and satisfactorily tested (ref. 2). This type of device is now routinely used in studies of harbours where set down makes a significant contribution to ship movement. The use of long crested waves, rather than the more naturally occurring short crested seas, is another aspect of physical modelling which is under active consideration. It is hoped that in the near future short crested wave makers, including set down compensation, will be in use in physical models. This will allow more realistic directional seas to be reproduced in harbour wave disturbance models.

8. The primary function of the breakwater is to offer protection to vessels at their moorings. In a physical model study a criterion will be required to judge if the wave conditions at a berth are acceptable. With a small harbour or marina it is often sufficient to establish a wave height which should only be exceeded at the berths during (say) a 1/50year return period event. For larger harbours a wave height criterion may not be adequate, and it will be recommended that a ship on its moorings should be modelled at the berth and its movements measured. This is because ship movement per unit wave height varies depending on the vessel's position in the harbour. Also where large ships respond at wave group periods the critical conditions may depend on the long wave height at the berth which is often not simply related to the storm waves at the berth. By measuring movements directly, accurate estimates of the time per year for which the berth is unusable can be made. It is therefore no longer necessary to rely on wave height measurements alone to judge the merits of a harbour development, where the breakwater may cost millions of pounds to construct.

9. In addition to representing ship movement, it is important that the correct reflection behaviour of the structures within the harbour is reproduced. Such structures may include those which are armoured with concrete units. Clearly in flume studies, where the stability and hydraulic performance of the armour units is being investigated accurate large scale model units will be required. However, in wave disturbance studies the cost of production of the large number of units which would be needed at relatively small scale generally prohibits their use. This can pose a problem since it is necessary in the model to reproduce reflection coefficients of the proposed breakwaters to ensure that wave activity in the lee of the breakwaters is correctly represented. To overcome this problem work is in progress to collate information on the reflection behaviour of various types of breakwater structure (ref. 3). This data will allow rubble breakwaters to be constructed in wave disturbance models which have reflection characteristics almost equivalent to breakwaters armoured with artificial

3

units. The performance of breakwaters constructed in this way, at scales of about 1:100, is adequate for the purpose of wave disturbance studies. It also allows greater flexibility in the number of alternative layouts which can be evaluated at a reasonable cost.

10. An important question which is frequently asked is how well do random wave physical models simulate the physical processes? For any investigation of harbour layout it is important that the modelling techniques which are used have been verified against field measurements. There are a number of published papers which describe the verification of physical models of wave disturbance (see for example ref. 4). For the physical modelling techniques used at Hydraulics Research such a verification was done using data collected at the Port of Acajutla, El Salvador (ref. 5). For this site a model study of proposed breakwater extension and reclamation was carried out in which moored vessels were represented. Measurements of wave height and ship movement were also carried out at the Port. Comparisons were made between wave spectra and ship motion spectra, the model and prototype values were found to be in good agreement. The results from this study indicated that random wave physical models do accurately represent the prototype, and do not suffer from any significant scaling problems.

Mathematical models

11. At present it should not be expected that a computational model of a port or harbour will provide as comprehensive a description of wave action as a physical model. The reason for this is that in a computational model any wave process has to be explicitly included and, in order for this to be possible, such a process must be both well understood and capable of being described mathematically. Non-linear effects such as wave breaking, flow separation and overtopping of breakwaters are mathematically complex and because of this are not, in general, included in computational models. However despite these shortcomings it is possible to model mathematically a number of the important mechanisms inside a harbour and, as a result, mathematical models have proved to be a useful tool for the engineer, particularly at the early stages in the design of a proposed harbour development.

12. There are a wide range of mathematical models presently available for use in harbour investigations. The various models are characterised by the governing equations which are solved, and the technique which is used to solve them. All models include certain fundamental processes, for example refraction and shoaling, and diffraction by breakwaters. Other effects such as non-linear wave interactions and partial reflections from structures may also be represented.

13. Much research in recent years has been directed towards modelling long period wind waves (15 to 30 seconds) and harbour resonance (wave periods of several minutes) using finite difference and finite element techniques (refs 6-10). These types of model are best suited to long period waves, but have certain drawbacks which become more evident for shorter wave periods. One of these is a technical difficulty in representing conditions at the boundaries of the model, particularly those which describe the radiated and reflected waves. The other main drawback of these methods is that they require a certain minimum number of grid points per wavelength to ensure accuracy of the numerical solution of the governing equations. These requirements can become excessive where the harbour area is large and/or the incident wavelengths are small.

14. In contrast to the finite difference and finite element models discussed in paragraph 13, ray techniques are particularly well suited to representing short period waves in harbours. The reasons for this are that they do not have the drawbacks of finite difference and finite element models as far as representation of the boundaries is concerned, and they are relatively inexpensive in terms of computing time and storage. This is because whilst the computing time in a ray model does increase for shorter periods waves, this increase is much less rapid than for finite difference or finite element models.

15. Recent research on mathematical models of harbour wave disturbance has been directed towards extending the range of physical processes which can be accurately represented. In particular, research is underway on including non-linear effects, such as set down beneath wave groups, in finite difference models. Preliminary results from this research (refs 12-13) appear to be promising, and will further extend the range of mathematical models which can be used with confidence in harbour wave disturbance studies.

16. As with physical models it is important when using a computational model in practical harbour design to be aware of how successfully the model represents a combination of certain physical processes. The verification of computational models is most frequently done by reference to physical model tests. Details of model studies carried out to verify various different mathematical harbour models may be found in, for example, refs 11, 14.

SELECTION OF APPROPRIATE MODELS

17. The range of applicability, and the limitations of the various models were discussed in the previous section.

We will now review how these factors affect the selection
of the type of model(s) to be used in a particular study.

18. In general, a mathematical model of harbour wave
disturbance can be used where a preliminary assessment of
wave conditions is required, or where a number of
different harbour layouts are to be compared. A
mathematical model will give information on wave activity
suitable for many engineering purposes quickly and at low
cost. The mathematical model approach will work
particularly well in cases where the bathymetry and
geometry of the harbour are fairly simple, and where the
wave processes occurring in the harbour are well
understood and represented in the mathematical model. For
example, a mathematical model can often be used in
feasibility studies, or to examine and compare a number of
breakwater layouts and their effect on wave disturbance.

19. For a more complex layout, where there are
processes which are not included in a mathematical model,
or where accurate quantitative information on wave
disturbance or ship movement is required, then a physical
model should be used. However, it will be appreciated
that a physical model investigation of a harbour will be a
more expensive undertaking than the corresponding
mathematical model study. It will also be understood that
as the complexity of a physical model increases it becomes
more time consuming to make layout changes so that all the
options can be tested.

20. The foregoing suggests that a joint approach, using
both physical and mathematical models, could expedite a
programme of wave disturbance tests. To do this a
mathematical model will be used to examine quickly and
relatively inexpensively various design options, with a
physical model being used to test a final design
extensively, perhaps with moored ships being included in
that model.   Another advantage of this approach is that
the mathematical model can be calibrated against the
physical model, and retained to test any future proposed
developments in the harbour, after the physical model has
been dismantled.

EXAMPLES OF APPLICATION
Physical model

21. An example of a site specific study where a
physical model was used to investigate breakwater layout
was that undertaken for Pittenweem Harbour, Scotland for
Fife Regional Council. The proposals for the harbour
involved enlarging the mooring area available for fishing
vessels with an entrance protected from wave action by a
new breakwater. Two breakwater designs were considered,
a vertically faced structure which would extend the
existing breakwater in a similar form, and a rubble mound
extension (see Fig 1). In addition to the proposed

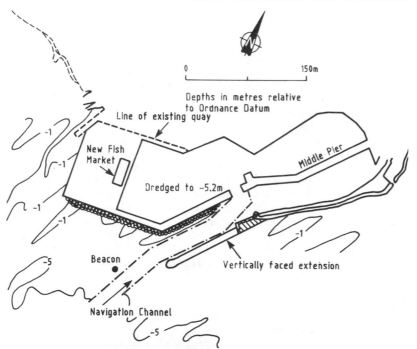

Fig. 1(a). Pittenweem – proposed vertically faced extension

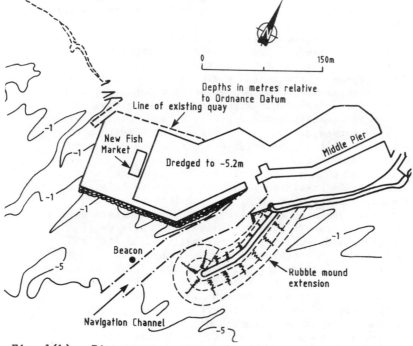

Fig. 1(b). Pittenween – proposed rubble mound extension

extension the existing scheme was also tested to obtain wave heights related to mooring problems, which had been reported from time to time.

22. For this study a mathematical model was not used because it was not considered to give an adequate representation of the wave breaking effects which were likely to occur in the shallow water at the harbour entrance. In addition, there were only two schemes to test and absolute values of wave heights were required. Therefore, a physical model study was considered to be appropriate.

23. The results from the study indicated that the rubble mound structure was the more effective of the two However, the vertically faced breakwater was considered to give conditions in the harbour which were acceptable, based on present experience, and at lower cost, since suitable rock is not readily available in this region. At the time of writing Fife Regional Council are proceeding with the design calculations for the vertically sided extension.

Mathematical model

24. A mathematical model study of wave disturbance was carried out for a proposed marina at the Town Quay, Southampton (ref. 11). The purpose of the study was to provide initial estimates of wave height inside the marina, which would be used to judge its suitability for the mooring of small craft. Six different designs for the marina layout were examined and compared for three incident wave directions. The marina was bounded by a timber piled quay to the west and a vertically faced quay to the east. It was intended that the breakwaters forming the entrance to the marina would be constructed from a piled A frame with a suspended skirt which would allow tidal flow underneath. Prior to commencing the mathematical model study a series of flume tests were carried out to determine the reflection behaviour of the skirt breakwaters, as this information would be required by the mathematical model.

25. For this study a mathematical model was used because only preliminary estimates of wave heights for the various schemes were required, and both time and cost were limited. The mathematical harbour ray model was used for this study (ref. 11) which includes the effects of diffraction by breakwaters, refraction, shoaling and partial or total reflections from harbour boundaries.

26. At all stages of this study the results from the mathematical model were used by the consulting engineer to refine the design of the marina layout. The best of those tested combined some of the better features of the earlier alternatives. It was recommended that if the marina project were to proceed further additional studies should

be carried out to examine the performance of final layout in more detail.

Joint physical and mathematical model study

27. A joint physical and mathematical model study of wave disturbance was carried out for a proposed marina in Torquay outer harbour (see Fig 2). The marina has subsequently been constructed. The purpose of the study was to establish the most cost effective means of providing sufficient shelter for boats moored to floating pontoons. Several different lengths and alignments of an extension to Haldon Pier were proposed as a means of reducing wave activity in the original harbour.

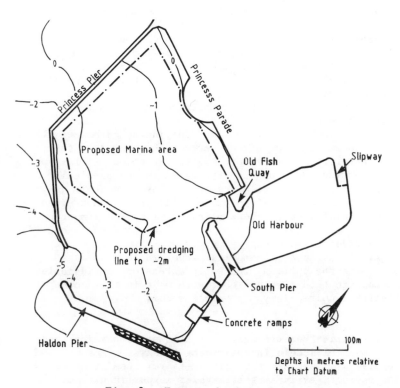

Fig. 2. Torquay harbour layout

28. The mathematical harbour ray model (ref. 11) was set up to represent the existing layout at Torquay. The model was then calibrated against the physical model results for the same layout for waves from the SE. The SE wave direction leads to the most severe wave conditions in the harbour. The agreement between the mathematical and physical model results was found to be good, particularly for positions in the marina area. The mathematical model

was then used to examine the effects of dredging the marina area, as shown on Fig 2. It was found that dredging the marina area would have no significant effect on wave disturbance inside the harbour. This result allowed the moulding of existing contours in the physical model to be removed at an early stage. All subsequent tests in both models were run with the dredged bathymetry.

29. The next stage was to examine methods for reducing wave activity in the harbour by extending Haldon Pier. The mathematical model was run for nine different extensions to Haldon Pier, all assumed to be vertically faced, for each of three sets of incident conditions. (1/50yr waves from SE, S, SSW). The conclusions which were drawn from these tests were that the optimum extension length to Haldon Pier was 40m with an orientation of between 250°N and 282°N. On the basis of the results from the mathematical model these two orientations of the extension were tested in the physical model for the SE and SSW directions. The results of the mathematical model also demonstrated that there was no need to test southerly waves in the physical model, which again reduced the number of tests conducted in the physical model.

30. Several tests were also done in the mathematical model to examine the effects of a rubble mound, rather than a vertically faced breakwater extension. This type of extension was subsequently tested in the physical model. The design which was eventually used at Torquay consisted of a 40m extension to Haldon Pier constructed from a reinforced concrete caisson, with rock armouring to protect the foundations from scour, and to absorb some of the incident wave energy.

31. By using a mathematical model in this study to determine the optimum breakwater extension considerable savings in both time and cost were made in the amount of testing required in the physical model.

CONCLUSIONS
A description has been given of the use of physical and mathematical models in optimising breakwater layout. Such models have been shown to provide a very useful means for the engineer evaluating breakwater layout in harbour design.

ACKNOWLEDGEMENTS
This paper describes work carried out in the Maritime Engineering Department of Hydraulics Research, Wallingford. The authors would like to thank collegues in the Department for their helpful discussions.

REFERENCES
1. BOWERS E.C. Long period disturbances due to wave groups. Proc. 17th Int. Conf. on Coastal Engineering, Sydney. ASCE, 1980, Vol 1, p610.
2. BOWERS E.C. and BERESFORD P.J. Set down beneath wave groups. Hydraulics Research, Wallingford, 1983, Report IT 243.
3. ALLSOP N.W.H. and HETTIARRACHCHI S.S.L. Design, construction and performance of wave absorbing structures: a review of literature and practice. Hydraulics Research, Wallingford, 1987, Report OD 89.
4. JENSEN O.J. and KIRKEGAARD J. Comparison of hydraulic models of port and marine structures with field measurements. Proc. Int. Conf. on Numerical and Hydraulic Modelling of Ports and Harbours, Birmingham. BHRA, 1985, Paper H3, p239-247.
5. SMALLMAN J.V. The use of physical and computational models in the hydraulic design of coastal harbours. PIANC Bulletin, 1986, No 53, p16-36.
6. ABBOTT M.B., PETERSEN H.M. and SKOVGAARD O. On the numerical modelling of short waves in shallow water. Journal of Hydraulics Research, 1978, Vol 16, No 3.
7. BETTESS P. and ZIENKIEWICZ O.C. Diffraction and refraction of surface waves using finite and infinite elements. Int. Journal for Numerical Methods in Engineering. 1977, Vol II.
8. ROTTMANN-SÖDE W., SCHAPER H. and ZIELKE W. Two numerical wave models for harbours. Proc. Int. Conf. on Numerical and Hydraulic Modelling of Ports and Harbours, Birmingham. BHRA, 1985, Paper K3, p285-293.
9. BERKOFF J.C.W. Mathematical models for simple harmonic linear water waves. Delft Hydraulics Laboratory, 1986, Report W154-IV.
10. COPELAND G.J.M. A practical alternative to the mild slope equation. Coastal Engineering, 1985, Vol 9, p125-149.
11. SMALLMAN J.V. The application of a computational model for the optimisation of harbour layout. Proc. 2nd Int. Conf. on Coastal and Port Eingeering in Developing Countries (COPEDEC). Beijing, China, 1987.
12. PRÜSER H-H., SCHAPER H. and ZIELKE W. Irregular wave transformation in a Boussinesq wave model. Proc. 20th Int. Conf. on Coastal Engineering, Taipei. ASCE, 1986.
13. SMALLMAN J.V., MATSOUKIS P.F., COOPER A.J. and BOWERS E.C. The development of a numerical model for the solution of the Boussinesq equations for shallow water waves. Hydraulics Research, Wallingford, 1987, Report SR 137.
14. JENSEN O.J. and WARREN I.R. Modelling of waves in harbours - review of physical and numerical methods. Dock and Harbour Authority, October 1986.

2. Results of model tests on 2-D breakwater structure

O. J. JENSEN and J. JUHL, Danish Hydraulic Institute, Hørsholm

SYNOPSIS. The authors have carried out a research project for investigation of the wave forces acting on breakwater armour units. The project included a literature study, hydraulic model testing and analysis of the test results. The hydraulic model testing was made in a wave flume on a 2-dimensional breakwater structure with an armour layer consisting of two rows of horizontal pipes.

INTRODUCTION

1. Many researchers have for many years looked into the question of wave forces on breakwater armour units. Most of the research has concentrated on studying the stability of breakwater slopes armoured with various types of armour units.

2. Only a few researchers have directly studied the wave forces on armour units by making measurements on idealized armour units in a hydraulic model (Sandström 1974, Ref. 2). The present research work is of the same nature.

THE MODEL
Wave Flume and Breakwater Model

3. The model tests were conducted in a 23 m long and 0.6 m wide wave flume. The flume set-up is shown in Fig. 1.

4. The model tests were conducted on a 2-D breakwater model with a slope of 1:2 of the armour layer. The crest height was chosen not to allow for wave overtopping. The armour layer consisted of two layers of horizontal pipes with diameter 50 mm to form an idealized and purely 2-dimensional representation of a breakwater armour layer. The porosity of the armour layer was selected to p = 0.40. Details of the model are shown in Fig. 2. A photo of the model exposed to waves is shown in Fig. 3. The wave forces on three of the pipes in the upper layer were measured simultaneously by use of strain gauge transducers giving two force components, i.e. vertically and horizontally. Nine of the pipes in the upper layer were prepared for force measurements.

5. The forces were measured on the 0.2 m wide middle sections of the horizontal pipes.

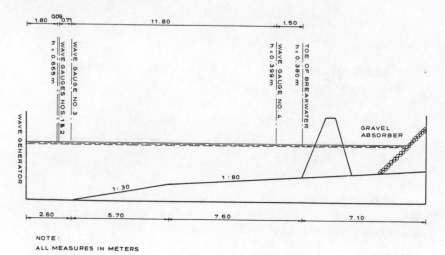

Fig. 1.   Set-up of wave flume.

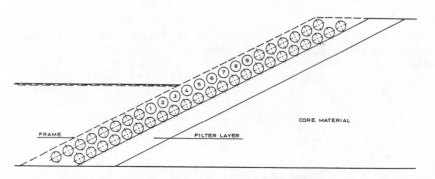

Fig. 2.   Details of test set-up.

Waves

6.   Sinusoidal, bichromatic and irregular waves were used for model testing. The wave combinations covered both non-breaking and breaking waves.

7.   <u>Sinusoidal waves</u>. Tests with sinusoidal waves were carried out for a large number of wave conditions (H, T).

8.   <u>Bichromatic waves</u>. Bichromatic waves are defined as a superposition of two sinusoidal waves with different frequencies (wave periods).

$$\eta_n = a_n \cdot \cos(\omega_n \cdot t - k_n \cdot x) + b_n \cdot \sin(\omega_n \cdot t - k_n \cdot x) \tag{1}$$

$$\eta_m = a_m \cdot \cos(\omega_m \cdot t - k_m \cdot x) + b_m \cdot \sin(\omega_m \cdot t - k_m \cdot x) \tag{2}$$

9.   The total surface elevation is then given by Eq. (3).

$$\eta = \eta_n + \eta_m = A_n \cdot \cos(\omega_n(t-t_o) - k_n \cdot x) + A_m \cdot \cos(\omega_m(t-t_1) - k_m \cdot x) \tag{3}$$

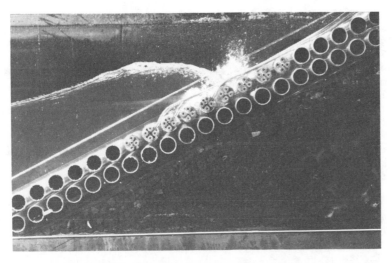

Fig. 3. Photo of the model exposed to sinusoidal waves
with a wave period of T = 2.0 s and a wave height of H =
0.15 m.

10. All the tests had $A_n = A_m$.
11. <u>Irregular waves</u>. Irregular waves with three different
energy spectra have been used as seen in Fig. 4.

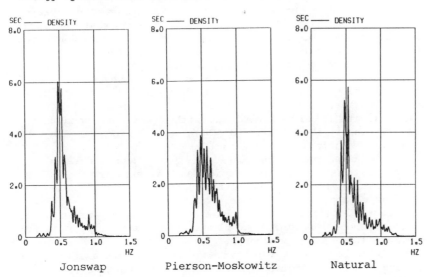

Fig. 4. Examples of measured spectra of the irregular
waves used for model testing (normalized spectra).

12. Based on the Pierson-Moskowitz and Jonswap spectra, artificial wave trains with the prescribed spectrum, but random
phases have been generated using a computer program.

15

13. The tests with irregular waves covered a range of wave combinations: Peak periods, T_p, from approx. 1.0 to 3.2 s and significant wave heights, H_s, from approx. 0.05 m to 0.20 m.

14. Examples of recorded time series of waves in front of the wave generator have shown that the Jonswap "artificial" computer generated waves give a significantly more smooth signal than the natural waves.

15. The waves in the flume were measured in 4 points by resistance type wave gauges.

16. The wave heights for the bichromatic and irregular waves were determined as $H_{rms} = 4 \times h_{rms}$. The wave height was measured 1.5 m in front of the breakwater, and has been used as reference wave height throughout the report.

17. Measurements of the run-up/run-down was carried out with a wave gauge placed parallel to the breakwater slope and in a distance of 50 mm (one pipe diameter).

18. <u>Force measurements</u>. Strain gauge transducers were used for measurements of two force components, i.e. vertically and horizontally.

19. The natural periods for the 0.2 m test section of the pipes fixed to the strain gauge transducers have been determined:

   (a) In air   100 Hz
   (b) In water   55 Hz
   (c) In half air/half water (pipe no. 5)   60 Hz

20. It should be noted that the transducers were calibrated to zero-force for still water level. This means that the buoyancy acting vertically upwards has been subtracted for pipes nos. 1-5 being either totally or partly submerged.

21. The buoyancy for a totally submerged pipe was 19.30 N/m and for the partly submerged pipe no. 5 the buoyancy was 9.65 N/m. This fact is important in the interpretation and comparison of the test results for the different pipes.

22. All tests were carried out with fixed wave conditions, i.e. stationary wave height (H, H_s), wave period (T, T_p), and water level during each test run.

23. The test runs with sinusoidal waves had a duration of 300 s, the test with bichromatic waves a duration of 600 s, while the tests with irregular waves had a duration corresponding to approx. 500 zero-crossing waves.

24. The signals from the wave gauges and the strain gauge transducers were recorded (with a logging frequency of 40 Hz) and stored by a micro computer.

25. Video recordings were made of all the tests allowing for subsequent identification of the form and characteristics of the water surface during run-up and run-down.

FORCES ON AN ARMOUR UNIT

26. Most stability formulae predict the required weight of the individual armour units to be proportional to the wave height in third power. Other stability formulae include a

linear dependency of the wave period (or parameters derived from this: wave length or wave steepness) and the square of the wave height, (ref. 2).

Wave forces on a two-dimensional breakwater

27. The determination of the forces acting on the idealized armour units can be compared with forces acting on a pipeline located on or close to the sea bed. For the present model set-up, the following should be considered:

(a) The flow pattern around the pipes is very complex, each individual pipe is influenced by the presence of neigh-bouring pipes

(b) Air entrainment occurs which makes the velocity field uncertain and decreases the density of the fluid

(c) The buoyancy varies with time as a result of run-up/run-down

(d) Wave breaking results in "Shock forces" or Impulse for-ces due to "Slamming".

28. By computer calculations the measured vertical and hori-zontal forces have been transformed to forces perpendicular and parallel to the slope.

29. Stability calculations including both force components have been carried out with the aim of making a simplified re-presentation of the acting forces.

30. It is assumed that the armour unit is supported in two contact points as shown in Fig. 5. Only symmetric contact points have been used for the calculations, i.e. $\theta = \theta_A = \theta_B$.

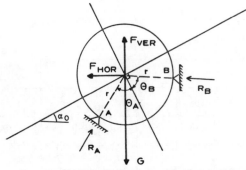

Fig. 5. Definition sketch of the used coordinate sys-tem, and simplified stability calculations.

31. An example of the measured run-up/run-down and measured horizontal and vertical forces is shown in Fig. 6 together with the calculated required weight for no movements. The re-quired effective weight to withstand roll-down, W'_d, is posi-tive whereas the required effective weight to withstand roll-up, W'_u, is plotted as negative values. The calculated re-quired weights presented throughout the paper have all been carried out with $\theta = 60$ deg.

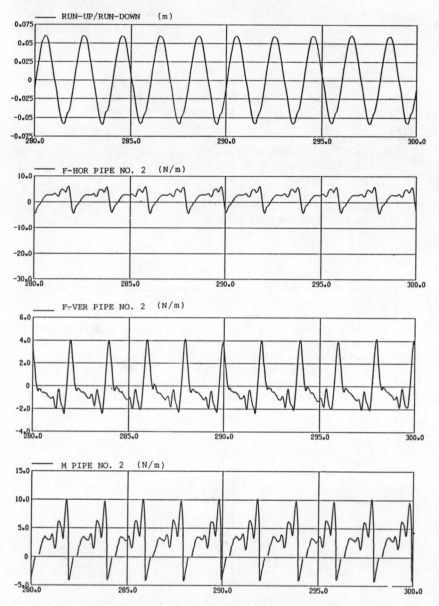

Fig. 6. Recorded and calculated time series for a sinusoidal wave with H = 0.07 m and T = 2.0 s.

32. Stability calculations considering rotation as determinant. The study has shown in agreement with Ref. 2 that the
necessary weights of armour units are expected to be proportional to the wave height in third power.

TEST RESULTS

33. <u>Sinusoidal waves</u>. A presentation of the force vector during one wave period starting at the moment when the run-up is zero (i.e. when passing the still water level) is shown in Fig. 7.

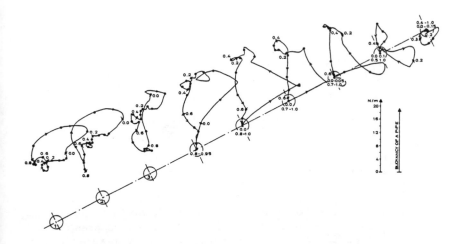

Fig. 7. Force vector during one wave period for wave conditions H = 0.11 m and T = 2.0 s.

34. Fig. 8 shows for sinusoidal waves the calculated required weights to withstand roll-up and roll-down respectively. W'_u and W'_d are presented for the nine pipes on which wave forces have been measured.

35. The calculated required weight to withstand roll-down shows that a maximum of W'_d is reached at pipe no. 2. Another interesting point is that a local maximum occurs at pipe no. 5 (located at still water level). Further, it should be mentioned that W'_d for pipe no. 2 is about twice W'_d for pipe no. 5.

36. For roll-up no maximum of W'_u is found at pipe no. 2, but still a local maximum at pipe no. 5. The test results show that the required calculated weights to withstand roll-up are generally less than the required weights to withstand roll-down. This is in good agreement with experience from physical model tests of rubble mound breakwaters, i.e. the major part of damage occurs during run-down for a slope of 1:2.0.

37. The tests have shown that the required weight, W'_d and W'_u is proportional to H in a power varying approximately from 1 to 2, which is less than predicted by the traditional stability formulae.

19

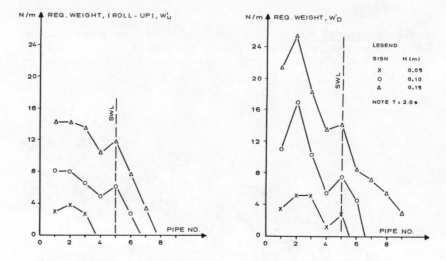

Fig. 8. Required weight to withstand roll-up/roll-down for the nine pipes. Wave period T = 2.0 s.

38. The results for pipe no. 2 and 5 have in Fig. 9 been plotted as a traditional stability number expressed by $H/(W'_d)^{1/3}$ as function of the surf similarity parameter $\xi = \tan\alpha/\sqrt{H/L_o}$.

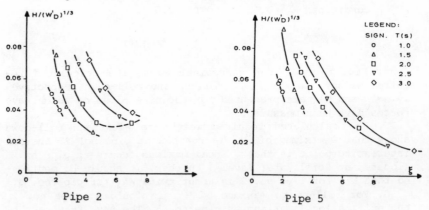

Pipe 2   Pipe 5

Fig. 9. Stability number $H/(W'_d)^{1/3}$ as function of surf similarity parameter, ξ.

39. It is seen that the stability number is decreasing with the surf similarity parameter but it is also seen that a curve was found for each wave period. However, it should be remembered that the duration of force impact has not been considered in the stability calculations.

40. Influence of wave period. The required weight to withstand roll-down and roll-up for pipes nos. 2, 5 and 8 are
shown in Fig. 10 as function of the wave period. It is observed that for pipe no. 2 the tests show a maximum in the required weight for T ∿ 1.5 s.

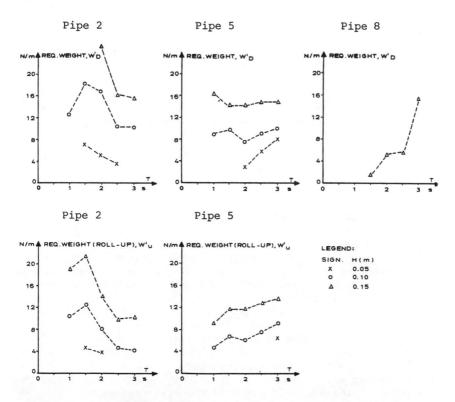

Fig. 10. Required weight as function of the wave period.
Sinusoidal waves.

41. This observation for the armour layer below SWL is in
agreement with Refs. 3, 4, 5 and 6 from observations of the
stability of rubble mound breakwaters. Further, it is found
that the wave period is determinant for which pipe is most
exposed. It is important to notice that at SWL, W'_d and W'_u
seems to be almost independent of the wave period. At pipe no.
8 above SWL, however, W'_d and W'_u increase with the wave period, which is due to increasing wave run-up by increasing
wave period.

42. These measurements therefore show that it is a rough
simplification to aim for one stability formula for description of the stability of the entire seaward armour layer of a
rubble mound breakwater.

43. Test results for bichromatic waves. The calculated required weights to withstand roll-down are shown in Fig. 11 as

21

function of the wave period. As for the regular waves, a maximum occurs for a wave period of approximately 1.5 s.

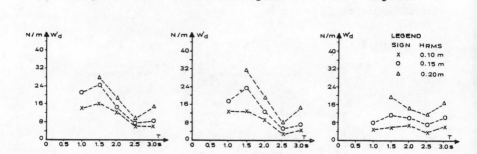

Fig. 11. Required weight, W'_d, to withstand roll-down for bichromatic waves as function of the wave period.

TESTS RESULTS FOR IRREGULAR WAVES

Jonswap spectrum

44. Influence of wave height. Generally, the test results show that the proportionality between the required weight and the wave height varies between the wave height and the square of the wave height.

45. Influence of wave period. In Fig. 12 the required weight necessary for stability is shown as function of the peak period. For pipe no. 2 it is seen that a maximum of the required weight occurs for a peak period of approximately T_p = 1.5 s and a minimum for approximately T_p = 2.5 s. For pipe no. 3 a maximum occurs for approximately T_p = 1.5 s while the results for pipe no. 5 do not show the same tendency.

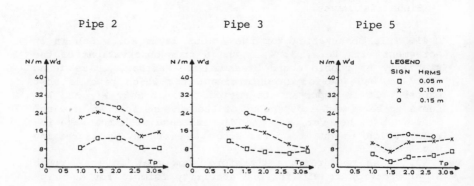

Fig. 12. Required weight, W'_d, as function of the peak period. Waves based on a Jonswap spectrum.

46. In comparison to the results for regular and bichromatic waves, it is seen that for irregular waves the influence of the wave period is of similar nature but less pronounced.

47. The test runs with a Pierson-Moskowitz spectrum and the natural wave train showed that the proportionality between the required weight and the wave train varied between the wave height and the square of the wave height.

48. <u>Comparison of results with irregular waves</u>. A comparison of the results obtained for irregular waves with the three different wave spectra is shown in Fig. 13.

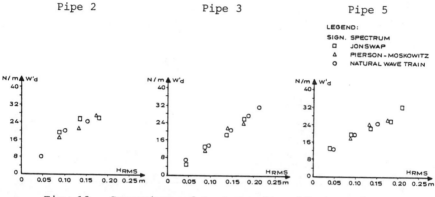

Fig. 13.   Comparison of test results with irregular waves.

49. The calculated required weight to withstand roll-up and roll-down are in Fig. 13 seen to be almost independent of the wave spectrum.

50. Further, from the test results it can be concluded that the shoaling characteristics (correlation between the wave height in front of the wave generator at water depth 0.665 m and the wave height in front of the breakwater at water depth 0.399 m) are similar for the three wave spectra.

51. Based on the test results, it seems reasonable to conclude that for spectra covering the same band of frequencies the form of the spectra is of minor importance for the stability of a breakwater.

52. <u>Wave slamming on armour units</u>. The analysis of the required weight to withstand roll-down and roll-up is only considering forces in the direction out of the armour layer, i.e. having a component in the direction perpendicular to the slope.

53. During run-up, however, large forces in the direction into the breakwater have been measured. Such forces were measured both for irregular, bichromatic and regular waves, but are generally more pronounced for irregular waves. These forces are characterized by a rapid growth and a short duration and occur when the water hits onto the units. Fig. 14 shows an example of the horizontal force on pipe no. 3 from test with

natural wave train (H_s = 0.19 m, T_p = 1.9 s). The time series are the wave gauge along the front, i.e. run-up/run-down (vertical). The run-up velocity of the wave causing the largest peak force is in the order of 1.12 m/s and the force corresponds to a C_s = 1.0 in the formula $F = \frac{1}{2} \rho \, C_s \cdot A \cdot u^2$. In this case, the force is seen to be about 1.6 times the buoyancy of the pipe. This type of force is not dangerous for the stability of the armour units but may instead be dangerous for the breakage of slender and fragile concrete armour units. For such units, it is possible that larger forces occur due to slamming than for the present breakwater consisting of circular pipes. The results of the present study will be further analysed in another paper to identify the relationship between the forces and the wave parameters.

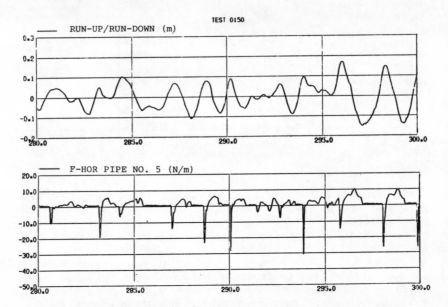

Fig. 14. Time series of run-up/run-down and horizontal force measured during testing with natural irregular waves (H_s = 0.19 m and T_p = 1.9 s).

54. The present paper presents a summary of the first analyses of the test results. In other papers the results of further analyses will be presented also including the important aspect of which individual waves or sequences of waves that cause the largest forces on armour units.

55. <u>Acknowledgement</u>. The research study of wave forces on armour units was partly financed by Statens Teknisk Videnskabelige Forskning (Danish Technical Research Council). The support is graceously acknowledged.

REFERENCES

1. Sandström, Aake, 1974. Vågkrafter på block i vågbrytarslän-ter (Wave Forces on Blocks of Rubble Mound Breakwaters), in Swedish with summary in English. Hydraulic Laboratory, Royal Ins. of Tech., Stockholm, Bulletin No. 83.

2. Jensen, O. Juul. A monograph on Rubble Mound Breakwater. Book published by Danish Hydraulic Institute, Dec., 1984.

3. Burcharth, H.F. Effect of waves on on-shore structures. Hydraulic Research Station Wallingford. Summary No. 40.

4. Bruun, P. and Günbak, A.R., 1977. Stability of Sloping Structures in Relation to $\zeta = \tan\alpha / \sqrt{H/L_0}$ Risk Criteria in Design. Coastal Eng. 1:287-322.

5. Ge, Wu and Jensen, O. Juul, 1983. Stability of Rubble Foundation for Composite Breakwaters. Conference on Coastal and Port Engineering in Developing Countries, Colombo, Sri Lanka, March 1983.

6. Mc Cartney, B., Ahrens, J. Wave Period Effect on the Stability of riprap. Civil Engineering in the Oceans, 1975, pp. 1019-1034.

3. Wave action on breakwater armour units

U. NASCIMENTO, Consultant on Soil Mechanics and Geotechnics, Lisbon, and C. PITA, Laboratorio Nacional de Engenharia Civil, Lisbon

SYNOPSIS. A new method for analysing the stability of rubble-mound breakwaters is presented. The method is based on the analysis of the equilibrium limit between the resultant of the drag and lift forces and the theoretical force needed to remove a block from the armour layer. These forces are proportional to a stability coefficient which depends on the Reynolds and Keulegan-Carpenter numbers. Values of this coefficient obtained using model tests are presented.

INTRODUCTION

1. Wave action on a rubble-mound breakwater can produce rocking or removal of the armour units and their breakage or deterioration.

2. Usual design methods consider only the removal of the units. Since the cause of all these effects is the same (wave action), a method of design must be established which, taking the wave characteristics and the behaviour of the structure into account, can evaluate all the aspects related to the stability of the structure.

3. This paper does not pretend to contain more than a first approach to a possible new method of evaluating the stability of this type of breakwaters.

4. Wave forces in a particular unit are dependent on the velocities and accelerations of the water flow when it enters or leaves the armour layer. These velocities and accelerations can be assumed as proportional to the velocities and accelerations of the water particles of the wave, before entering the breakwater.

5. The removal of a particular unit occurs when the wave force exceeds its resistance to being removed. When these two forces are equal, an equilibrium limit is achieved. We are going to try

to evaluate the resistance, so that the wave force can be calibrated.

6. The removal of a certain unit can be considered as a particular case of an erosion problem in rivers or channels.

7. The resistance to erosion can be expressed by the critical drag stress, τ_{Wc}, which can be evaluated by Lane Equation, 1952 (see ref. [1]):

$$\tau_{Wc} \, (kgf/m2) = d75 \, (m) \qquad (1)$$

in which d75 is the 75% quantile of unit diameters. For non-horizontal slopes, this equation can be generalized as follows:

$$\tau_{Wc} = d75 + b \qquad (2)$$

where b is a constant, which is lower than zero for the run-down and positive for the run-up. This equation is derived for block diameters between 0.5 and 10 cm. Analysing this equation it was concluded that it will be valid also for greater diameters since the shape of the blocks is similar and since the flow is still turbulent.

8. The computation of the critical drag stress can be made using model test results, by evaluating the hydraulic buoyancy, the wave drag and lift forces.

WAVE FORCES

9. A fixed body, in a fluid which flows with variable velocity U(t), is acted by drag forces, Fd(t), lift forces, Fl(t), and mass forces, Fm(t). All those forces are variating with time. Drag forces can be evaluated by:

$$Fd(t) = 0.5 \, Cd(t) \, \rho \, A \, U(t)|U(t)| \qquad (3)$$

in which ρ is the specific mass of the body, A is the exposed area of the body to the flow, U(t) is the velocity of the flow at the instant t. As it is known, when the flow is produced by wave propagation, U(t) depends on the depth.

10. Cd(t) is a non-dimensional drag coefficient, which depends on the shape of the unit, on smoothness of its surface, on Reynolds number of the flow, Re:

$$Re = \frac{U \, ea}{\nu} \qquad (4)$$

and on Keulegan-Carpenter number, Kc:

$$Kc = \frac{U(t)T}{ea} \qquad (5)$$

ν is the kinematic viscosity of the fluid, T is the period of the flow and ea is a characteristic dimension of the unit.

10. Drag forces have the direction of the flow.

11. Lift forces can be evaluated by:

$$Fl(t) = 0.5 \ Cl(t) \ \rho \ A \ U^2(t) \qquad (6)$$

$Cl(t)$ being a non-dimensional lift coefficient, dependent on the same variables as $Cd(t)$.

12. Lift forces are normal to the flow.

13. Mass forces can be evaluated by:

$$Fm(t) = Cm(t) \ \rho \ V \ \frac{dU(t)}{dt} \qquad (7)$$

in which $Cm(t)$ is a non-dimensional mass coefficient, function of the Reynolds number of the flow, Re, of the flow frequency and the shape of the unit. V is the volume of the unit. $\frac{dU(t)}{dt}$ is the acceleration of the flow.

14. Mass forces have the direction of the acceleration.

15. It is assumed that water velocities and accelerations are proportional to the wave particle velocities and accelerations near the structure, which may be evaluated by the wave theories. It is also assumed that wave forces are larger for the run-up than for the run-down.

16. As Blevins, 1977 [2], showed (fig. 1), the instant of maximum wave forces in a submerged cylinder occurs at an instant at which horizontal acceleration (and consequently the mass forces) is approximately zero. At this moment the horizontal velocity is maximum and vertical velocity is zero.

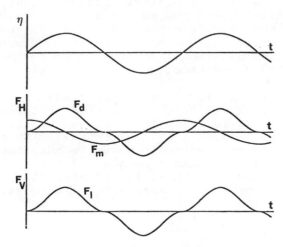

Fig.1 - Variation of lift and drag forces with time

17. If the vertical acceleration is neglected, the maximum of the resultant of these forces can be expressed:

$$F = \sqrt{Fd^2 + Fl^2} = 0.5 \, C \, \rho \, A \, U^2 \tag{8}$$

in which C (square rooth of the sum of the square of Cd and the square of Cl) is a constant dependent on the Reynolds Number, the Keulegan-Carpenter Number, the shape of the unit and the smoothness of its surface.

18. Water particle velocities can be evaluated by the proper Wave Theory for the conditions under study, using for instance Le Mèhaute limits ([3]).

ARMOUR RESISTANCE
Erosion Mechanism

19. As said before, Lane (1952) determined empirically the critical drag pressure in rivers and channels with horizontal beds. He found that for coarse materials above 5 mm of diameter τ_{Wc} is proportional to its size and can be expressed by formula (1).

20. When the drag pressure reaches the critical value, τ_{Wc}, this means that it becomes equal to the shear strength developed between the surface of the particles and the underlying material and then the erosion begins. This shear strength, given by Coulomb´s law, is in cohesionless material obtained by the product of the normal stress in the shear plane and the corresponding friction coefficient, $\tan \phi$.

21. If the flow is plainly laminar, that normal stress will equal the weight of the solid phase deducted from the corresponding hydraulic buoyancy.

22. It occurs, however, that the values of shear strength thus obtained are 5 to 10 times higher than those obtained empirically by Lane and given by his formula.

23. This great difference results from the influence of the lift forces induced by turbulence.

24. In order to evaluate this, we use the concept of turbulence zone in the rockfill (Nascimento, 1983). Pressure fluctuations due to turbulence are considered to decrease from a value $\pm\Delta p$ to zero at a depth z inside the rockfill. There will thus be a hydrodynamic pressure gradient, γ_t, whose main value is given by the relation:

$$\gamma_t = \frac{\pm \Delta p}{z} \tag{9}$$

25.   By generalizing Archimedes´ principle we may
show that this hydrodynamic pressure gradient within
the rockfill, by apparent volume unit, will be equal
to the product of solid volume (1 - P) to the
hydrodynamic pressure gradient, due to turbulence
or, in simple terms, turbulent pressure gradient,
γ_t .
26.   The turbulent force by surface unit, ft, will
be:

$$ft = (1 - P) \gamma_t d \qquad (10)$$

in which P is the rockfill porosity and d the
diameter of the unit. Thus the shear strength s will
be given by:

$$s = (1 - P)(\gamma_s - \gamma_w - \gamma_t) \ d \ \tan\phi \qquad (11)$$

27.   This force is random (in space and time) and
can be positive or negative.   Nevertheless only the
lift force will be considered, since this is the
most dangerous for the structure stability.
28.   The erosive force will act with an angle w
defined by:

$$\tan w = \frac{\tau_w}{ft} \qquad (12)$$

29. Erosion occurs when the drag stress equals the
shear stress, s.
30.   The comparison between τ_{wc} with s (computed
by (11)) permits to compatibilize Coulomb´s law with
Lane´s   equation and would provide an indirect
approach to evaluate the order of magnitude of γ_{tc} and
w_c through the following expressions:

$$\gamma_{tc} = [\ 1 - \frac{0.06}{(1 - P) \tan\phi}](\gamma_s - \gamma_w) \qquad (13)$$

$$\tan w_t = \frac{\tan\phi}{\dfrac{(1 - P) \tan\phi}{0.06} - 1} \qquad (14)$$

Run-up
------

31. Following the mechanism already described, the
critical strength of a layer of armour units in a
rubble-mound breakwater by unit area of slope will
be given by:

$$\tau_c' = D_c' \ \gamma_w \ d \qquad (15)$$

where D_c' is the critical drag coefficient given by:

$$D'c = \cfrac{(1 - P)(Sr - 1) \cos \alpha \, (\tan \phi' + \tan \alpha)}{1 + \cfrac{\tan \phi'}{\tan w_c}} -$$

$$- \cfrac{i'_w \cos(\alpha - \beta')[1 - \tan \phi' \tan (\alpha - \beta')]}{1 + \cfrac{\tan \phi'}{\tan w_c}} \qquad (16)$$

where P is the porosity of rockfill, $Sr = V_r / V_w$ the relative density of unit material to the sea water, α the slope of the breakwater, ϕ' the angle of internal friction and

$$i'_w = \cfrac{\sin (\alpha'_w + \alpha) \cos \alpha'_w - \sin \alpha}{\cos (\alpha - \beta')} \qquad (17)$$

is the hydraulic gradient of percolation that enters the rockfill. α'_w is the inclination of the surface of water uprushing on the slope, β' the inclination of the flux lines of percolation and w_c is the angle between the force and the slope.

32. There are, however, two ways of instabilizing an armour unit, depending on whether the normal stress that the unit exerts against the underlying units is cancelled or not. If it is not cancelled out, the armour unit is displaced by dragging and the corresponding drag force is given by (15). If it is cancelled out, the armour unit is raised and moves without any friction resistance of the underlying units. In this case coefficient Dc of expression (16) is replaced by the coefficient D given by:

$$D'_L = \tan w_c \, [(1 - P)(Sr - 1) \cos \alpha + i'_w \sin (\alpha - \beta')]$$
$$(18)$$

33. The drag force will obviously corresponds to the lowest value of either D'c or D'_L.

34. For each armour unit, the critical drag resistance, Rc, will be:

$$Rc = \tau_{Wc} \, A \qquad (19)$$

A being the area of the slope corresponding to each armour unit.

Run-down
--- ----

35. The critical resistance of a layer of armour units is given by:

$$\tau_c = Dc \, V_w \, d \qquad (20)$$

being:

$$Dc = \cfrac{(1-P)(Sr-1)\cos \alpha \; (\tan \phi - \tan \alpha)}{1 + \cfrac{\tan \phi}{\tan w_c}} -$$

$$- \cfrac{i \quad \cos(\alpha - \beta)[1 + \tan \phi \tan(\alpha - \beta)]}{1 + \cfrac{\tan \phi}{\tan w_c}} \qquad (21)$$

The different symbols have the meanings already indicated in expression (16), duly adjusted to run-down conditions.

Critical slope
-------- -----

36. As can be deduced from (20), when the slope of the rubble-mound reaches a value that cancels out the numerator of that expression the critical drag pressure vanishes. For any slope above that value, the rubble-mound submitted to percolation is not any more stable. Thus this slope can be said the critical slope of the rubble-mound, α . If $\beta = \alpha_w = a$, the critical slope α_c will be given by:

$$\tan (\alpha_c)_{\beta = \alpha} = \frac{V_r - V_w}{V_r} \tan \phi \qquad (22)$$

This expression was deduced by Taylor (1948) for saturated soil slopes.

STABILITY OF ARMOUR UNITS
Ultimate Equilibrium Condition
-------- ----------- ---------

37. The ultimate equilibrium of an armour unit is achived when the acting force, F, equals the critical resistance:

$$F = Rc \qquad (23)$$

F being given by relation (8) and Rc by relation (19). From the equality of (8) and (19), this results at the instant of maximum horizontal velocity:

$$ea = \frac{U^2 c}{2 \, Dc \, g} \qquad (24)$$

38. Taking into account that the volume can be computed by:

$$\cdot \; V = \left(\frac{ea}{K_\Delta} \right)^3 \qquad (25)$$

K_Δ being a shape coefficient. The weight of the armour unit will be obtained:

$$W = \frac{V_r\ U^6\ C^3}{8\ K_\Delta^3\ Dc^3\ g^3} \qquad (26)$$

where V_r is the specific weight of the armour unit material.

39. The preceding analysis makes use of the expressions deduced for wave run-up. Although an identical analysis can be made for wave run-down, the number of variables (since no simplifications can be made in that case) would make its pratical applications difficult. Without affecting the possibility of future development to that approach, the hypothesis of equal effects of wave run-up and run-down, early accepted, leads us to consider expression (26) valid for both cases.

APPLICATIONS

40. A certain number of tests with tetrapods, cubes, and dolosse was performed using regular waves to evaluate C coefficient in relation (26). The results obtained assuming $\alpha = \alpha_w = \beta$ and for the wave conditions at which the first block was removed are presented at figures 2, 3 and 4.

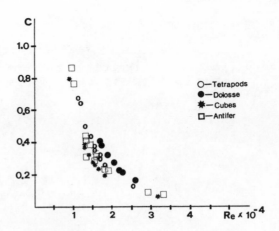

Fig. 2 - Stability coefficient, C, in terms of Reynolds number, Re.

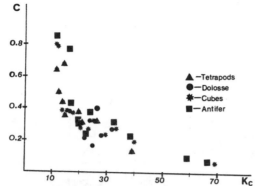

Fig. 3 - Stability coefficient, C, in terms of
Keulegan-Carpenter number, Kc.

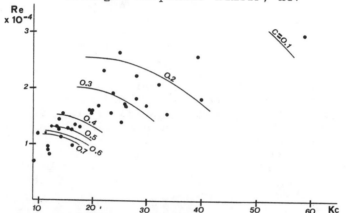

Fig. 4 - Stability coefficient, C, in terms of
Reynolds (Re) and Keulegan-Carpenter (Kc)
numbers.

41. The results obtained seem quite good, although
the number and type of test results disponible so
far are still insufficient for design purposes.

42. The stability of different types of armour
units can be described using only their friction
coefficient, porosity and shape coefficient.

43. Other interesting result is the inexistence of
dispersion of stability coefficient for similar Re
and Kc. According to the results obtained, the
dispersion of test results can be explained by the
increasing of coefficient C for lower Re and Kc.

44. Assuming that the effect on a breakwater of a
real sea state is the sum of the effects of
individual waves, this method can be applied to
irregular waves. An example of this application is
presented in Table I. The results present also a
good aggreement with empirical conclusions about pa

35

relations between regular and irregular waves.

TABLE I

TETRAPODES   cot   = 1.5   = 2.5 ton/m3
JONSWAP Spectrum   Hs = 10 m
Breakwater depth of 12.5 m

Hudson formula using SPM coeffi- cients	Breaking wave	H = H1/10	W = 109 tons
		H = H1/3	W = 53 tons
	Non- breaking wave	H = H1/10	W = 95 tons
		H = H1/3	W = 47 tons
Method presented	Stability coefficient obtained from fig. 4 for individual wave heights and periods		W = 85 tons

44.   Additional studies are being performed in order to investigate hydrodynamics effects neglected until now, to evaluate the actual wave velocities and to obtain stability coefficients for large Reynolds and Keulegan-Carpenter numbers.
45.   As soon as C coefficients are calibrated for removal of blocks, the force F can be computed for other conditions and the method can also be used to evalute rocking and internal stresses in the blocks produced by wave action.

REFERENCES
1.   NASCIMENTO, U., 1985 - "Estabilidade de enrocamentos em quebra-mares e barragens galgáveis", LNEC, Dezembro de 1985.
2.   BLEVINS, H., 1977 - "Flow induced vibrations", Nostrand Reinhold Company, New York, 1977.
3.   CERC, 1984 - "Shore Protection Manual" (SPM), Fourth Edition, CERC, Washington, 1984.

Discussion on Papers 1 and 2

Paper 1

MR G. CRAWLEY, Fairclough Howard Marine, Chatham
Paper 1 refers to cost-saving achieved by the use of
mathematical models compared with physical models. Could the
Authors provide model costs for the Torquay Harbour example
for
   (1) mathematical model – set-up and cost per run
   (2) physical model   – set-up and cost per run.

MR J. BERRY, Bertlin and Partners, Reigate
In the case of the Torquay Harbour, severe wave conditions may
occur when a swell from the east, resulting from several days
of strong easterly winds, is combined with the effects of
local waves.
   Is it possible to combine the effects of a swell with that
of locally generated waves by mathematical modelling, in order
to determine wave conditions inside a harbour?

MR P. HUNTER, Sir Alexander Gibb and Partners, Reading
Is it not normal practice to include pontoons or moored boats
in physical models of marinas? The pontoons are of shallow
draft and may not be important, but boats will act as a
floating breakwater to damp out resonance and to reduce wave
heights within the marina. Possibly this leads to
conservative conclusions being drawn from model tests.

MR P. LACEY, Ove Arup and Partners, London
What tolerances do the Authors use to establish a good fit
between physical test results and computational test results?
   After construction and monitoring of prototype projects does
Hydraulics Research amend the computational models to increase
accuracy and sensitivity, as comparison with physical models
is still comparing best guesses with best guesses.

MR H. LIGTERINGEN, Fredric R. Harris (Holland), The Hague
It is well-known that a major problem related to random-wave
generation in port models is the reflection of the long waves
(set-down), in particular. While the importance of simulating
these long waves is recognized, one must be careful that
reflections from model boundaries, which are open in nature,
do not exaggerate long wave energy. How does Hydraulics
Research solve this problem?

MR J. D. METTAM, Bertlin and Partners, Reigate
As a consulting engineer, I find that my clients are rarely
prepared to allocate either time or money to physical
modelling. I therefore hope that, with experience,
mathematical modellers will become sufficiently confident to
recommend the use of their work on its own without recourse
having to be made to physical models to confirm their
conclusions. Could the Authors please say whether we are now
reaching that degree of confidence for normal, straightforward
wave penetration testing?
   The Authors discussed set-down as found in random wave
testing in physical models. Could they say to what extent the
existence of set-down has been confirmed in the prototype?
Long period waves are most important if they are sufficiently
regular to set up resonance inside harbours. With the long-
crested waves generated in models, within a small basin with
many reflecting surfaces, is there not a danger of
exaggerating both the regularity and magnitude of the set-down
effects compared with the short-crested multi-directional seas
more normally encountered in the protopype?

MR W. A. PRICE, Rendel, Palmer and Tritton, London
Could the Authors say whether the predictions they made,
especially of wave heights, were realistic? Are field
measurements planned to check these predictions? Would this
be a productive research project?

MR H. L. WAKELING, Consulting Engineer, Esher
With reference to the use of mathematical models, the Authors
have said that a change in alignment of a breakwater can be
modelled easily. Could they say to what accuracy the
breakwater can be modelled, as the examples they have quoted
appear to imply that relatively small changes have been
examined. Does this mean that the Authors have had to use a
small grid size in the model, with consequent increase in
computer costs?

DR SMALLMAN (Paper 1)
The contributions of Messrs Crawley, Lacey, Mettam and Price
raise issues concerning the relative costs and performance of

mathematical and physical models. At present, prices studies of the type carried out for a harbour of a similar size to Torquay would cost around £10 000 for a mathematical model and £30 000 for a physical model. For the Torquay study, both types of model were used, with the mathematical model allowing several layout options to be compared for many different wave conditions before testing of the best layout in the physical model. This approach is frequently used in wave disturbance studies, as the use of the mathematical model allows considerable cost and time savings to be made in the amount of testing required in the physical model, while retaining the accuracy of the absolute wave height values measured in the physical model.

In many joint computational and physical model studies, the mathematical model is at some stage calibrated against the physical model. This may be done either before the mathematical model tests, where time is available and/or the harbour layout is complex, or following the completion of the physical model tests. In the latter case, the computational model can then be retained and used to examine any further harbour developments once the physical model has been removed.

In comparing physical and computational model results, an agreement to within 10% of the measured wave heights is usually considered to be reasonable. In many instances, the agreement between wave heights for the two types of model is 5% or better. It is clearly more usual to calibrate mathematical model results with those from a physical model, as prototype data wave disturbance at many locations throughout the harbour - which are required for a full comparison - are rarely available. It would be a productive research project to carry out a comparison between field data and mathematical harbour model results, and we are hopeful that it may be possible to make such comparisons in the future.

With regard to using physical model results to calibrate mathematical models, we are confident that the physical model does represent accurately the mechanisms occurring in a real harbour. This is borne out by the results of the Acajutla study which is referred to in the Paper.

Finally, two of the contributors ask about the accuracy of the mathematical models. At the present time, we are sufficiently confident that the results from mathematical models are accurate enough for most harbour layouts, for feasibility studies or where comparisons are to be made between a number of schemes, and that they will allow the most effective scheme to be chosen. However, where absolute values of wave heights are required for final design or where the harbour is more complex, a physical model is still the most appropriate type of study. In almost all cases, the use of a hydraulic model in a harbour study will be cost-effective in terms of savings made in the construction costs of the harbour. An example of such savings is given in Paper 7.

In reply to Messrs Berry and Hunter, vessels at berth in

harbours and marinas during storm events will be subject to
the combined influence of waves entering the harbour and those
generated inside the harbour by virtue of the wind blowing
across the very local fetch. (They may also be affected by
the wind in exposed locations.)

In respect of storm waves entering the harbour, wavelengths
are large in terms of the dimensions of the vessels, and
therefore the vessels will have very little effect in
dissipating the incident wave energy. Model tests have shown
this to be true, even for fairly large floating vessels.

However, short-period, locally generated waves may reflect
and be dissipated to some extent, since wavelengths are in the
order of a few metres and comparable to the typical dimensions
of length (or even breadth) of the moored vessels.

Unfortunately, it is not usually possible to generate such
short periods in normal physical wave disturbance models where
the upper limit of the wave frequency capable of being
generated is around 2.2 Hz. At typical model scales in the
range 1:50 to 1:100, this frequency will result in maximum
periods in the model of the order 3.5 s to 5.0 s respectively,
which are too long to represent locally generated waves.

Therefore, as the short waves are not able to be generated
and the longer period waves are little affected by small craft
there is little value in representing the craft or the
pontoons in the model.

However, an allowance for local waves can be made in order
to judge the acceptability of any harbour layout. This is
achieved by using a mathematical wind-wave prediction model to
calculate the local storm wave height (H_{SLOC}) generated in the
harbour. This height may be combined with the storm or swell
wave height (H_{SS}), measured in the physical model or
calculated in a numerical model, to produce a resultant wave
height that can be compared with wave height criteria for the
vessels in question. If the individual wave height components
are considered to be uncorrelated, their energies may be added
and the total wave height (H_{ST}) may be calculated from

$$H_{ST} = (H_{SS}^2 + H_{SLCO}^2)^{1/2}$$

This method can be used to combine wave effects measured in
physical models, or those calculated in mathematical models,
or those derived from a combination of both.

In reply to Mr Ligteringen, is is well known that long waves
constitute a problem in physical model wave studies because
they can be reflected from boundaries, thus increasing long
wave energy in the basin. If spurious second order waves are
minimized, by the use of suitable methods, then it is
basically the set-down waves which give rise to these boundary
effects. Wave set-down disappears during the breaking of
short period waves when the boundary is gentle and breaking
takes place over a certain distance, i.e. as spilling
breakers.

If the beach slope is too steep then wave set-down will be

reflected as free long waves, known as surf beats. While both
these effects are natural phenomena, it is the re-reflection
of their energy from other boundaries in the basin which is a
potential source of error.

Models are constructed in this way in order to represent, as
far as possible, the natural harbour and coastal boundaries
adjacent to the area of interest. All other boundaries are
made up of shallow shingle slopes which are very good
absorbers of wave energy. This leaves the re-reflection of
waves from the paddle as the only major possible source of
spurious long-period energy. If the wave paddle is placed
many wave lengths away from the study area, then this will
help to moderate any re-reflections propagating towards the
site through diffraction, refraction and bed friction.

In numerical models, similar problems can potentially occur
in representing open boundaries. At Hydraulics Research, we
have found that the most effective absorber of long-period
energy in numerical models is the 'sponge layer' technique as
put forward by Larsen and Dancy (reference 1). This is used
in our numerical models at open boundaries where set-down is
represented; it has been found to work well (reference 2) and
to minimize unwanted reflected long-period wave enegy.

In reply to Mr Mettam (part 2 of question), the three points
raised should be answered individually since they relate in
turn to the existence of set down in nature, the problems with
its generation in physical models and, finally, the degree to
which set-down in models is representative of conditions in
nature.

In answer to the first question, the phenomenon of set-down
beneath wave groups in the real sea was first described by
Longuet-Higgins and Stewart in 1964 (reference 3). However,
the discovery of its significance for harbour design had to
await the use of random waves in physical models of harbours
in the early 1970s. Since then, many researchers in this
field have recorded the existence of the phenomenon through
measurements in nature. One example in particular involves
the analysis of waves recorded offshore and inside Port
Talbot.

The results from Port Talbot showed that, where the primary
waves were large, long waves with periods in the range 30 s to
many minutes occurred naturally in the sea, and that most of
the energy could be explained by set-down. When the primary
waves were small, most of the long wave energy appeared to be
due to surf beats generated on the beaches to the south east
of Port Talbot.

The second point raises the question of re-reflections
within the wave basin (obviously re-reflections of long waves
within harbour boudaries is a real effect). As described
previously, there will always be some unwanted long wave
energy in the basin, but if care is taken to provide
sufficient spending material along the sides of the basin and
the wave generator is located at a reasonable distance from
the study area then these problems can be minimized.

41

Finally, there is the question of set-down in models and in nature. Research carried out at Hydraulics Research has shown the importance of compensating for set-down at the wave maker. If this is not done, spurious long waves are produced. Set-down compensation is now used on a regular basis in physical models of harbours employing long-crested random waves, and it has led to a more realistic representation of vessels ranging at wave group periods.

As waves propagate into coastal waters, wave reflection tends to reduce the directional spead of wave energy. This has been used in the past as an argument to justify the use of long-crested waves in harbour models, because it was thought that the spread in direction remaining at the harbour entrance was small enough not to cause a significant effect on results obtained with long-crested waves. However, research has shown that directional spread can have significant effects on harbour and ship motion responses. In particular, when using long-crested wave results as a basis for comparison, set-down is some 50% smaller even for wave conditions with a relatively narrow directional spread.

Naturally, the next step is to provide short-crested wave makers, with set-down compensation, in random wave models. This is expected to improve the accuracy of physical models to a significant degree.

In reply to Mr Wakeling, it is possible to change both breakwater length and alignment fairly easily in the mathematical model (harbour ray model) used in the Torquay study. In this model, the harbour boundaries are represented as straight line segments in the model grid squares. The position of the line segment is independent of the mesh used to resolve the bathymetry. This allows the position of the harbour boundaries to be specified accurately and avoids the 'saw tooth' effect which occurs in finite difference models.

This method of representing the boundaries means that relatively small changes (say 10 m) in the breakwater length can be examined in the model without having to use a small grid size, with the consequent increase in computer costs.

Another advantage of the method used is that each individual line segment has its own specified reflection coefficient. it is therefore relatively easy to change the reflection behaviour of a boundary, and this enables different possible structures (e.g. vertical walls or rubble slopes) to be examined.

References

1. LARSEN J. and DANCY H. Open boundaries in short wave simulation - a new approach. Coastal Engineering, 1983, 7, pp 285-297.

2. SMALLMAN J. V., MATSOUKIS P. F., COOPER A. J. and BOWERS E. C. The development of a numerical model for the

solution of the Boussinesq equations for shallow water
waves - coastal depth case. Hydraulics Research,
Wallingford, 1987, Research Report SR 137.

3. LONGUET-HIGGINS M. S. and STEWART R. W. Radiation
stresses in water waves: a physical discussion with
applications. Deep Sea Research, 1964, Vol II, p.529.

Paper 2

DR S. S. L. HETTIARACHCHI, University of Moratuwa, Sri Lanka
A series of tests was performed on instrumented Shed and Cob
armour units which were not in touch with their adjacent
members or with the underlayer (see Paper P3 by Hettiarachchi
and Holmes in this volume). It was observed that wave induced
hydraulic impact loads were acting in the directions parallel
and perpendicular to the slope, corresponding to the point of
wave impact. These forces, which acted over a very small time
interval (approximately $0.01 \sim 0.001$ seconds), were
superimposed on the gradually varying type of force (details
in reference 1). The characteristics of the measured force
records were dependent on the relative position of the
instrumented armour unit, the incident wave climate, the
dynamic response of the transducer (reference 1) and the
method of data acquisition.
    Could the Authors comment on the above observations and
state whether the method they adopted enabled them to record
the peak forces acting on the cylindrical members?
    Do the Authors feel that in the analysis of results it is
important to differentiate clearly between the effects of
impact types of load and the gradually varying types of load?

MR H. LIGTERINGEN, Fredric R. Harris (Holland), The Hague
In reply to the Authors' request for suggestions on how to use
the available test results, I would like to give the following
ideas.
    In his analysis of wave forces on armour units, Iribarren
used a rather simple expression for drag, lift and inertia
forces. It may be useful to go back to this early work for
comparison of theory and experimental results.
    With reference to wave slamming effects on concrete armour
units, Cyril Galvin published the results of a theoretical
study some years ago, and these may provide a basis for
evaluating the present results.

DR G. VAN DER MEER, Delft Hydraulics
The Authors used monochromatic, bichromatic and random waves.
Monochromatic waves can give an idea of the scatter and
repeatability of the phenomenon (wave forces). Random waves

are close to prototype conditions. Why, however, did the Authors use bichromatic waves, which bear no relation to nature and which were not used to investigate, for example, grouping effects. What did they expect to do with these results?

Only maximum forces were used in the analysis of the random wave test. The maximum force depends on the number of waves used. Furthermore, the maximum force is not very reliable and a 1% or 2% value from the exceedence curve will probably give less scatter. Is it good practice to analyse the data in this way, and do the Authors indeed expect less scatter?

DR O. J. SAYAO, F. J. Reinders and Associates Canada Limited, Brampton, Ontario
Has the Danish Hydraulic Institute considered a test set-up with a single layer of pipes, as this would give some insight into the importance of the contact forces?

MR J. D. SIMM, Rendel, Palmer and Tritton, London
The Authors have commented on the variation of stable armour weight with position on slope. I get the impression that this effect is less with irregular than with regular waves. Would they comment further on this?

Wave slam appears to be a most interesting effect. It would be useful if the Authors' results could be evaluated in some way, in order to determine the loosening effect which wave slam has on armour layers.

MR R. V. STEPHENS, Hydraulics Research Limited, Wallingford
Since January 1988, Hydraulics Reasearch have been conducting a series of field measurements of wave-induced pressures on breakwater armour units. The breakwater which has been instrumented is in an area of very high tidal range. It is clear from the results so far that the nature of wave loading on an individual unit is strongly dependent upon its position relative to static water level. As the tide level drops, a transition is observed between a smooth, quasi-hydrostatic loading pattern and a sharply peaked impact loading due to wave breaking action. Although the frequency of occurrence of such sharp wave loadings diminishes with falling tidal level, the magnitudes of impact peaks increase in some cases. It would seem very important to attempt to characterise the different types of wave loading pattern and to study the influences which static water level changes relative to an armour unit position have on the transition between smooth loadings and peaked impacts.

MR J. G. BERRY, Bertlin and Partners, Reigate.
It would seem from Paper 2 that the effects of air entrainment

on armour units are so variable that it will never be possible
to find a formula for armour stability.

Is the air entrainment shown in Fig. 3 of Paper 2, and in
similar photographs presented by the Authors, representative
of what occurs in the prototype?

MR JENSEN and MR JUHL (Paper 2)
In reply to Dr Hettiarachchi, the wave forces and impacts on
the cylinders were recorded by use of strain gauge
transducers. The set-up was made as stiff as possible, the
frequencies of oscillation were 100 Hz in air and 55 Hz in
water. The fastest recorded rising time of the peak pressures
was in the order of 0.05 s, which proves the adequacy of the
set-up for recording of peak pressures.

The Authors feel that their test set-up is quite different
from the one used in the UK for prototype measurements on Shed
and Cob units, as the pressure transducers used on the UK
recorded local pressures due to wave impact/slamming, while
the Danish Hydraulic Institute model included measurements of
the total force on the entire armour unit.

As presented in the Paper, the Authors are of the opinion
that it is necessary to analyse their own test results with
respect to duration of wave loading, i.e. the wave impulse.
With respect to the Shed and Cob prototype measurements, the
situation is different (as presented above), and the Authors
agree that for these measurements a clear differentiation
between the two types of loading appears appropriate.

The Authors will try to follow the suggestions made by Mr
Ligteringen in his contribution.

In reply to Dr van der Meer, the Authors decided to study
both monochromatic, bichromatic and random waves following the
research project of Hydraulic Research Wallingford and
published by Burcharth (reference 2).

The Authors had no prior expectation, but were curious to
observe the differences for the various types of wave.

The Paper presents the first analyses of the test results;
therefore, for simplicity, the maximum recorded values were
used. The Authors agree with Dr van der Meer that, for
example, the 1% value of wave forces would give more reliable
or consistent values when comparing the results for various
test series. The Authors are presently expanding their
analysis in the proposed way.

In reply to Dr Sayao, the Authors have not considered a test
set-up with a single layer of pipes, as they set out to make a
purely 2D-structure with the highest degree of resemblance
with a traditional rubble mound breakwater, including two
layers of armour blocks, a filter layer and a quarry-run core.
The Authors do not understand how the tests could give an
insight into contact forces, as they were conducted with fixed
cylinders separated from each other.

In reply to Mr Simm, it is correct that the tests have shown
that for irregular waves the variation in weight requirements

as function of the position on the slope is less than for regular waves. The same tendancy was observed for the influence of the wave period. This is as expected, because a train of irregular waves contains a wide range of wave heights and periods. Therefore, the results for irregular waves would be, say, the integrated results of all the wave heights and periods and, therefore, necessarily more 'smoothed' and with less variation than for regular waves.

The Authors agree that the observed wave slamming is an interesting and important phenomenon. However, they do not understand how these impacts with a direction into the mound of material should have a loosening effect on an armour layer of randomly placed armour units. Perhaps the effect on the armour layer as a whole would rather be a compaction.

It is the Authors' opinion, as expressed in the Paper, that the major effect of the slamming would be the risk of breakage of slender armour units. The Authors are at present analysing their results to extract the maximum information on the observation phenomenon of wave slamming.

Mr Stephens should refer to the answer given to Dr Hettiarachchi.

The Authors agree with Mr Berry's statement that air entrainment in the upper part of the water wedge of the run-up causes considerable scatter and uncertainty in armour stability. As the Authors have pointed out, the variation and characteristics of the hydrodynamic loading on armour units vary as a function of the position on the slope. It appears difficult to imagine, therefore, that one single formula should ever be able to embrace these variations in an acceptable way.

However, engineering science is also faced with the problem of air entrainment in flowing water - for example, on a spillway - and in the process of wave breaking, when waves are breaking on a gently sloping seabed where they are not meeting a structure. Consequently, the physical process of air entrainment is likely to be an important object of study in the near future, in order to improve the general understanding of this important phenomenon, but also because a mathematical formulation of the air entrainment process will be required for any mathematical modelling of all the phenomena mentioned above.

It is general knowledge - for example, from physical models and prototype observations of breakwaters and spill-ways, etc. - that the air entrainment in a model will always be on the low side of prototype conditions, but how much lower is not yet fully understood. In connection with the basic research of air entrainment, this is an important problem to investigate, as it has a bearing on many disciplines of physical scale modelling. In the Authors' opinion, this is, in fact, one of the most important outstanding problems in hydraulic modelling.

References

1. BIGGS J. M. Introduction to structural dynamics. McGraw Hill, New York.
2. JENSEN O. J. and JUHL J. The effect of wave grouping on on-shore structures. Coastal Engineering, 1979, 2, pp 189-199.

4. Modelling interior process in a breakwater

Dr F. B. J. BARENDS and P. HÖLSCHER, Delft Geotechnics

SYNOPSIS. The stability of a rubble mound structure depends for an important part on interior processes related to the motion of pore water in the structure. Complex transient phenomena are observed. In the conference Breakwaters '85 a general view on geotechnical aspects has been given focusing on mechanical behaviour [1]. Here, the attention is drawn to the modelling of local stability of boundary elements in a rubble mound slope subjected to waves.

THE ART OF MODELLING

1. To predict the behaviour of a structure models are used. Models reflect reality only partly. The selection of a suitable model is not simple. This is outlined by considering the various steps in a modelling procedure:
- definition of the problem
- decision on essential phenomena (conceptual model)
- selection of a simulation model
- calibration of model parameters (model tuning)
- model prediction and evaluation

2. First of all a well-defined problem description is required. For a successful model application one should, in fact, have already some idea of the outcome. A correct problem definition helps in this respect.

3. Understanding of relevant phenomena, their physical background and mathematical formulation, and practical experience form requisite keystones for the definition of a relevant conceptual model, which can be purely empirical or which can be constructed using conservation principles and constitutive laws. Precise knowledge of the processes for the composition of a sound conceptual model is rarely available, and one has sometimes to accept even rigorous assumptions which may be difficult to verify.

4. For the investigation and solution of a problem using a simulation model one may select a physical model or a mathematical model.

5. The application of a physical (scale) model requires good comprehension of scale rules. The philosophy behind scaling is not founded on similitude only, but on dynamic equilibrium of the actual forces in the model and in

reality. The general length, an intrinsic length (grain size), time, gravity, and certain material properties (viscosity, strength, roughness, weight) may be used to scale forces due to friction, viscosity, deformation, inertia and surface tension correctly.

6. The problem is there is no single correct way. Forces for each of these phenomena have to be scaled in their own characteristic manner (Froude, Weber, Reynolds, Mosoni-Kovacs, etcetera). Due to non-linearity, changing direction and varying intensity of forces the scale is, in fact, time-dependent, boundary-value-dependent, and phenomenon-dependent (elasic/plastic, laminar/turbulent). It is impossible to meet all the requirements in a single scale model. It is the art to recognise dominant features in a problem, and to find the best model suited to simulate those in a realistic fashion.

7. Mathematical models are mathematical representations of reality, usually defined in terms of partial differential equations, the elaboration of which is achieved by analytic or numerical methods. Sometimes it is a simple formula based on empirical concepts, particularly useful when it is well-calibrated. Extrapolation outside the scope of tested applications may, however, be hazardous.

8. Analytical and numerical methods go hand in hand. The first provides parametric insight, the second a numeric answer. There are complications. The mathematical character (elliptic, parabolic, hyperbolic) may alter requiring a different algorithm, not necessarily known beforehand.

9. Model tuning is best achieved by comparison with alternative models. The comparison should be based on a representative example, a bench-mark test.

10. In conclusion one may state that at one hand physical modelling is objectionable because of scaling and measuring problems, and at the other hand mathematical modelling suffers from incomplete concepts and algorithms. If both types of modelling can be applied complementaryly, we may improve the way of our engineering.

11. In the following discussion this approach is applied to the investigation of the influence of turbulent porous flow to the local stability of individual boundary blocks.

MODELLING LOCAL STABILITY OF A BOUNDARY BLOCK

12. For the assessment of the local stability of an individual block subjected to porous flow a numerical simulation model and an empirical formula based on physical experiments will be applied.

13. The macroscopic porous flow field in a rubble mound structure is characterized by average parameters. For the evaluation of the stability of a single boundary block the local inhomogeneity is essential. The block may move out of position due to a local drag force by porous flow, but this motion changes the drag force. The question is whether this phenomenon is progressive or self-healing.

14.   For the investigation a well-conditioned situation is considered: a steady turbulent porous flow through a rubble mound sill under a caisson-type breakwater (Fig. 1a). The procedure consists of two steps: 1) the determination of the influence of local variation of the (turbulent) permeability (due to block motion) on the local gradient (driving force), and 2) the determination of the actual local drag force on that block.

15.   Since average properties and general formulas are used the result provides, however, a qualitative view. At present it does not seem possible either to perform adequate local measurement nor to compose a microscopic conceptual description with a satisfactory accuracy for a quantitative view.

16.   The local inhomogeneity-induced influence of the permeability in a steady turbulent porous flow field is determined iteratively by a numerical finite element model. A particular element containing the considered boundary block is selected (Fig. 1b).

17.   In this element the permeability is varied and the corresponding local gradient normal to the boundary is calculated. It includes the effect of flow concentration due to the permeability variation. Regression analysis yields: $708K=i^{-1.6}$ (Fig. 2a). Since per definition: $q^2=Ki$, one obtains: $q^2=0.0195K^{0.4}$.

18.   The local filter velocity can, from its definition, be conceived as the local absolute velocity in the absence of the boundary block. In this respect a well-known empirical formula for the drag force R on a block in a free turbulent flow in a pipe can be applied: $R=\alpha D^2 \rho Cq^2$, where αD^2 represents the effective surface and C is the drag coefficient. Inserting the expression for q yields: $R=0.0195\alpha D^2 \rho CK^{0.4}$.

19.   The turbulent permeability is related to the local geometry. Many empirical formulas exist. Here, $K=2gn^5 D/C$ is used [2], which yields: $R=0.025\alpha D^{2.4} g^{0.4} C^{0.6} n^2$.

20.   For the evaluation of this expression several remarks are due. The factor D represents the diameter of the object in a free pipe flow. However, in the local pore geometry the "pipe" containing the observed block is narrow. When the block moves the pipe geometry changes (Fig. 2b). If the motion of the block is expressed by $\delta=d/D$, one can state that D is a decreasing function of δ, see Fig. 2c.

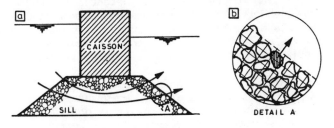

Fig. 1.   Turbulent flow through a rubble mound sill.

21. The factor C represents the drag coefficient related
to the shape and roughness of the object and its orientation
with respect to the main flow direction. When the boundary
block moves, it may, by rotation, choose a more favourable
position as to oppose less resistance. Therefore, C can be
conceived as a function of δ, such as sketched in Fig. 2c.

22. The factor n represents the effect of the changes in
permeability due to the block motion. It can be conceived as
a function of δ, such as sketched in Fig. 2c. The initial
effect is an increasing drag force, as the block moves out,
which is attributed to flow concentration. If the block has
moved significantly, this effect remains about constant.

23. The corresponding drag force R (Fig. 2c) as function
of δ increases during initial motion. A precise graph cannot
be provided without accurate geometric analysis and a better
formula for the drag force (see e.g. [3]). The drag force
tends to decrease rapidly when the block is lifted out in
the order of 50%.

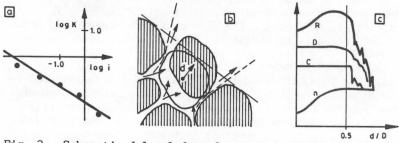

Fig. 2.   Schematized local drag force on a moving block.

24. This analysis shows the effect of porous flow to the
local stability of an individual block using numerical and
physical models. The lift has a progressive nature in the
initial stage, and a strong self-healing potential when the
block is lifted more than about 50%. If no other conveying
mechanism occurs, the block will remain at position.
However, the flow field is usually unsteady, and, in the
case of a breakwater subjected to wave loading, strong
surface currents occur, which may give a final' kick to move
the block completely out of place.

25. For the stability of a block in a porous structure the
interblock friction forces play an important role, particularly at the boundary. This aspect has not been included.
If, however, placement of surface blocks allows one or few
blocks to be relatively free, this analysis applies, and
there erosion usually starts.

MODELLING INTERIOR PROCESS IN A BREAKWATER SLOPE

26. In the previous section it was shown that the porous
efflux may contribute to erosion, but it is not the main
cause. In this section the influence of core permeability,
storativity, and wave period is considered. A numerical

TABLE I   Energy levels of wave-induced pore water motion

grain size m	Epot Nm	Ekin Nm	Ekin/Epot -
0.32	506	496	0.980
0.08	90	5	0.056
0.02	41	1	0.002

model is applied. It is tuned and verified by measurements
in specially instrumented physical model tests. The maximum
parallel porous velocity in the slope may be a more significant cause of erosion.

27. The wave-induced porous flow in a rubble mound is
characterized by dynamic wave transmission counteracted by
internal friction in the porous structure. Usually the
kinetic energy in an unsteady porous flow field is not taken
into account. This effect is small in dense and fine pores,
but not in coarse.

28. An example of a numerical simulation elucidating this
aspect is shown in Fig. 3, in which a vector plot of the
water motion during a harmonic wave loading is presented,
showing the motion at the very moment that accelerated pore
water from the rising water (left) and the steady water
(right) meet inside the core causing a vertical jet in the
phreatic water body in the rubble mound. For this particular
moment the kinetic energy is not negligible.

29. Results of similar simulations performed for various
grain sizes are compiled in Table I. Acceleration and
deceleration forces are important in unsteady flow in coarse
porous media. A small scale test may grossly underestimate
inertia. This aspect can be accounted for in a numerical
model by adopting a proper porous flow law, i.e. which
includes unsteadiness. This can be achieved by considering
dynamic seepage forces and virtual mass effects in the
dynamic equilibrium equation.

31. A 1D numerical Lax-finite-difference algorithm for the
determination of the dynamic internal water table yields acceptable results for the investigation of local phenomena.
It is conditonally stable, but it may cause some numerical
dispersion at small time steps [4].

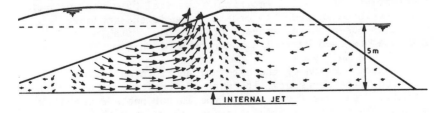

Fig. 3.   Unsteadiness in dynamic porous flow.

53

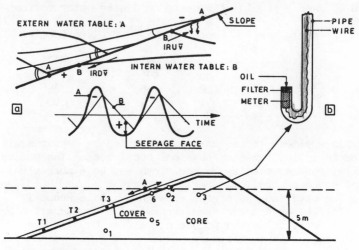

Fig. 4. Lay-out of investigated breakwater.

32. The 2D dynamic porous flow field is simulated by a 2D finite element approach. The boundary conditions consist of the external and internal water table and the pressures at the submerged slope. In physical tests performed by the Delft Hydraulic Laboratory the external water table has been measured by a wave-height meter parallel to the slope. The internal water table is determined with the Lax-scheme.

33. Pressures at the slope are measured by three meters along the down slope (Fig. 4). To avoid turbulent pressure deviations occurring at the slope the meters have been placed just inside the top layer. Another six pressure meters are placed in the core at strategic positions for verification purposes. Water pressure meters function well in a saturated environment. Therefore, they are installed in a special way (Fig. 4b) in order to avert malfunctioning.

34. Tests have been run for random waves, for regular waves and for different core materials. In the early stage of the numerical verification it appeared that the local boundary condition near the fluctuating water table is of essential importance and that the local measurements were insufficient. A special empirical numerical preprocessor has been developed in order to create a consistent boundary condition based on the sparse information, suitable to simulate observed data. The background of this preprocessor is outlined in Fig. 4a.

35. Disconnection of the external and internal water table appeared to be necessary. It causes a "positive" and a "negative" seepage face; across the latter one air intrusion occurs. The numerical simulation has shown that the porous flow field is rather sensitive to these details.

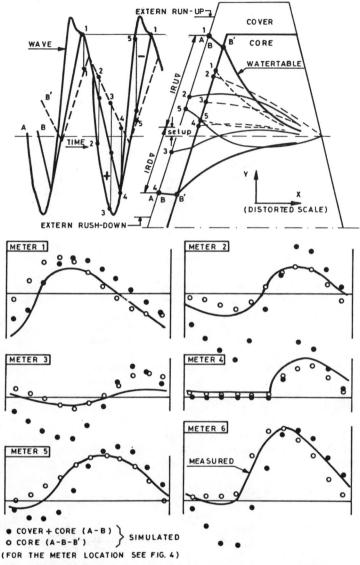

Fig. 5. Mathematically simulated and measured response.

In fact, extensive measurement was needed. As this was not possible, verification by numerical simulation has been worked out so far for regular waves only, i.e. for three different wave periods and three different types of core material. The characteristic data is compiled in Table II.

36. The hydraulic data contain a factor for internal run-up (IRU) and one for internal rush-down (IRD). The actual water velocity of the internal water table is limited theoretically by a (gravity-controlled) gradient of 100%.

TABLE II Characteristic data

Hydraulic data	period s	height m	IRU -	IRD -
wave1	3.5	1.00	1.5	1.0
wave2	4.5	1.00	1.4	1.0
wave3	7.0	1.00	1.2	1.0
Material data	porosity -	diameter m	roughness -	shape -
cover layer	0.40	0.25	2.0	1.4
core1	0.40	0.25	1.5	1.5
core2	0.40	0.05	1.5	1.5
core3	0.40	0.01	1.5	1.5

Effects by semi-saturation, inertia and friction may cause deviatons. The actual internal run-up and rush-down, disconnected from the external water table, is defined by the theoretical maximum velocity multiplied with IRU and IRD, respectively. In this way observed internal set-up and unsaturated flow can be accounted for to some extent.

37. The material data has been selected based on experience and it is used in a turbulent flow law [2]. These data have not been altered during the evaluation.

38. In Fig. 5 some results (●) are presented for a structure composed of a cover and core2 subjected to wave3 loading. The agreement between simulated and measured pressure history is rather poor. Similar results were obtained for other loading conditions: wave1 and wave2. The most obvious missing in the simulation can be a disconnection of the internal water table at the separation of cover and core.

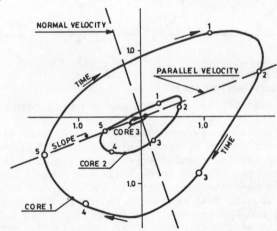

Fig. 6. Hodograph at the water table in the slope.

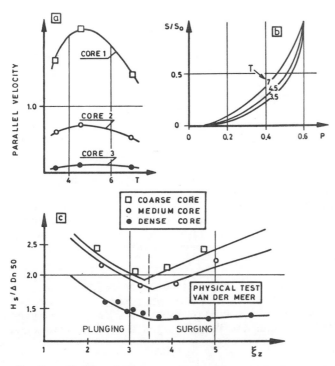

Fig. 7. Correlation between porous flow and observed damage.

This has been checked by a simulation without cover using the preprocessor for the local boundary condition directly at the core slope.

39. The corresponding numerical results (o) show a better agreement with measurements (Fig. 5). One may observe that disconnection may cause initial storage (internal set-up), delay and amplitude decay. The differences between observed and simulated response still present are due to effects of air intrusion, which is not included in the simulation.

40. Next, the influence of the core density has been investigated. The influence is best observed by focusing at the generated velocity of porous flow at the internal water table near the slope by means of a hodograph such as shown for wave2 in Fig. 6.

41. From hodographs for all waves the maximum velocities parallel and normal to the slope are obtained. The maximum parallel velocity is extreme at an intermediate wave period, for all cores (Fig. 7a). Probably, the storage is little for shorter waves, whereas the rush-down is slow for longer waves. Actual lift out may be upward, because the upward parallel velocity is larger due to the wave form during rush-up.

57

42. The normal velocity is a measure for the cyclic storage in the structure. Longer waves and higher permeabilities give larger storage. Van der Meer [5] suggested an extended Hudson formula including the effect of core density by a permeability factor P, which can be related to the cyclic storage (Fig. 7b). The ratio of cyclic storage S for the considered structure and the one for a homogeneous coarse structure S_0 determines the actual value of P.

43. The relation between wave and core characteristics for a fixed damage level, presented by Van der Meer (Fig. 7c), shows a critical wave marked by the transition from plunging to surging type. With regard to the phenomenon found for the parallel velocity (Fig. 7a) it is conceivable that this critical wave depends not only on the wave type, but also on the cyclic storage capacity. An explicit answer may be found after a thorough investigation including well-instrumented physical tests and adequate numerical simulations, such as has been done for regularly paved revetments [6].

ACKNOWLEDGEMENT
The work presented is performed by Delft Geotechnics in cooperation with Delft Hydraulic Laboratory and supported by the Dutch Department of Public Works. The inspiring suggestions by Mr. J.W. van der Meer are gratefully acknowledged.

REFERENCES
1. Barends,F.B.J. Geotechnical aspects in rubble mound breakwater design, Proc. IC Developments in Breakwaters, Inst. Civil Eng., Thomas Telford PC, London, 1986.
2. Hannoura,A.A. and Barends,F.B.J. Non-Darcy flow; a state of the art, Proc. Euromech 143 Flow and Transport in Porous Media, Balkema PC, Rotterdam, 1981.
3. Willets,B.B. and Naddeh,K.F. Measurements of the lift on spheres fixed in low Reynolds numbers flow, J Hydraulic Res., vol.24(5), 1986.
4. Hölsher,P. and Barends,F.B.J. Transport in porous media; numerical modelling of turbulent dynamic unsteady phreatic porous flow, Delft Geotechnics, Rp 690511, 1986.
5. Van der Meer,J.W. Stability of rubble mound revetments and breakwaters under random wave attack, Proc. IC Developments in Breakwaters, Inst. Civil Eng., Thomas Telford PC, London, 1986.
6. Hjortnaes-Pedersen,A.G.I., Bezuijen,A. and Best,H. Non stationary flow under revetments using FEM, Proc. IX ECSMFE, vol.1, 1987, Dublin.

5. Pore pressure response and stability of rubble-mound breakwaters

J. D. SIMM, Rendel Palmer & Tritton, London, and T. S. HEDGES, University of Liverpool

SYNOPSIS. Regular wave tests on a scale model of a typical rubble-mound structure are described in which pore pressure response and tetrapod armour stability were observed. The tests confirm that armour stability increases with mound permeability and indicate that a regular pattern of tetrapods is generally more stable than a random configuration. The pore pressure measurements are evaluated and are used in simple Bishop's method static stability calculations to provide some insight into the changes in the geotechnical stability of the mound caused by wave action. Some practical consequences of these investigations are indicated.

INTRODUCTION

1. The stability of rubble-mound breakwaters is a subject of ongoing interest. Discussion continues as to the most appropriate form of armouring. In addition, the failure of the Sines Breakwater highlighted the importance of the geotechnical stability of a breakwater (ref.1). Previously the geotechnical stability of a breakwater mound, as distinct from its foundation, was largely ignored in design calculations. However, as Hedges (ref.2) states the armour should not be assumed to act as a retaining skin to the core and underlayers.

2. This paper reports on some flume tests on a 1:50 scale model of a breakwater, the prototype of which was recently completed at Ras Lanuf, Libya. In these tests, the movements under wave action of tetrapods placed in different patterns were observed. The pore pressure response in the mound was also measured and evaluated, particularly in relation to the mound stability.

3. The Ras Lanuf main breakwater was selected as the prototype for these tests because it was typical of many breakwaters constructed in the last 10 or 20 years. Its design and construction was fully described during the previous Breakwaters Conference (ref.3). Fig. 1 shows a typical section through the model and the following points should be noted:

(i)   The model water depth was about 300mm, representing a prototype water depth of 15m, very typical of many rubble-mound breakwaters.

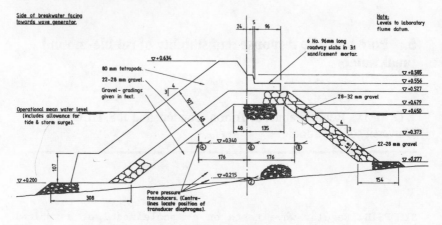

Fig. 1. Section through model breakwater

(ii)   The breakwater side slopes were 3 in 4, an angle often
        chosen for ease of construction, being approximately
        equal to the natural angle of repose of rockfill. This
        angle has been criticised as being too steep when
        considering geotechnical stability and it was hoped that
        measurement of pore pressures might provide some
        illumination on this matter.

(iii)   The armour units used were tetrapods – very popular
        until recently –and in the prototype were placed as near
        as possible to a regular pattern and orientation.   The
        tests sought to examine the value of this placing
        technique when compared with the normal approach of
        placing with random orientation.

DESCRIPTION OF TESTS

4. The model breakwater (Fig.1) was tested using regular
waves in the flume indicated in Fig.2. Regular waves were used
because of limitations in the paddle performance.

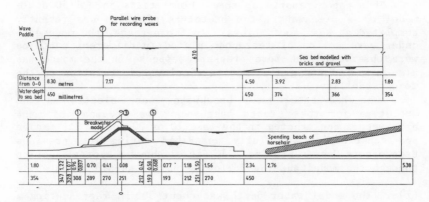

Fig. 2. Longitudinal section through test flume

5. For each test carried out, the model breakwater was subjected to a storm composed of four successive regular wave conditions of increasing severity. The storm parameters are given in Table 1.

Table 1   Equivalent prototype wave conditions to those used in model tests

Wave Set	A	B	C	D
Wave height H (m)	4.9	6.1	7.0	6.9
Wave period T (s)	8.5	9.0	9.1	10.4
Duration (hours)	3	3	3	3

6. Four tests were carried out using the above storm. The distinguishing features of each of these tests are listed in Table 2.

Table 2   Distinguising features of main breakwater tests

Test No.	Model breakwater core grading	Tetrapod armour placing pattern
1	Normal D_{50}=9mm	Random
2	Normal D_{50}=9mm	Regular
3	Coarse D_{50}=15mm	Random
4	Coarse D_{50}=15mm	Regular

7. The model breakwater core gradings noted in Table 2 are shown in Fig.3 along with their prototype equivalents. The intention was that the 'normal' core grading (3) should represent linear (Froudian) scaling of the prototype. In fact, information received after the tests were underway indicated that the grading was a little coarse. The second or 'coarse' core grading (4), was intended to match that required to counteract Reynolds number scale effects on the flow within the mound. The grading required was initially assessed using the methods of Le Mehaute, Hudson/Keulegan, and Kogami (refs. 4 and 5). The average of their results gave curve (5). However, strict adherence to this curve at the upper end was not possible as this would have led to unrealistic modelling

61

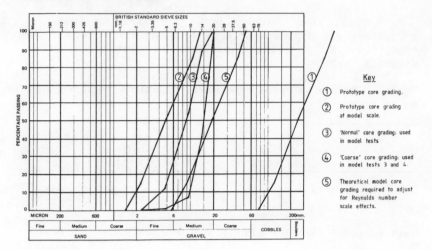

Fig. 3. Gradings of core material

of the core/underlayer interface. It was, in any case, unnecessary because it is the bottom end of the grading curve which mainly governs core permeability and thus influences scale effects.

8. The tetrapod armour arrangements referred to in Table 2 were achieved as follows:

(i)   Random: Units were lowered into position using a cotton thread (simulating crane placing) at predetermined grid positions. However, no restraint was applied to the units to determine their final orientation.

(ii)   Regular: Units were lowered as in (i) but their orientation was constrained to follow the pattern described in ref.3.

The resulting arrangements are shown for comparison in Fig.4. It should be noted that the regular placing pattern is only possible if the underlayer size is in the W/10 to W/20 range recommended by Sogreah rather than in the W/5 to W/10 range suggested by ref. 6 (W is the primary armour unit weight).

ARMOUR STABILITY

9. The amount of damage to the armour units during the tests is summarised in Table 3. The final column of this table gives the percentage of units 'damaged' by total removal from the armour slope, by displacement over a distance of more than one tetrapod spacing, or by rocking.

10. The four tests show the regular armour placing pattern to be more stable than the random armour placing pattern, both in terms of total damage and in percentages of units rocking during the tests. (The reduction in units rocking is particularly important when large units are being employed).

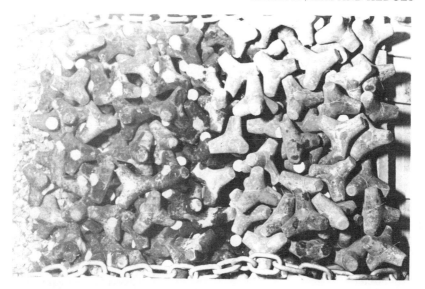

(a) Random Pattern

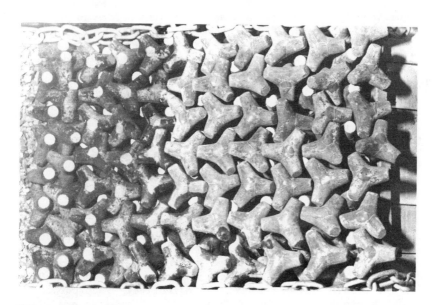

(b) Regular Pattern

Fig. 4. Random and regular tetrapod placing patterns

Table 3 Percentage damage to tetrapod armour

Test No.	Tetrapod placing pattern	Model core grading	Percentage of tetrapods rocking during test	Percentage of tetrapods 'damaged' during test (including rocking tetrapods)
1	Random	Normal	4	5
2	Regular	Normal	2	2
3	Random	Coarse	1	2
4	Regular	Coarse	NIL	NIL

There appear to be two contributory factors involved in the increased stability with the regular placing system.

11. The first factor is that armour stability appears to be increased as a result of the smoother surface profile presented to the uprushing wave. With the regular pattern there are no isolated tetrapods (or parts of tetrapods) which are much higher than the general surface level. There is, therefore, no opportunity for any one unit to be exposed to significantly higher than average hydraulic forces during wave uprush (and downrush). This concept is supported by Bruun and Johannesson (ref. 7) who state that "on the basis of analyses of coefficients of inertia and drag, it may be concluded that a streamlined shape of the exposed side of an armour unit increases stability".

12. It is important to note that this first factor also implies that the resistance to wave uprush is reduced. This was confirmed by measurements of wave run-up on the face of the breakwater.

13. The second factor involved in the increased stability of the regularly placed armour is the increased interlocking which is initially achieved. This presumably delays the 'loosening up' of the armour layer which is required before units can be removed. However, it should be noted that once movement does commence, the loss of mutual support between units could lead to fairly rapid breakdown of the whole armour layer.

14. In relation to the above factors, it is pertinent to note that the successors to the Tetrapod, such as the Haro (ref.8) or the Accropode (ref.9), effectively incorporate in their systems both the improved stability factors just described.

15. Finally, Table 3 shows the increased stability of the armour units when the coarser (distorted scale) core material was used. This confirms the significant effect of mound permeability on the stability of armour units already noted by Bruun and Johannesson (ref.7) and suggests that conventional linear scale hydraulic models may sometimes underestimate armour layer stability.

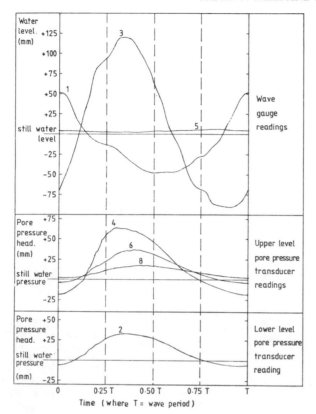

Fig. 5. Simultaneous wave gauge and pore pressure transducer measurements, Test 2, Wave Set D.

PORE PRESSURE RESPONSE

16. Pore pressures were measured during all the wave tests listed in Table 2. A typical set of results is shown in Fig.5. The numbers against the traces in this figure relate to the gauge positions indicated in Figs. 1 and 2. A number of general observations may be made from these traces:

(i) The smoothing and attenuation of the wave form as it passes through the mound is very clear.

(ii) The pressures at the centre of the mound appear to vary approximately hydrostatically (compare the output from gauges 2 and 6). At the front face, pressure changes are more influenced by dynamic effects and variations in pressure head are less than corresponding changes in water level.

(iii) There is a set-up in mean water level (or residual water pressure head) within the breakwater mound during wave attack. The magnitude of this set-up varies according to position in the mound, but for any one

65

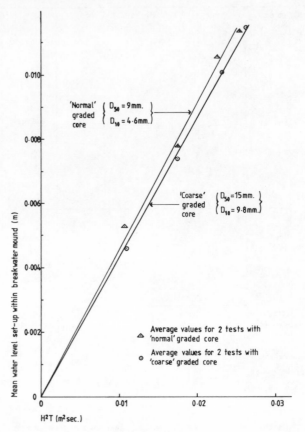

Fig 6. Variation of set-up of mean water level in breakwater mound with H^2T

position is proportional to the square of the wave height times the wave period (and hence is approximately proportional to the power of the incident waves). An example of this proportionality is seen in Fig.6 which relates to data measured at gauge 2. This figure also confirms the findings of Bruun and Gunbak (ref.10) in that the residual water pressure head is larger for the more impermeable core. Since this implies larger pressure gradients across the armour layer during downrush of waves, it has been advanced as an explanation for the reduced armour stability already noted in paragraph 15.

GEOTECHNICAL STABILITY ANALYSIS

17. It was of interest to the authors to pursue the approach of Barends (ref.1) who had used Bishop's method of slip circle analysis to assess the geotechnical stability of the Sines West Breakwater. The resources available for the

present work were rather modest but some preliminary calculations were carried out using a standard computer package and the pore pressures measured at maximum run-up and run-down during test 2, wave set D (Fig.5). A number of simplifying assumptions had to be made:

(i)   Pore pressures within the mound were assumed equal to the hydrostatic head under the piezometric line.

(ii) The water outside the breakwater was assumed to be a 'soil' of zero shear strength.

(iii) The apparent angle of internal friction, $\emptyset'$, for the core material was assumed to be 40°.

(iv) The interlock effect of the tetrapod armour was modelled by assuming for this layer that not only was $\emptyset'$ = 45° (as for the rock underlayer) but that there was also an apparent cohesion $c' = 30$ kN/m^2.

18.   Analysis showed that at the time of maximum run-up of the incident waves, the geotechnical factor of safety of the front face of the mound was lower than for still water conditions. This result is probably attributable in large part to the additional weight of water on the upper levels of the slope which increases the disturbing moment about potential slip centres.

19.   At the time of maximum run-down, there appeared to be an increase in the geotechnical factor of safety of the front slope. However, this effect was probably unreal. Assuming a hydrostatic pressure distribution beneath the water surface at the toe of the mound almost certainly resulted in predicted pore pressures in this region which were lower than those which actually existed (see paragraph 16). As a result, the shear strength of the mound would have been overpredicted. In reality, rapid drawdown of the water level in front of the slope would be expected to produce a reduction in geotechnical stability relative to the still water condition (see refs. 2 and 11).

20.   Under still water conditions, the critical slip circle for the front face of the breakwater was a deep slip passing through the toe of the mound. In contrast, at the time of maximum run-up the critical slip was shallower and emerged through the armour layer higher up the face of the slope. Thus, whilst the armour layer may not be designed to act as a retaining skin, the enhanced shear strength which these layers often possess may well provide a valuable contribution to the geotechnical stability of the breakwater.

21.   From the above it may be postulated that one mechanism for overall failure of the front face of a breakwater could be where failure:

(i)   commences with a shallow geotechnical slip of the material at the top of the breakwater under the low factor of safety arising at maximum run-up, and

(ii) progresses to a hydraulic failure as the destabilised material is pulled out of, and down, the front face of the structure.

PRACTICAL CONSEQUENCES

22. Geotechnical analysis of breakwater mounds has historically been neglected. Recent studies have indicated that for breakwaters in deep water (say 30–50 metres), this neglect can be serious. However, many successful breakwaters have been constructed in water depths of the order of 15–20 metres with the relatively steep face slope of 3 in 4. This slope is popular as it represents the angle to which the rockfill of the core naturally falls and therefore reduces construction costs. However, it has been difficult to show why this form of construction has not led to geotechnical failures. It now appears that one important aspect of the explanation may lie in the high shear strength of interlocking armour layers when 'prestressed' under their own weight at these steep slopes. Furthermore, a thick, free-draining armour layer also tends to increase the shear strength of the structure by increasing the effective stress on potential failure planes within the core (see ref. 11). In deep water, where breakwaters are much larger, this retaining effect of the armour skin is proportionately reduced.

23. Use of interlocking or other types of armour unit whose stability is assessed using formulae such as those of Hudson (ref 12) or Van der Meer (refs 9 and 13) will continue to be popular for deep water breakwaters until such time as armour layers of hollow blocks (ref.14) can be more widely used. Whilst continuing to use such interlocking units, it is wise to ensure

(i)   that the surface profile presented to the uprushing wave is 'smooth' enough to prevent any one unit being subjected to excessively high hydraulic forces, and

(ii)   that the interlocking effect is maximised, making use of the prestressing effect of down-slope forces where permitted by geotechnical considerations.

24. Production of the 'smooth' armour layer suggested in paragraph 23 will be assisted if the following procedures are adopted:

(i)   Use of an underlayer stone weight no greater than about W/10, where W is the armour weight. Note, however, that the size of the underlayer stones must still be large enough to prevent their removal through the voids in the armouring and to ensure adequate shear resistance between the armour and the underlayer (ref 2).

(ii)   Careful site control (profiles, diving inspections, etc) to ensure that design leading dimensions and tolerances are achieved. The suggested tolerances for the outer surface of the underlayer are from minus zero to plus one half of the minimum stone size above the design profile, but without any sudden steps, hollows or bumps.

REFERENCES

1. BARENDS, F.B.J., VAN DER KOGEL, H. UIJTTEWAAL, F.J. and HAGENAAR, J. West Breakwater - Sines. Dynamic-geotechnical stability of breakwaters. Proc. Conf. Coastal Structures '83, Am. Soc. Civ. Engrs., 1983, 31-44.

2. HEDGES, T.S. The core and underlayers of a rubble-mound structure. Proc. Conf. on Breakwaters - design and construction, Instn. Civ. Engrs., 1983, 99-106.

3. HOOKWAY, D.W. and BRINSON, A.G. Construction of rubble mound breakwaters at Ras Lanuf, Libya. Developments in Breakwaters: Proc. Conf. Breakwaters '85, Instn. Civ. Engrs, 1985, 175-189.

4. HUDSON, R.Y. Stability of coastal structures. In R.Y. Hudson et al. Coastal Hydraulic Models, Special Report No. 5 for U.S. Army Coastal Engng. Res. Center, U.S. Govt. Printing Office, Washington, 1979, 314-452.

5. KOGAMI, Y. Researches on stability of rubble-mound breakwaters. Coastal Engng. in Japan, 1978, vol.21, 75-93.

6. U.S. ARMY COASTAL ENGINEERING RESEARCH CENTER. Shore Protection Manual, 4th edition. U.S. Govt. Printing Office, Washington, 1984.

7. BRUUN, P. and JOHANNESSON, P. Parameters affecting stability of rubblemounds. J. Waterways, Harbors and Coastal Engng. Div., Am. Soc. Civ. Engrs., 1976, vol. 102, No. WW2, 141-164.

8. DE ROUCK, J., WENS, F., VAN DAMME, L. and LEMMERS, J. Investigations into the merits of the Haro breakwater armour unit. Int. Conf. Coastal and Port Engng. in Developing Countries, Peking, 1987, vol.1, 1054-1068.

9. VAN DER MEER, J.W. Stability of tubes, tetrapods and accropode. Proc. Conf. Breakwaters '88, Instn. Civ. Engrs, 1988.

10. BRUUN, P. and GUNBAK, A.R. Stability of sloping structures in relation to $\xi = \tan \alpha / \sqrt{H/Lo}$ risk criteria in design. Coastal Engng., 1977, vol. 1, 287-322.

11. HEDGES, T.S. Geotechnics. Developments in breakwaters: Proc. Conf. Breakwaters '85, Instn. Civ. Engrs., 1985, 301-308.

12. HUDSON, R.Y. Laboratory investigation of rubble-mound breakwaters. J. Waterways and Harbors Div., Am. Soc. Civ. Engrs., 1959, vol. 85, No WW3, 93-121.

13. VAN DER MEER, J.W. Stability of breakwater armour layers - design formulae. Coastal Engng., 1987, vol. 11, 219-239.

14. WILKINSON, A.R. and ALLSOP, N.W.H. Hollow block breakwater armour units. Proc. Conf. Coastal Structures '83, Am. Soc. Civ. Engrs., 1983, 208-221.

6. Stability of cubes, tetrapods and accropode

J. W. VAN DER MEER, Delft Hydraulics Laboratory

SYNOPSIS. Results of an extensive research program on stability of rubble mound revetments and breakwaters were presented in recent years. In fact new stability formulae were introduced for an armour layer consisting of rock. In addition to this research Delft Hydraulics has performed basic model tests on breakwaters armoured with Cubes, Tetrapods and Accropode(R). The stability of these artificial units under random wave attack is the subject for the present paper. Finally a comparison of stability between rock and the mentioned artificial units is made.

ROCK STRUCTURES
Background
1. New practical design formulae have been developed which describe the stability of rubble mound revetments and breakwaters consisting of rock under random wave attack. The formulae were based upon a series of more than two hundred and fifty model tests. The work of Thompson and Shuttler (ref. 1) were used as a starting point. First results were published at the Breakwaters '85 Conference (ref. 2) and final results were published in ref. 3. The application of the formulae in a deterministic and probabilistic design were given in ref. 4.

Formulae for rock
2. The final formulae established for rock structures will be summarized first as it gives the basis for the investigation on stability of artificial units. The stability formulae derived are (ref. 3):

$$H_s/\Delta D_{n50} * \sqrt{\xi_z} = 6.2 \ P^{0.18} \ (S/\sqrt{N})^{0.2} \tag{1}$$

for plunging waves, and

$$H_s/\Delta D_{n50} = 1.0 \ P^{-0.13} \ (S/\sqrt{N})^{0.2} \ \sqrt{\cot\alpha} \ \xi_z^P \tag{2}$$

for surging waves

where:
H_s = significant wave height at toe of structure   (m)

ξ_z = surf similarity parameter, $\xi_z = \tan\alpha/\sqrt{s_z}$ (-)
s_z = wave steepness = $2\pi H_s/gT_z^2$ (-)
T_z = zero up-crossing wave period   (s)
α = slope angle   (degrees)
Δ = relative mass density of stone, $\Delta = \rho_a/\rho - 1$ (-)
ρ_a = mass density of stone or unit (kg/m^3)
ρ = mass density of water (kg/m^3)
D_{n50} = nominal diameter of stone, $D_{n50} = (W_{50}/\rho_a)^{1/3}$ (m)
W_{50} = 50% value of mass distribution curve   (kg)
P = permeability coefficient of the structure   (-)
S = damage level, $S = A/D_{n50}^2$ (-)
A = erosion area in a cross-section (m^2)
N = number of waves (storm duration)   (-)

3. The influence of dimensionless wave height, wave period and damage level on stability, computed with equations (1) and (2), are shown in Fig. 1 for a breakwater with $\cot\alpha = 1.5$, $P = 0.5$ (permeable structure) and $N = 3000$. The curves on the left side of Fig. 1 are given by equation (1) and on the right side by equation (2). Collapsing waves are present at the transition from plunging to surging waves.

4. Curves are shown for two damage levels, $S = 2$ for "start of damage" and $S = 8$ for "failure" (filter layer visible).

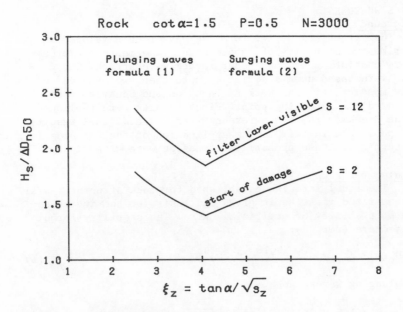

Fig. 1 Stability of rock slopes

Dimensionless governing variables

5. For armour layers consisting of rock the following conclusions were derived (ref. 3):

72

- Stability was determined in a dimensionless form, using:
  the significant wave height parameter: $H_s/\Delta D_{n50}$
  the surf similarity parameter: ξ_z
  the slope angle: $\cot\alpha$
  the damage as a function of the number of waves: $S/\sqrt{N}$
  the permeability of the structure: P
- Within the conditions tested the following parameters did
  not influence the stability:
  the grading of the armour
  the spectrum shape and groupiness of waves.

SET-UP OF RESEARCH

6. Tests on breakwaters with artificial armour units were
based on above mentioned conclusions. The research was limited to only one cross-section (slope angle and permeability)
for each armour unit. Therefore the slope angle, $\cot\alpha$, and
consequently the surf similarity parameter, ξ_z, will not be
present in a stability formula to be developed on the results
of the research. The same yields for the permeability coefficient, P.

7. Breakwaters with armour layers of interlocking units
are generally built with steep slopes in the order of 1:1.5.
Therefore this slope angle was chosen for tests on Cubes and
Tetrapods. Accropode(R) are generally built on a slope of
1:4/3, and this slope was used for tests on Accropode(R).
Cubes were chosen as these elements are bulky units which
have good resistance against impact forces. Tetrapods are
widely used all over the world and have a fair degree of interlocking. Accropode(R) were chosen as these units can be
regarded as the latest development, showing high interlocking, strong elements and a one layer system.

8. A uniform 1:30 foreshore was applied for all tests.
Waves were generated at a water depth of 0.90 m and the water
depth at the structure amounted to 0.40 m. Each complete tests
consisted of a pre-test sounding, a test of 1000 waves, an
inter-mediate sounding, a test of 2000 more waves, a final
sounding. Sometimes a test was extended with another 2000
waves. After each complete test the armour layer was removed
and rebuilt. Fig. 2 gives the cross-sections tested.

9. A tests series consisted generally of five tests with
the same wave period, but different significant wave heights.
Wave heights ranged from 0.10 to 0.25 m and the periods applied were: T_z = 1.4, 1.7, 2.2 and 2.9 s, covering a large
part of wave steepnesses found in nature. Generally 20 tests
were performed on each different armour unit which resulted
in a total of about 60 tests.

10. Damage to rock structures is usually measured by means
of a surface profiler. Damage, S, is then defined by the erosion area related to the nominal diameter (see Section 2).
Damage to artificial armour units is often measured as the
number of units displaced more than one diameter. Although
damage is often given as a percentage, this definition has a
lot of shortcomings. It is dependent on the slope angle and

the total number of units in the armour layer. Therefore, different investigations can hardly be compared.

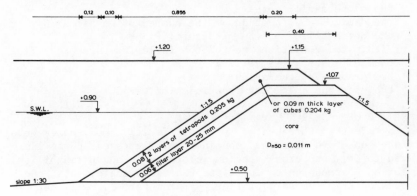

a) Cross-section for Cubes and Tetrapods

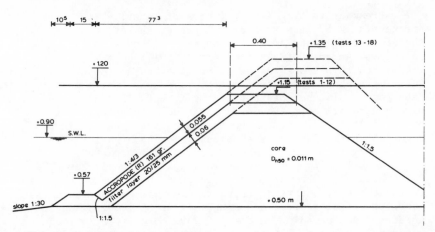

b) Cross-section for Accropode (R)

Fig. 2   Tested cross-sections

11.   Another definition is suggested for damage to artificial armour units. Damage here is defined as the relative damage, N_0, which is the actual number of displaced units related to a width (along the longitudinal axis of the breakwater) of one nominal diameter, D_n. The nominal diameter is defined by:

$$D_n = (W/\rho_a)^{1/3}, \text{ where: } W = \text{mass of armour unit.} \tag{3}$$

For Cubes D_n is the side of the cube, for Tetrapods $D_n = 0.65h$ where h is the height of the unit and for Accropode(R) $D_n = 0.7h$. The definition of the relative damage, N_0, is comparable with the definition of S, although S includes displace-

ment and settlement, but does not take into account the porosity of the armour layer. Generally S is about two times N_0.
12. As only one slope angle was investigated, the influence
of the wave period should not be given in formulae including
ξ_z, as this parameter includes both wave period (steepness)
and slope angle. The influence of wave period, therefore, will
be given by the wave steepness $s_z = gT_z^2/2\pi H_s$.

13. Governing variables
Sections 6-12 have reviewed the governing variables for stability of artificial armour units on the basis of the set of
variables for rock structures, given in Section 5. The final
governing variables are given by:
- the wave height parameter: $H_s/\Delta D_n$
- the wave steepness: s_z
- the relative damage: N_0
- the number of waves (storm duration): N

RESULTS
14. Damage curves were drawn for each period and each storm
duration. An example of such damage curves is shown in Fig. 3.
From these damage curves $H_s/\Delta D_n$ and ξ_z values were taken for
several damage levels, according to the procedure described
for rock slopes (ref. 2 and 3). These values were plotted in
so-called $H_s/\Delta D_n$-ξ_z plots, showing the influence of the wave
period, storm duration and damage level, as was already given
in Fig. 1 for rock structures. The $H_s/\Delta D_n$-ξ_z plots for Cubes,
Tetrapods and Accorpode(R) are shown in Figs. 4-6, for N =
3000 and for two damage levels: $N_0 = 0$ (start of damage) and
$N_0 = 1-2$ (severe damage, the actual number depends on the
unit considered). Results of the units will be described se-
perately.

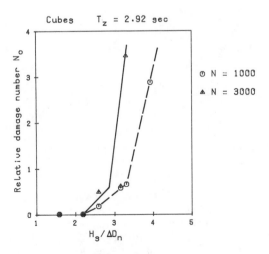

Fig. 3 Example of damage curves for Cubes

Stability of Cubes

15. Fig. 3 gives the damage curves for one wave period. From the analysis of this figure it follows that the influence of the storm duration (number of waves) is negligible for the no-damage criterion, $N_0 = 0$. This can also be expected: if 1000 waves do not displace any unit it can be expected that another 1000 or 2000 waves are not able to displace more units. When some damage is considered, the damage becomes a function of the storm duration.

16. Fig. 4 shows the results for Cubes. This figure shows a slight influence of the wave period. Longer wave periods (large ξ_z values) increase the stability which is according to rock slopes, Fig. 1. No transition is found between plunging and surging waves which is probably due to the steep slope considered.

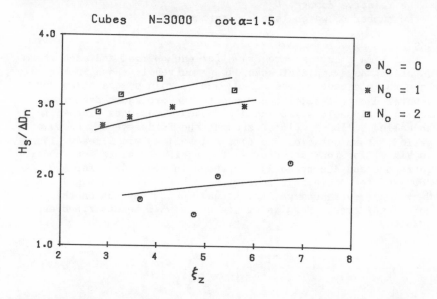

Fig. 4 Stability of Cubes

17. The final formula for stability of Cubes includes the relative damage level, N_0, the number of waves, N, and the wave steepness, s_z, and is given by:

$$H_s/\Delta D_n = (6.7\ N_0^{0.4}/N^{0.3} + 1.0)\ s_z^{-0.1} \tag{4}$$

Stability of Tetrapods

18. Figure 5 shows the $H_s/\Delta D_n - \xi_z$ plot for Tetrapods. The influence of wave period on stability is more pronounced for Tetrapods than for Cubes (Fig. 4). The same conclusion of the influence of storm duration was found, however.

19. A similar formula as (4) was found for Tetrapods:

$$H_s/\Delta D_n = (3.75\ N_0^{0.5}/N^{0.25} + 0.85)\ s_z^{-0.2} \tag{5}$$

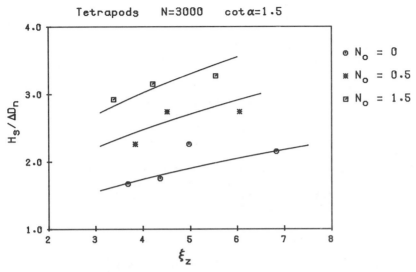

Fig. 5   Stability of Tetrapods

Stability of Accropode(R)

20.   Accropode(R) are placed in a one layer system. The Accropode(R) were placed according to the specifications given by SOGREAH and described in ref. 5. The cross-sections tested are shown in Fig. 2. Both partly overtopping (10–40%) and non-overtopping (< 10%) structures were tested.

21.   The results are shown in Fig. 6 for no damage ($N_0 = 0$) and severe damage ($N_0 > 0.5$). No influence of the storm duration was found. Furthermore, no influence of the wave period was found, as the curves in Fig. 6 are horizontal.

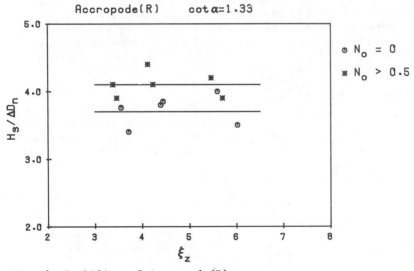

Fig. 6   Stability of Accropode(R)

22. From Fig. 6 two more and important conclusions can be drawn. The stability for start of damage is very high compared to Cubes and Tetrapods (Figs. 4 and 5). This is caused by settlement of the steep slope (cotα = 4/3) during the bedding in test with low waves. After settlement the armour layer acts as a "blanket" where each unit contacts several neighbours. Start of damage (N_o = 0) and severe damage or failure, given by N_o > 0.5 are very close, however. This means that the initial stability of Accropode(R) is very high, but that the structure fails in a progressive way. The results found for start of damage should not be used as design values, therefore.

23. As storm duration and wave period have no influence on the stability of Accropode(R) and as the "no damage" and "failure" criteria are very close, the stability can be described by two simple formulae:

Start of damage, N_o = 0: $\qquad H_s/\Delta D_n = 3.7 \qquad\qquad$ (6)

Failure, N_o > 0.5: $\qquad H_s/\Delta D_n = 4.1 \qquad\qquad$ (7)

RELIABILITY OF FORMULAE

24. In ref. 4 the formulae for rock were used in a probabilistic design, considering also the reliability of the formulae itself. This reliability (scatter) consists of a part due to random behaviour of a rubble mound structure and a part due to curve fitting. The coefficients 6.2 and 1.0 in equations 1 and 2 were treated as stochastic variables, having a normal distribution, an average equal to the values 6.2 and 1.0 respectively, and a standard deviation of 0.4 and 0.08 respectively.

25. A similar procedure can be followed for the formulae of artificial units. The coefficients 3.7 and 4.1 in equations 6 and 7 for Accropode(R) can be considered as stochastic variables. From analysis it followed that the standard deviation (assuming a normal distribution) amounted to σ = 0.2. The procedure for equations 4 and 5 is more complicated. Assume a relationship:

$$H_s/\Delta D_n = a * f (N_o, N, s_z) \qquad\qquad (8)$$

The function $f(N_o, N, s_z)$ is given in equations 4 and 5. The coefficient, a, can be regarded as a stochastic variable with an average of 1.0 and a standard deviation. From analysis it followed that this standard deviation is σ = 0.10 for both formulae on Cubes and Tetrapods.

COMPARISON OF STABILITY

26. Equations (1), (2) and (4)-(7) describe the stability of rock, Cubes, Tetrapods and Accropode(R). A comparison of stability is made in Fig. 7 were for all units curves are shown for two damage levels: "start of damage" (S = 2 for rock and N_o = 0 for artificial units) and "failure" (S = 8

for rock, $N_O = 2$ for Cubes, $N_O = 1.5$ for Tetrapods and $N_O >$ 0.5 for Accropode(R)). The curves are drawn for N = 3000 and are given as $H_s/\Delta D_n$ versus the wave steepness, s_z.

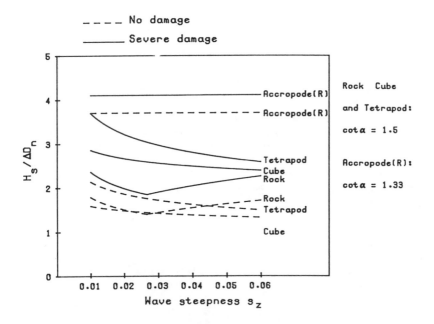

Fig. 7   Comparison of stability

27.   From Fig. 7 the following conclusions can be drawn:
- Start of damage for rock and Cubes is almost the same. This is partly due to a more stringent definition of "no damage" for Cubes ($N_O = 0$). The damage level S = 2 for rock means that a little displacement is allowed (according to Hudson's criterion of "no damage", however).
- The initial stability of Tetrapods is higher than for rock and Cubes and the initial stability of Accropode(R) is much higher.
- Failure of the slope is reached first for rock, than Cubes, Tetrapods and Accropode(R). The stability at failure (in terms of $H_s/\Delta D_n$ values) is closer for Tetrapods and Accropode(R) than at the initial damage stage.
28.   The complete investigation is described in refs. 6 and 7 for Cubes and Tetrapods and in ref. 8 for Accropode(R).

REFERENCES
1.   THOMPSON D.M. and SHUTTLER R.M. Riprap design for wind wave attack. A laboratory study in random waves. HRS, Wallingford, 1975, Report EX 707.
2.   VAN DER MEER J.W. Stability of rubble mound revetments and breakwaters under random wave attack. Developments in

Breakwaters, ICE, Proc. Breakwaters '85 Conference, 1985, London, Chap. 5.

3. VAN DER MEER J.W. Stability of breakwater armour layers - Design formulae. Coastal Eng., 11, 1987, pp. 219-239.

4. VAN DER MEER J.W. Deterministic and probabilistic design of breakwater armour layers. Proc. ASCE, Journal of WPC and OE, 1988, Vol. 114, No. 1.

5. VINCENT G.E. Rubble Mound Breakwaters - Twenty Applications of the ACCROPODE(R) Technique during its first six years of existance, 1987. SOGREAH Consulting Engineers.

6. DELFT HYDRAULICS. Stability of rubble mound breakwaters. Stability formulae for breakwaters armoured with Cubes, 1986. Report on basic research, S467 Volume VI. (Confidential).

7. DELFT HYDRAULICS. Stability of rubble mound breakwaters. Stability formula for breakwaters armoured with Tetrapods, 1987. Report on basic research, H462 Volume II. (Confidential).

8. DELFT HYDRAULICS. Stability of rubble mound breakwaters. Stability formulae for breakwaters armoured with Accropode(R), 1987. Report on basic research, H546.

Discussion on Papers 4-6

PAPER 4

MR N. W. H. ALLSOP, Hydraulics Research Limited, Wallingford
How does the Author couple external flows to internal flows
without a physical model? Can one continue to use the
continuum assumption?

Laboratory experiments appear to show very large hydraulic
gradients. Is field or laboratory information available on
the flow performance (and its variability) in real breakwater
cores under these very steep gradients?

DR S. S. L. HETTIARACHCHI, University of Moratuwa, Sri Lanka
Could the Author state the form of the expression adopted in
the model for the hydraulic gradient for flow through porous
media? It is presumed that both drag and inertia terms were
included.

How were the respective coefficients (drag and inertia) in
the expression for the hydraulic gradient determined? In
particular, were the drag coefficients determined under steady
or unsteady flow conditions?

When applying the mathematical model to a typical prototype
situation, how would the Author determine the above
coefficients for the prototype material? Is it necessary to
perform permeability tests at a large scale?

The model described in Paper 4 accounts for the movement of
the outcrop point. Was the analysis of the Author similar to
that given in reference 1? If not, could he explain very
briefly the method adopted?

DR J. W. VAN DER MEER, Delft Hydraulics
In Paper 4, Fig. 7(a) shows that a coarser core gives higher
parallel velocities than a dense core. Fig. 7(c) shows that
the armour layer with a coarse core underneath is more stable
than with a dense core. This looks contradictory.

As stated in paragraph 43 of the Paper, the cyclic storage
capacity is important, and indeed it is. The higher storage
capacity of a coarse core gives a great deal of inflow during

wave run-up. This decreases both the run-up and run-down forces in the armour layer, and it makes the armour layer with a coarse core more stable.

MR W. A. PRICE, Rendel, Palmer and Tritton, London
The Authors show that velocities peak at particular wave periods. Presumably this result was obtained for a particular slope action. Would not the value of the period corresponding to a peak velocity be expected to vary with slope angle?

DR BARENDS (Paper 4)
In reply to Mr Allsop, the approach uses measured dynamic wave pressures at the breakwater slope obtained in physical (scaled) models. From these pressures, boundary conditions are derived for the simulation of the internal processes using numerical models. Additional pressure meters installed in the core of a physical model are used to verify and/or calibrate the numerical model results. Little is known about the real behaviour under high and dynamic gradients. Few real breakwaters are well instrumented at present, and data are not available yet.

In reply to Dr Hettiarachchi, the flow law used in the simulation includes non-steadiness and porous turbulency, covering inertial effects. The corresponding virtual mass coefficient has been measured in a few scaled unsteady tests. Large variations have been observed. In the simulation, values between 1 and 2 are usually chosen. More investigation is required, preferably large-scale permeability unsteady tests. In reality, induced anisotropic effects may be important (two-dimensional turbulent flow). The approach for the movement of the outcrop point is shown in Fig. 4 of the Paper. The internal flow is limited by the maximum theoretical gravitational vertical flow. The simulation model HADEER is partly based on the studies of Nasser and Hannoura.

In reply to Dr van der Meer, higher velocities are to be expected in coarser cores. The drag force depends on the ratio velocity/permeability. The permeability in coarser cores usually increases more significantly than the velocity; hence, the stability of coarser structures can be greater irrespective of higher velocities. There is no contradiction in this.

The effect of storage capacity on armour stability for coarser cores, as described by Dr van der Meer, sounds logical and consistent.

In reply to Mr Price, the slope angle is one of many factors which affect the velocity patter in the core. For numerical analysis, measured pressures at the slope are required. These pressures will alter with the slope geometry, and so will the internal velocities will. In Paper 4, no other slope angles have been considered, which restrict the generality of the results presented.

Additional discussion on Paper 4 (from the Authors)

DR BARENDS and MR HÖLSCHER

Paper 4 outlines the importance of disconnection of the
internal water table at interfaces between layers of
different material, such as between armour and filter, or
filter and core. The presented analysis is partly based on
the application of a numerical model, the Hadeer code, which
is suited to simulate dynamic surface wave propagation in
porous structures. A facility which adequately models the
disconnection (and also the internal wave reflection) at
interfaces is not included (yet). Therefore, the analysis
has been performed for a homogeneous core body. The cover
(armour plus filter), which was relatively thin in the
considered situation (viz. Fig.4), could not be included for
the assessment of the disconneted water table.
The velocity visualized in Fig.6 and Fig.7 of paper 4
corresponds to the flow field in the core body. In this
respect it represents the filter flow inside the core at a
position near the slope. The parallel velocity mentioned in
Fig.7 does not reflect the parallel velocity in the cover.
In Fig.6 of paper 4 a so-called hodograph (ὸδοσ=path) is
used to present the velocity. Similar hodographs like Fig.6
for the other two wave conditions (wave1 and wave2) are
presented. Some irregular behaviour appears at the moment of

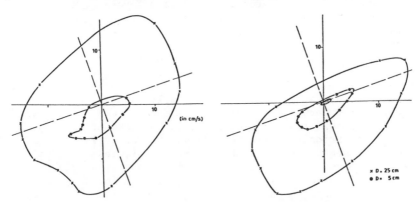

Figure 1. Hodographs for wave3 and wave1 (paper 4)

the wave through at the slope. The armour permeability
allowing air intrusion at wave recession provides a
reservoir effects for runup which strongly influences the
local behaviour, particularly for short waves [1]. The
assessment of this condition is difficult due to steep runup
wave pressures and the seepage face partly present.

1. Burcharth, H.F. Stability of armour units in
oscillatory flow, Proc. Coastal Structures '83, Washington
D.C., 1983

Reference
1. NASSER M. S. The outcrop point movement in non-Darcy flow. Proceedings of the Sixth Canadian Congress of Applied Mechanics, 1977.

PAPER 5

DR S. S. L. HETTIARACHCHI, University of Moratuwa, Sri Lanka
The tests performed by the Authors have indicated that the regular armour placing pattern is more stable than the random armour placing, in terms of both total damage and in percentage of units rocking.

On the basis of these test results and of results from similar tests performed by others (reference 1), would the Authors recommend the use of these units placed to a predetermined layout? In particular, because, even though the initial stability of the armour layer may be high, the structure may fail in a progressive way soon afterwards.

DR J. P. LATHAM, Queen Mary College, London
I would like to comment on the suggestion that exposure of individual blocks to extensive hydraulic forces could be prevented by producing a 'smooth' armour surface. It may be possible to produce a smoother surface by careful placement than by random placement, but the advantage achieved in increasing stability may be partially offset by the consequences of higher run-up. To assess the extent to which a 'smoothing' armour placement operation is advantageous with respect to hydraulic efficiency, it will be necessary to consider the detailed description of the different profiles that can result from careful placement. These detailed or 'high resolution' profile methods and the analysis of roughness are described in Poster Paper 2 (Latham and Poole) in this volume. The implicit assumption that a single layer of randomly placed armour units is generally rougher than a double layer was not borne out for dolosse units, the most hydraulically efficient of the armour units that we looked at in our static models.

DR J. W. VAN DER MEER, Delft Hydraulics
The upper part of Fig. 5 of Paper 5 shows that the transmitted wave height has a mean signal above still water level (set-up). Could the Authors explain this set-up behind the breakwater and the influence of it on the measured set-up inside of the breakwater.

DR F.B.J. BARENDS, Delft Geotechnics

1. The slope of a breakwater subjected to waves, swell or tides represents a geometric nonlinear effect to the generated porous flow field in the structure. The inflow surface along the slope at the moment of a high level is larger than the outflow surface at the moment of a low water level. Moreover, the average flowpath for inflow is shorter than the outflow path. Hence, during cyclic water level changes more water will enter the structure than will leave. Consequently, there will be an state, in which the outflow of the surplus of water is realised by an average internal setup of the waterlevel inside the structure.

2. It is possible to determine a simple practical formula which expresses the magnitude of this internal setup.

3. The penetration length λ of the cyclic water level at the slope into the porous structure can be approximated by:

$$\lambda \approx 0.5 \sqrt{(KDT/n)}$$

where K is the average permeability [m/s], D the still water depth [m], T the period of the cyclic loading [s], and n the porosity. This expression is based on the linear solution of the penetration of cyclic water level in a porous structure.

4. The slope steepness (angle γ) causes an extra storage of water volume every cycle in comparison to a linear situation with a vertical slope (dashed area, Figure 1). This extra volume can be represented by an average infiltration, expressed by the following formula:

$$I = I_0 \exp(-x/\lambda) \qquad \text{with} \qquad I_0 = 2cdf/T$$

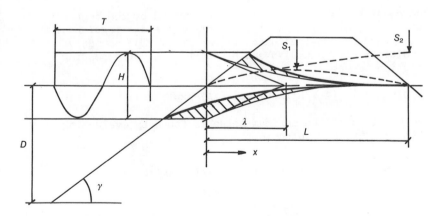

Fig. 1. Wave penetration into a slope

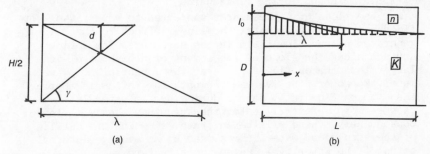

Fig. 2. Schematization

Here, d is related to the inflow/outflow surface, and f is related to the in/out flowpath. Geometric consideration yields (Figure 2a): $d = 0.5\ H/(1+(2\lambda/H)\tan\gamma)$. The flowpath factor can be approximated by: $f = (H/(2\tan\gamma) + \lambda)/\lambda$. Finally, the intensity becomes:

$$I_0 = cH^2/(2\lambda T\tan\gamma)$$

The factor c is related to effects by runup/rushdown, air intrusion and inhomogeneity to be determined in experiments.

5. The average internal setup can be assessed by solving a schematized problem (Figure 2b). The corresponding flow field is described by:

$$(h^2)_{,xx} = -2I/K$$

The fundamental solution reads:

$$h^2 = Ax + B - \xi D^2 \exp(-x/\lambda) \quad \text{with} \quad \xi \approx 0.1\ cH^2/(n\lambda D\tan\gamma)$$

The constants A and B are to be determined by the boundary conditions, of which two typical cases are distinguished.

6. Case I: constant harbour waterlevel. The boundary conditions are: $h(x=0)=D$ and $h(x=L)=D$. The corresponding solution becomes:

$$(h/D)^2 = 1 + \xi(1-x/L+(x/L)\exp(-L/\lambda)-\exp(-x/\lambda))$$

The position of the maximum setup is found by evaluating $(h^2)_{,x}=0$, which results in:

$$x_{max} = bL, \quad \text{with} \quad b = (\lambda/L)\ln((L/\lambda)/(1-\exp(-L/\lambda)))$$

and:

$$(h_{max}/D)^2 = 1 + \xi(1-b+b\exp(-L/\lambda)-\exp(-bL/\lambda)) = 1 + \xi F_I$$

7. Case II: impervious inside. The boundary conditions are: $h(x=0)=D$ and $h(x=L)_{,x}=0$. The corresponding solution becomes:

$$(h/D)^2 = 1 + \xi(1-(x/\lambda)\exp(-L/\lambda)-\exp(-x/\lambda))$$

The position of the maximum setup is found at the impervious boundary at $x=L$, which results in:

$$(h_{max}/D)^2 = 1 + \xi(1-\exp(-L/\lambda)(1+L/\lambda)) = 1 + \xi F_{II}$$

8. The above given expressions can be presented in a practical form. The internal maximum average setup becomes:

$$s/D = \sqrt{(1+\xi F)} - 1 \quad \text{with} \quad \xi \approx 0.1 \ cH^2/(n\lambda D\tan\gamma)$$

The function F is related to the actual case (open or closed inner side), and presented in Figure 3. The factor b denotes the relative position of the maximum setup in the open case.

9. The formula for the internal setup does not include the waterdepth effect, which in itself will cause a nonlinear effect, important when the cyclic amplitude of the free water table is large in comparison to the still waterdepth.

10. The given formula is valid for the internal setup which may occur after several cycles.

11. The actual watertable inside the core is determined by the internal setup plus the attenuated fluctuating water table. If the core is dense the fluctuation is low, but the internal setup is high.

12. As a practical example a particular model test mentioned in paper 5 (Simm and Hedges) is worked out. From the given grainsize distribution it may be assumed that $K \approx 0.05 m/s$, using an empirical relation between intrinsic permeability κ and D_{10}: $\kappa/D_{10}^2 \approx (6\pm2)_{10}-4$. The corresponding

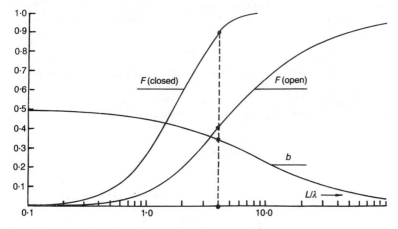

Fig. 3. Graphs for determination of internal setup

penetration length λ becomes with D=0.24, T=8.5/$\sqrt{50}$=1.2 (scaled) and n=0.3: λ=0.5$\sqrt{(KDT/n)}$≈0.10. The situation corresponds to a case I. With L≈0.40 we obtain from Fig.3: L/λ=4, b≈0.35 F≈0.41. With tanγ=0.75, c=1.25, and H=4.9/50=0.098 (scaled): ξ=0.1H^2/(nλDtanγ)≈0.22. The maximum setup becomes: s≈D($\sqrt{(1+\xi F)}$-1)=0.0108.

13. The value reported (their Fig. 6) is about 0.012. Probably, the filterlayer overlaying the core causes some extra infiltration (probably c=1.35 is better).

14. If the model breakwater has a backfill the evaluation becomes with F≈0.9 (Figure 3): s≈D($\sqrt{(1+\xi F)}$-1)=0.023, significantly larger than with an open lee side.

15. A breakwater with a backfill exposed to waves on a slope may develop an internal setup eventually completely saturating the core. No damping or attenuation of wave pressures in the core is then possible, and the backfill structure will be subjected to strong wave forces. The stability of the innerside of a breakwater with a backfill needs special attention.

MR SIMM and MR HEDGES (Paper 5)
By laying model tetrapods in the regular pattern shown in Fig. 4(b) of Paper 5, it was possible to improve their stability over that obtained by random placement. The Authors recommend the use of this pattern, where appropriate, for the reasons given in the Paper. However, it would be wrong to assume that all regular placing patterns improve armour stability. Carver and Davidson's tests on dolos-armoured breakwaters, referred to by Dr Hettiarachchi, showed that of the three patterns which they selected for testing, only one proved to be more stable than random placement. Stability was also shown to be strongly influenced by the armour packing density.

While recommending the regular pattern shown in the Paper for use with tetrapods, it is clear that other units, developed since the tetrapod, may now offer more attractive armouring solutions in appropriate conditions. Examples include the following

(a) single layer random systems, such as that incorporating the accropode (see paragraph 14 of Paper 5) in which the orientation of the individual units is not critical

(b) single layer regular voided armour units, such as the Cob, Shed and Diode, which have proved their worth in situations where the breakwater toe is exposed at low tide, allowing ready construction of a suitable toe beam.

On the matter of wave run-up, to which Dr Latham refers, this was increased when the regular placing pattern was used, as noted in paragraph 12 of the Paper. The Authors do not believe that it is sensible to increase the slope roughness by exposing some of the armour units to much higher than average hydraulic forces, which is what tends to happen with random placement. If run-up is a problem then consideration should be given to alternative means of controlling it, such as employing larger armour units or changing the profile of the breakwater face. However, the problem of increased run-up should not be exaggerated and, in this respect, the following points should be noted.

(i) For the cases tested, the increases in the run-up/ wave height ratio for the regular pattern, when compared with the random pattern, was of the order of only 25%.

(ii) With the random tetrapod pattern, experience shows that ridges and gullies tend to form during placement. The gullies collect more than average amounts of water, and run-up can 'shoot' out of the top of them, considerably increasing overtopping rates. With the regular placing pattern, the ridge and gully phenomenon is considerably reduced. Hence, the apparently greater run-up with the regular tetrapod pattern may not be reflected in an increase in the overtopping rate and, indeed, this rate may even be reduced.

Although not shown in Fig. 2 of Paper 5, the test flume was constructed within a tank. The water was free to flow between the flume and the surrounding tank through connections behind the permeable spending beach and beneath the wave paddle. Therefore, in the absence of waves, the water levels in the flume and the tank were identical. However, as Dr van der Meer points out, under the action of waves there was a small set-up in the mean-water level behind the model breakwater. This was associated with the mass transport of water through the structure induced by wave action on its seaward slope. The set-up would have been sufficient to drive a return flow from behind the breakwater, by way of the surrounding tank, to the seaward side of the structure. A similar flow would be induced in the prototype, with the water returning seaward past the end of the breakwater. It is not known whether the model set-up was in scale with that experienced on site, but, in any case, set-up on site could be expected to vary around the structure.

Clearly, the set-up within the rubble mound is affected to some degree by the water level behind it, because the difference between these two levels will influence the drainage of water to the lee side of the structure. However, we did not measure this influence. Dr Barends, who provides a means of estimating the set-up within a rubble mound, illustrates the significant difference between cases when water is allowed to drain through the lee side of the structure and when drainage is prevented.

Reference

1. CARVER R. D. and DAVIDSON D. Dolos-armoured breakwaters. Proceedings of the 16th Conference on Coastal Engineering, 1978, chapter 136.

PAPER 6

PROFESSOR M. A. LOSADA, University of Cantabria, Santander
I agree with Dr van der Meer that the influence of Iribarren's number on the stability of cubes is small. However, this is true for Iribarren's number higher than a certain value, which can be clearly seen in Fig. 1 (taken from reference 1). In the same Paper, the third conclusion was that porosity and roughness dampen the influence of Iribarren's number on stability, and thus the influence of the wave period on stability. For almost impermeable slopes, the influence of Iribarren's number on stability is very clear, showing a remarkable peak, as can be seen in Fig. 2 (taken from reference 2).

Could the Author explain how he drew the curves in the Figures in his Paper. It is important to know the · criterion used in order to fix the confidence levels. My co-author and I (reference 2) used the best fit curves fitted by linear regression. The confidence levels were established assuming a gaussian distribution of the fit variable (see Fig. 3).

On the basis of his laboratory experience, could the Author explain what happens with an accropode slope when damage is initiated. The cube slope develops an S profile and achieves an additional stability, but what happens with the accropode?

MR J. D. METTAM, Bertlin and Partners, Reigate
In the Author's new formulae, the influence of breakwater slope is taken into account only within the surf similarity parameter ξ_z.

I would be interested to know whether the Author has attempted to fit a formula to his test results which includes Hudson's $\cot\alpha$, or perhaps even better Iribarren's slope functions which include an angle of repose as well as a slope of face. I would have more confidence in a formula which included these factors.

MR J. READ, G. Maunsell and Partners, London
Dr van der Meer's Paper has enlarged our knowledge of armour behaviour and takes us nearer to having a true design formula for precast armour units.

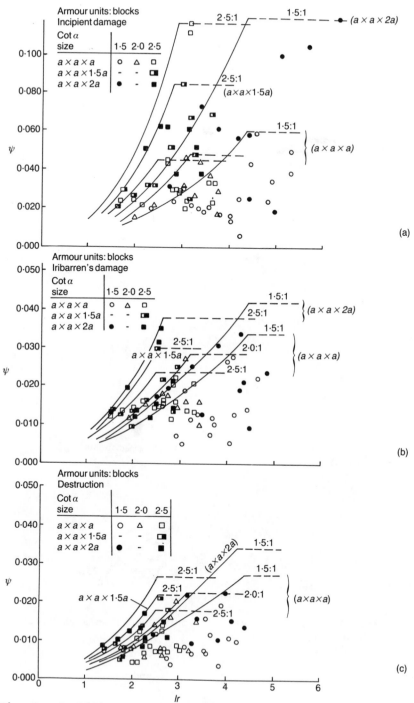

Fig. 1. Stability function for blocks: (a) incipient damage;
(b) Iribarren's damage; (c) destruction

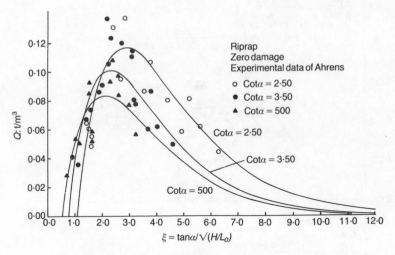

Fig. 2. The parameter Q after Ahrens and McCartney's tests (rip-rap)

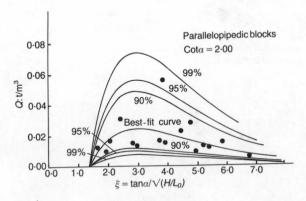

Fig. 3. Control curves for confidence levels of 90%, 95% and 99%

The work so far has necessarily been limited in extent, and it may be of interest to see how data from flume tests at Hydraulics Research (reference 3) fit the proposed formulae. The present formulae relate only to one slope for each of the three armour units studied and none of the HRS data was measured for the same slopes. The abscissa of Fig. 4 of the Paper is, however, the surf similarity factor, so that it was possible to deduce an equation of the same form as equation (4) but incorporating the surf similarity factor instead of the wave steepness. It is as follows.

$$\frac{H_s}{\Delta D_{n50}} = \left(9.1 \frac{N_o^{0.4}}{N^{0.3}} + 1.36\right) \xi_z^{0.2}$$

It was then possible to calculate the cumulative damage for each test, each stage of the tests being plotted against the observed number of extractions. Nine tests are plotted, one with a slope of 1:2.0 and the rest at 1:2.5. The permeability was the same for all tests and was assessed at 0.4 on the scale suggested in the earlier paper for rock. The result is shown in Fig.4.

A little algebraic rearrangement of Van der Meer's original equation for rock (reference 4) leads to the following expression for the damage coefficient for plunging waves. (A similar expression can be written for surging waves.)

$$S = \sqrt{N} \left(\frac{H_s \sqrt{\xi_z}}{6.2 p^{0.18} \Delta D_{n50}}\right)^5$$

It follows from this that, for any particular set of conditions, the damage is inversely proportional to the fifth power of the equivalent cube size. If the fifth roots of the observed and calculated damages are compared they will be found to bear a constant ratio to each other, as Figs 5 and 6 illustrate. Fig.5 shows the results for the same nine tests plotted in Fig.4, and Fig.6 shows a similar plot for Stabits. (The Stabits were tested at a slope of 1:1.5.)

This suggests that in each of these two cases an effective equivalent cube size could be substituted for the actual value, and the formula would then predict the correct degree of damage. For cubes, an effective size 1.77 times greater than the actual is indicated by the plot, whereas for Stabits the ratio would be 2.43.

As the observed damage was measured by the number of extractions rather than the area of erosion, it is to be expected that the actual damage coefficients recommended for use in design will differ from those suggested for rock. For

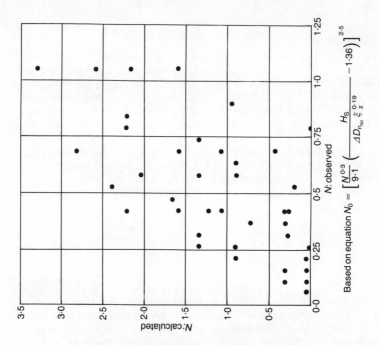

Based on equation $N_0 = \left[\dfrac{N^{0.3}}{9.1}\left(\dfrac{H_S}{\Delta D_{n_{50}}\,\xi_z^{0.19}}-1.36\right)\right]^{2.5}$

Fig. 4. Comparison of calculated and observed extractions for cubes

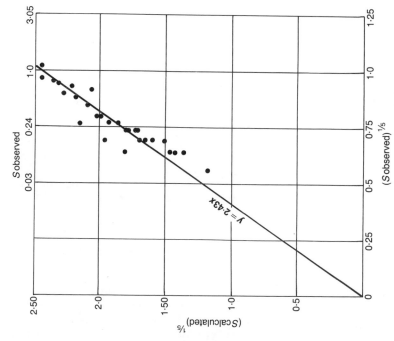

Fig. 6. Comparison of calculated and observed extractions for Stabits

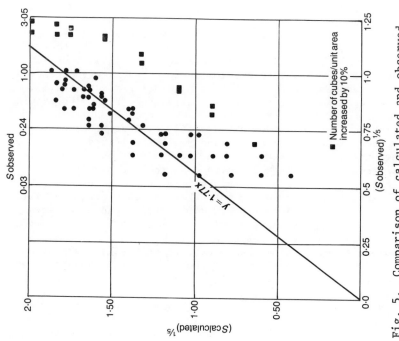

Fig. 5. Comparison of calculated and observed damage for cubes

both these cases, a value of 'S' of 1 represented about 3.9%
extractions, so that a value of 0.75 might be used for design.

A great deal more needs to be done before this method can be
recommended for general use. Three tests were carried out
with cubes on a slope of 1:2. One of these tests has been
included with the other tests done on a slope of 1:2.5, as it
appears to fit with that population, but the other two had
approximately 10% more armour units than the first and showed
a much higher amount of damage. They are shown separately in
Fig.5. Therefore, tests to establish the influence of the
placing density of units are needed, as well as tests covering
a range of slopes and permeabilities. Tests also need to be
extended to examine the effects of surging waves.

Nevertheless, this method of analysis appears promising and
it would be interesting to know if the Author has considered
such an approach and if he has any data which might confirm
its appropriateness.

MR J. D. SIMM, Rendel, Palmer and Tritton
Would Dr van der Meer like to comment on the level of the
factor of safety used for accropode design. This seems to be
very conservative, even when one takes account of the rapid
breakdown of the armour layer once damage is initiated. Could
a somewhat lower factor of safety be used?

MR L. STADLER, Ports Authority, Israel
I would like to present test results on tetrapods and Antifer
cubes from 2D and 3D models. Firstly, 2D 16 m^3 tetrapods,
built to a 1:1.67 slope, showing extrations and extractions
plus rocking damage figures, plotted for both percentage
against wave height and N_o against $H_s/\Delta D_n$ (Fig.7). On Van der
Meer's Fig.5, the above results are plotted, for the relevant
surf parameter $\xi = 3.37$ for both damage definitions (Fig.8).

Firstly, I am of the opinion that damage definition should
include rocking, which has been shown to cause breakage.
Secondly, Figs 9-11 show extensive model test results of a
proposed extension to the Ashdod Breakwater (Israel). Fig.9
shows the 1150 m breakwater, including location of the five
test sections. Damage is defined as rocking over two-thirds
of the time, and displacements of more than one cube height
are extracted. Fig.10 presents the results plotted for all
five stations, as N_o against $H_s/\Delta D_n$; considerable variations
are shown. Fig.11 shows results for Station 5, where a wide
band of damage levels was obtained.

Summarising, I think that a more comprehensive definition of
damage should be used and this should certainly not be based
on extractions (out of profile units) only, which lead to low,
unrealistic damage figures.

I should also like the Author's comments on the following:
the influence of wave attack angle on N_o values; the influence
of different toe (berm) arrangements; and on whether any

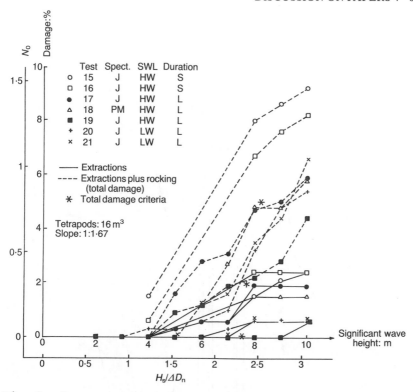

Fig. 7. Damage curves for 2-D model

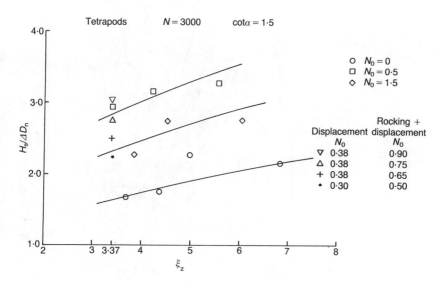

Fig. 8. Stability of tetrapods

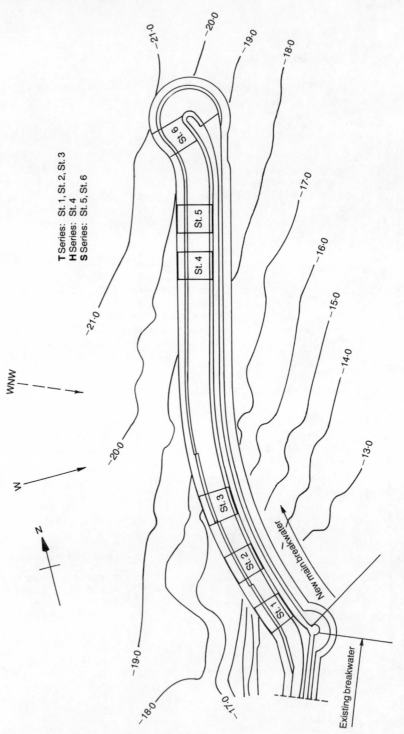

T Series: St. 1, St. 2, St. 3
H Series: St. 4
S Series: St. 5, St. 6

Fig. 9. Test section location

repeat tests were conducted.

DR VAN DER MEER (Paper 6)

In reply to Professor Losada, the ψ number in Fig. 1 is the
same as $(H/\Delta D_n)^{-3}$. Fig. 1 shows a large increase in stability
for increasing ξ number, but only up to a certain value where
the curve becomes horizontal, indicating that for surging
waves no influence on stability of the wave period is present.
The reason for a horizontal line is not very clear, regarding
the scatter of the data. The trend is, however, more or less
similar, as found by the Author for rock (see Fig. 1 in the
Paper), because plunging waves stability increases with
increasing ξ number until the region of collapsing waves is
reached, where a minimum of stability is found. Increasing
the ξ number again, the stability increases with longer
surging waves.

That this peak of minimum stability was not found in the
present tests might be due to the fact that the slope was so
steep that only collapsing waves were present for the shortest
wave period and that plunging waves were never present. The
transition from plunging to surging waves for random waves
(Losada's tests were all performed with monochromatic waves)
and for a 1:1.5 slope will be found somewhere between ξ = 3-4.
If more gentle slopes had been tested (with consequently
smaller ξ values), I am convinced that a minimum in stability
would have been found.

With regard to porosity (and permeability of the core) the
Author's results were different from those of Professor Losada
(for rock) (see also reference 4 of the Paper and reference
5). With increasing permeability of the core, more uprushing
water can penetrate into the core, which causes a less severe
downrush and hence a higher stability. The curves for an
impermeable core are almost horizontal (for surging) waves;
whereas for a permeable core, the stability increases
considerably with increasing ξ values (see also Fig. 1 of the
Paper).

Stability curves were derived by using regression analysis.
The analysis procedure was more or less similar to that for
rock, described in reference 5. First the relationship
between $H_s / \Delta D_n$ and the wave steepness s was established;
then the threshold limit for N_o =0 was obtained. Whereupon,
the relationship between $H_s/\Delta D_n$ and the damage level N_o was
established, and finally the relationship between the damage
after 1000 and 3000 waves. The reliability of the formulae
was described in the Paper (sections 24 and 25). With the
described standard deviations (assuming a normal
distribution), it is easy to draw a damage curve with
confidence levels: Fig. 12 gives an example.

The greatest damage measured on a cube slope was S = 12
(with N_o in the order of 6, the number of displaced units was
more than 100); the profile of this test is shown in Fig. 13.
A similar test for accropode is shown in Fig. 14; the damage

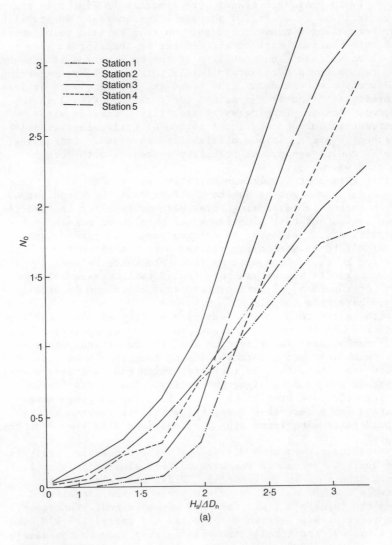

Fig. 10 (above and facing). Mean total damage curves
for main breakwater trunk: (a) West; (b) West North West

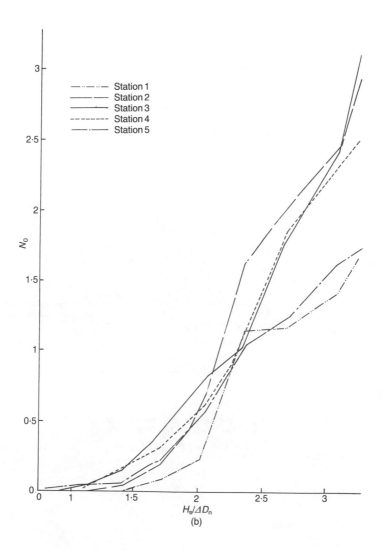

Fig. 10 – continued

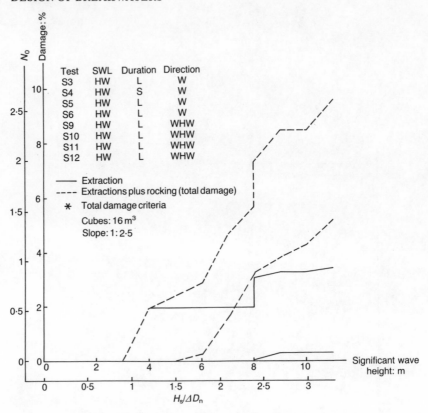

Fig. 11. Damage curves for 3-D model

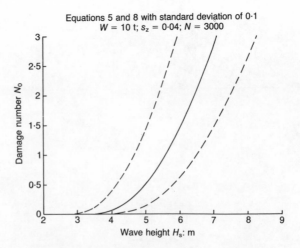

Fig. 12. Stability curve for tetrapods
with 90% confidence levels

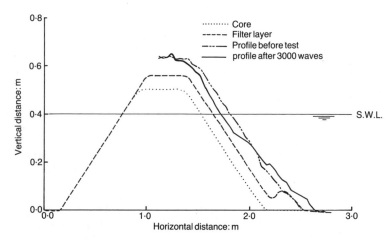

Fig. 13.   Profile of test on cubes; N_o = 6 and S = 12

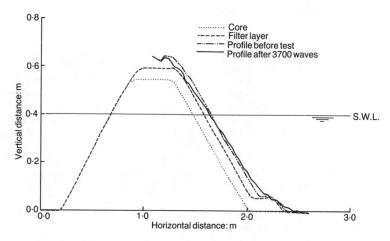

Fig. 14.   Profile of test on accropode; N_o = 2.5 and S = 12

was N_o = 2.5 and S = 6.2. The damaged profiles are very similar. It should be noted that these damage levels are beyond any conventional design level and probably between what Professor Losada calls 'Iribarren's damage' and failure.

I agree completely with Mr Mettam that a formula with the $\cot\alpha$ included would be a better formula. The fact is that only one slope angle was tested for each unit. Therefore, it was simply impossible to include the slope angle in he formula with only the present test results. Although some figures in the Paper were given with the ξ_z number on the horizontal axis, the results were always presented for one slope angle. Therefore, only the wave steepness was used in the formula.

In my opinion, a more gentle slope with bulky units (such as cubes) will be more stable than the 1:1.5 slope tested. Maybe a factor $\cot\alpha^{1/3}$ can be used as a first guess in that case. For units with more interlocking (such as tetrapods, but especially accropode), a steep slope gives more initial settlement and, therefore, more interlocking than on a gentle slope. The difference between a 1:1.5 and a 1:4/3 or a 1:2 slope will be small for these units, smaller than a factor $\cot\alpha^{1/3}$ would suggest.

The equation proposed by Mr Read, rewriting equation (4) with the surf similarity parameter instead of the wave steepness holds only for the slope angle tested. The use of this proposed equation for other slope angles leads to erroneous results, Equations (4) and (5) give a lower stability for a smaller ξ_z value (see Figs 4 and 5 of the Paper). If the surf similarity parameter is used to introduce the effect of the slope angle, this results in smaller ξ_z values for a gentler slope (for the same wave steepness) and hence in a lower stability for a gentler slope; and it is, of course, not true that a gentler slope will be (much) less stable that a steeper slope. Therefore, the scatter in Fig. 4 is related to a wrong interpretation and wrong transformation of the original formulae.

On the other hand, data of other slope angles than the tested 1:1.5 can give valuable information on the influence of the slope angle on stability (see Mr Mettam's question). Re-analysis of Mr Read's data in the same way as the presented results might lead to correct interpretation of the influence of the slope angle.

Mr Read rewrites the formula for rock and then compares the equation with results on cubes and stabits. Probably it is acceptable to do this for the Hudson formula where only a coefficient and a $\cot\alpha$ factor play a role besides the dimensionless wave height $H_s / \Delta D_n$. However, it is not acceptable for the new formulae on rock: too many parameters are involved which do not have the same influence on stability for different unit shapes. This means that the proposed method by Mr Read cannot directly be recommended for general use.

The relationship between counted displaced units N_o and the profiled damage S was established in the tests (both methods were applied, but the results with N_o were presented). In these tests, the relationship was found to be $S = 1.8N_o + 0.4$ for cubes, and $S = 2N_o + 1$ for tetrapods and accropode. A design value of $S = 0.75$, suggested by Mr Read, is just between the values found with the above equation, using $N_o = 0$.

In reply to Mr Simm, the safety factor for accropode was determined using the K_D value suggested by Sogreah (reference 5 in the Paper). The safety factor is 1.5 on the required nominal diameter, and therefore a factor 3.4 on the required weight. However, even with this safety factor, the unit weight is smaller than for tetrapods, and the saving in concrete is already 40% as only a single amour layer is applied. Probably a smaller safety factor is possible, but the saving in costs will be small; and after the breakage of large units on the well-known breakwaters it is perhaps preferable to construct an armour layer with a unit which has a substantial safety factor.

Mr Stadler shows a great deal of data in a compressed form. It is almost impossible to re-analyse the data without having the original test reports. Fig. 7 shows seven damage curves with, in total, considerable variation. Without further explanation the figure suggests that the tests were comparable and the scatter is due only to the random behaviour of the units. Probably two differences in analysis procedure can be found with that described in the Paper. The first difference is that Mr Stadler gives cumulative damage: the tests were started with a low wave height and the wave height was increased in steps. The test results described in the Paper (see Fig. 3) were independent. The slope was rebuilt after each attack with 3000 waves. The effect of cumulative damage was described in the reply to Mr Ackers in the discussion at the Breakwaters '85 Conference (reference 2 of the Paper, pp 195-197). A second (possible) difference is the wave height which has been used. Fig. 7 shows that a lower water level gives less damage, which implies that the offshore wave height is possibly used in the $H_s / \Delta D_n$. The wave height to be used should, however, always be the wave height in front of the structure.

The above considerations show that it is not easy to use test results of other investigations. It is possible, but it takes a lot of careful analysis.

I agree completely with Mr Stadler that not only are extractions important, but rocking (maybe even various categories of rocking) is also of importance. For a first design with formulae, rocking is less important. Large breakwaters should always be tested in a model, including the determination of rocking.

I have only basic experience with 3D testing. My impression is that, in most cases, perpendicular wave attack is worse.

A high and long berm certainly has an effect on stability. Most designs of the rehabilitated breakwaters have such a berm, which is more like an S-shape. Stability of a

105

breakwater can be increased considerably by such a berm or an S-shape.

Repeat tests were conducted for rock (reference 5). As the slope was rebuilt after each wave attack with one wave height, the data points in a damage plot already give a good idea of reliability and scatter (see, for instance, Fig. 3 of the Paper). Scatter and reliability were also discussed at the Breakwaters '85 Conference (reference 2 of the Paper, p. 199).

Designers are always interested in a good design formula. The reliability, however, should be proven by independent data. It is clear that Mr Read and Mr Stadler have this independent information. Also, Mr Jensen of the Danish Hydraulic Institute has information on a number of tests with cubes (personal communication). Furthermore, Delft Hydraulics has performed numerous investigations with armour units, and these tests are independent of the tests described in the Paper. It would be worthwhile to select a set of data of each of the above-mentioned investigations and to re-analyse the data in the same way as presented in the Paper. If the discussers were willing to co-operate, I would perform this analysis.

References
1. LOSADA M. A. et al. Stability of blocks as breakwater armour units. J. Structural Engineering, 1986, 112, No. 11, 2392–2401.

2. LOSADA M. A. and GIMENEZ-CURTO L. A. The joint effect of the wave height and period on the stability of rubble mound breakwaters using Iribarren's number. Coastal Engineering, 1979, 3, 77–96.

3. ALLSOP N. W. H. Concrete armour units for rubble mound breakwaters and seawalls: recent progress. Hydraulics Research, Wallingford, March 1988, Report SR 100.

4. VAN DER MEER J. W. Stability of breakwater armour layers – design formulae. Coastal Engineering, 1987, 11, No. 3, 219–239.

5. VAN DER MEER J. W. Rock slopes and gravel beaches under wave attack. Delft University of Technology, Doctoral thesis. Also, Delft Hydraulics Communication No. 396.

7.   Crumbles Harbour Village Marina, Eastbourne

P. LACEY and R. N. WOOD, Ove Arup & Partners, London

SYNOPSIS. The paper describes the various hydraulic
modelling procedures used to test harbour and breakwater
layouts for a marina at Eastbourne between the years 1973 to
1987. The difference in techniques available in the early
1970's which consisted mainly of physical models to the now
familiar mathematical computation modelling systems which are
used in conjunction or in their own right in the later 1980's
are described. The paper will discuss and compare, as a case
history, the original tests conducted by the British
Hovercraft Corporation Ltd, with the tests recently
undertaken by Hydraulics Research Ltd, Wallingford, both
being testing stations in the United Kingdom.

INTRODUCTION
1.   Since 1973 the idea of a Harbour for Eastbourne has
been in the Eastbourne Borough's thinking, a proposal
embraced in principle in the East Sussex County Structure
Plan from 1975 to date.
2.   Town growth with additional housing, a choice of local
jobs and leisure opportunities has been consistently
advocated over the same period. Keel boats and larger craft
have at present no place at Eastbourne though on the sea
front in-shore fishermen's boats are pulled up the shelving
beach.
3.   In 1975 there were ten places between Dover and Lands
End providing major marina facilities and a survey carried
out by the National Yacht Harbour Association indicated a
growth in marinas. Now there are at least twenty two places
indicating a major boom in leisure activities coupled with
large scale developments around water.
4.   The Crumbles at Eastbourne is considered a good
location for a marina as can be seen from Figure 1. The
existing site is shown in more detail in Figure 2.
5.   The marina constructed at Brighton, with a planned
capacity of 2,300 boats, catered for a great number of craft
unable to find a berth in the Solent in the late 1970's but
the eighty five miles of coastline between Newhaven and
Ramsgate has now attracted many marinas.

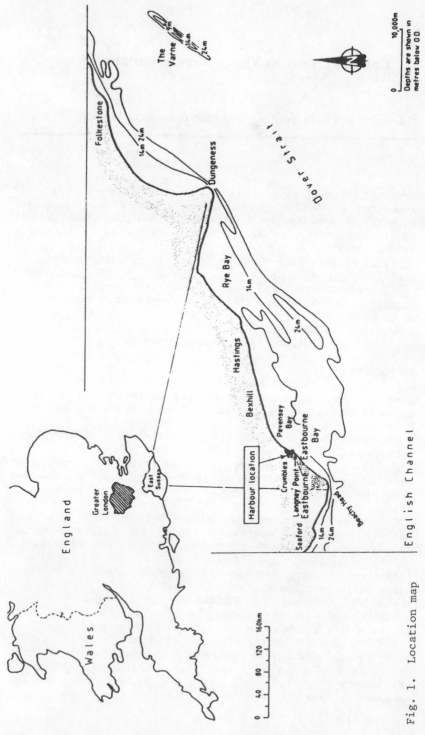

Fig. 1. Location map

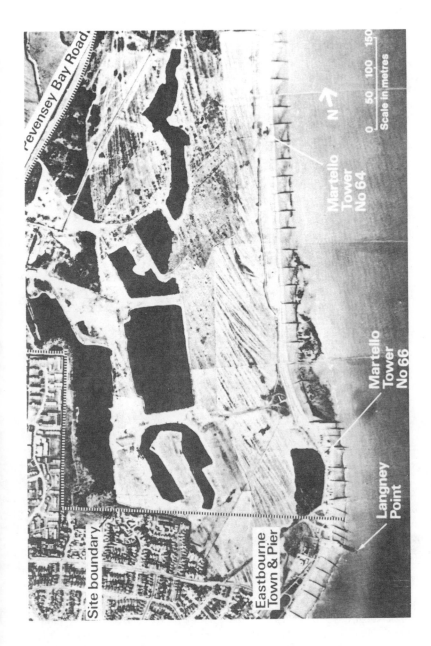

Fig. 2. Existing site

6. The Trustees of the Chatsworth Estates have been the champion of a marina over many years and in 1985 Tarmac/Enterprise Zone Development approached them with an offer to develop the site.

The 1975 Eastbourne Harbour Scheme.

7. The term "Harbour" included three component parts; the Outer Tidal Harbour; the Inner Tidal Harbour, and the Locked Harbour. These are shown in Figure 3. The tidal parts of the harbour were to secure safe quiet water as a transition between conditions in the open sea and the Locked Harbour which is reached by navigating via a lock.

8. The location of a harbour on this length of coast is generally favourable in that Beachy Head affords protection from the prevailing south-westerly seas. Because of this, and the refraction of waves approaching from various directions in shoaling water, most of the heavy seas that do reach the harbour approach from the south to south-west quarter.

The Outer Tidal Harbour

9. The following design criteria have dictated the form of this:

(a) To minimise the entry of wave energy, thus providing a protected passage for boats to the Inner Tidal Harbour and quiet water conditions at the locks.

(b) To secure entry and passage to the locks at all states of the tide while minimising the effect on vessels of adverse wind and sea combinations during critical manoeuvres.

(c) To minimise adverse effects on the coastal regime caused by the interruption of littoral drift, and the deposit of sediment at the harbour mouth.

The 1985 Crumbles Harbour Village, Eastbourne

10. The general layout of the present scheme shown on Figure 4 is very similar to that of 1975. The breakwater arms have been reduced in length and the outer harbour of refuge has disappeared.

11. The reduction in size of the breakwaters was made due to the need to economise on capital outlay. It is realised that although the design criteria for both the Schemes are still the same, a perfectly acceptable but smaller protection can be provided. The pedestrian access featured in the 1975 Scheme has also been removed for reasons of economy.

12. Local concern over environmental problems such as beach erosion, siltation, flooding and pollution has come to the very forefront since 1975. A scheme of this size has attracted local pressure groups who demanded answers to questions on the effect on the environment especially flooding and beach erosion.

THE PROBLEMS

13.   The problems posed by the scheme are not new to coastal works as practically every marina situated on the coast faces them.

14.   The cutting of an entrance channel through the existing beach profile and tidal harbour behind the existing sea defences means that wave action and water movement must be assessed and an engineering review of the necessary provisions against surge and inland flooding made.

15.   The breakwaters have been designed to give protection from storm waves attacking the harbour and must be tested to ensure that residual wave height within the harbour are reduced to acceptable limits for yacht berths and that the proposed construction is sturdy enough to survive such storms.

16.   The intrusion on the coastline regime caused by constructing breakwater arms needs special attention to detect, again by testing, the effect on littoral drift and accretion or erosion patterns. The movement of beach material and the effects of such movement on the entrance channel were considered of prime importance by the client and the Southern Water Authority as the existing nett drift is from south to north moving towards Dungeness.

HYDRAULIC MODELLING

17.   The modelling of this harbour has spanned a period of fourteen years encompassing two separate physical models and a mathematical model discussed later in this paper.

1975 Scheme

18.   The 1974 hydraulic modelling was undertaken at the British Hovercraft Experimental and Electronic Laboratories and consisted of tests to evaluate the hydraulic design and configuration. Many studies were carried out and are listed in Appendix A before the tests began. Wave rider buoy readings were taken between May 1973 and July 1974, to provide data to allow the offshore wave climate to be assessed.

19.   A model of the Tidal Harbour with adjoining beaches and seabed surfaces extending over an area of more than 250 hectares, was built at a scale of 1:60. Initial tests investigated the effects of the harbour works on the tidal regime and in particular sought to establish a form for the Outer Harbour which interrupted the transport of seabed material as little as possible. Later tests investigated the effects of stormwaves from the east, south-east and south using fixed tide levels and regular waves of different heights and periods. The layout was developed so that the wave energy entering the Inner Tidal Harbour was reduced to acceptable levels.

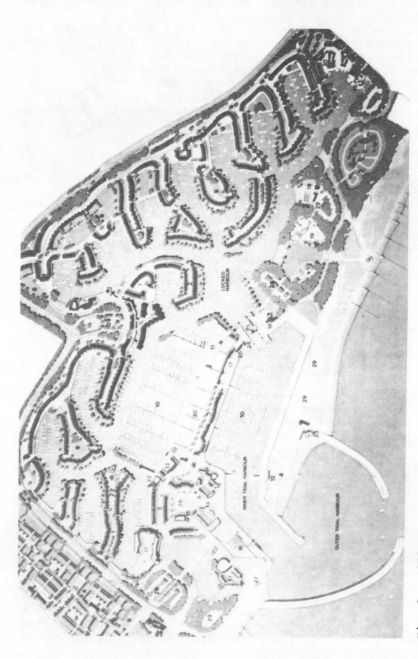

Fig. 3. 1975 scheme

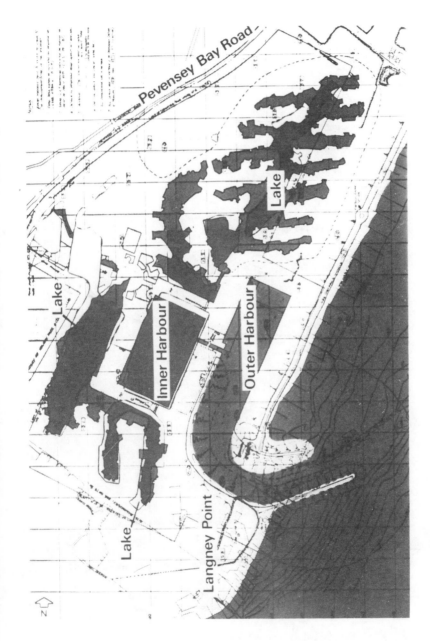

Fig. 4. 1985 scheme

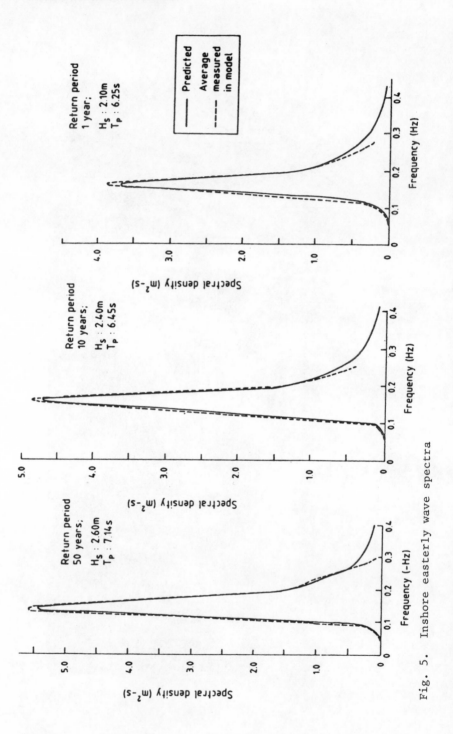

Fig. 5. Inshore easterly wave spectra

20. The harbour is enclosed by two unequal rubble mound breakwaters, curved in plan and extending beyond the -7.0 metres Ordnance Datum (mOD) seabed contour (3.3 metres below Chart Datum). To determine the detailed design of these models, of the trunk and head of the south-west breakwater, were constructed at 1:43.5 scale. The tests investigated in detail the performance of the breakwater under varying conditions of wave attack.

21. The harbour mouth faces east-north-east and the entry to the Inner Tidal Harbour is sited in the lee of the south-west breakwater, facing south-east and flanked by spending beaches.

22. The curved breakwaters and the form of the harbour mouth were designed to encourage the smooth flow of the tidal currents across the harbour mouth, minimising the deposition of sediment. The form of the harbour mouth, without an overlap of the breakwaters, leaves the harbour open to the entry of wave energy. The harbour mouth has therefore been aligned to face the relatively quiet east-north-east sector. Thus the Outer Tidal Harbour will provide protected anchorage from all wave directions except on the rare occasions when there are heavy seas from the south-east and east. The breakwaters were parapeted for functional reasons and, in normal weather conditions, the public will be able to walk along them to the harbour mouth.

23. The entrance to the Inner Tidal Harbour has been designed to achieve the maximum reduction of wave energy by careful siting and the introduction of a short inner breakwater.

1985 Scheme

24. As can be seen from Figure 4 the appearance of the schemes is very similar. The 1985 Scheme however, excluded the harbour of refuge with the breakwater arms pulled back close to the shore. They do not now extend past the -6.0 mOD sea bed contour.

25. This configuration has been selected to protect the Outer Tidal Harbour from severe storm wave conditions providing calm waters for yacht mooring and to minimise the effect of the breakwater arms on the littoral drift.

26. In most other respects, the hydraulic modelling test requirements were the same. The methods of attaining the requirements however were somewhat different. Computational models were used to predict extreme offshore wave conditions, refraction at the harbour entrance and wave conditions inside the harbour. The shingle movement was also assessed using historical data and a mathematical model.

27. The physical model that was constructed represented a sea bed area of 2900 m x 1800 m and was to a scale of 1 to 100 and used a random sea wave generator to test the layout and outer tidal basin.

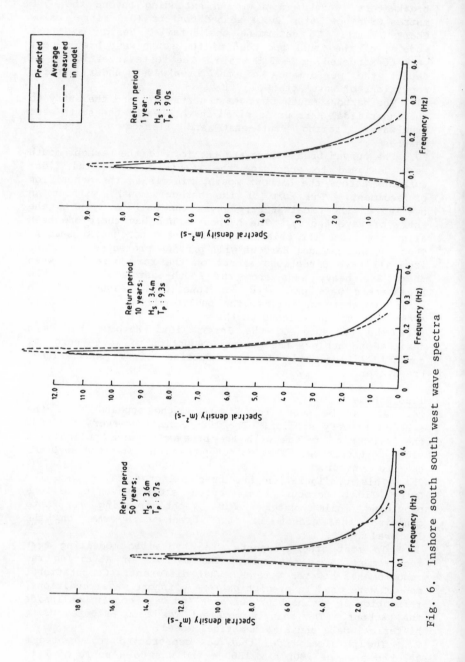

Fig. 6. Inshore south south west wave spectra

Fig. 7. Proposed harbour layout – 1975

117

28. The effect of the breakwater arms on the long shore drift under wave attack from the two worst directions, easterly and south-south-westerly was examined by the introduction of crushed anthracite at certain groyne positions and at the Langney Point outfall.

29. The original wave inputs used for the 1973 Scheme, were used by Hydraulics Research Ltd, Wallingford, where all the model testing for the 1985 Scheme was carried out. The wind fetch lengths and the predicted extreme offshore significant wave heights are given as Appendix B and C as produced by the studies.

COMPUTATIONAL MODELLING

30. As has been stated previously, the main difference in the model testing of the two schemes was the added input provided by the new computational techniques.

31. Before the Trustees of the Chatsworth Estates and Tarmac EZD would commit funds to another series of physical models they wished the consulting engineer to give professional advice on the likely hydraulic performance of the revised 1985 scheme.

32. Hydraulics Research Ltd, Wallingford, was commissioned in February 1986 to carry out a preliminary study of wave effects. The study was to provide information suitable for the feasibility of the scheme to be assessed. This was done using a wave hindcasting model to predict extreme offshore wave conditions using the Hindwave model, a back tracking refraction model and a short wave model to determine wave conditions inside the new harbour. A mathematical model of the 1975 Scheme was set up and found to give good agreement with the original physical model results obtained. A model of the new proposed harbour was then produced.

33. The recommendations of this preliminary study were to proceed with a physical model similar to the 1975 Scheme to give a full investigation of the tidal harbour and also give an indication of drift and erosion patterns.

34. In addition, a study was also undertaken by Hydraulics Research Ltd, Wallingford, on the effects of the breakwater construction on beach transport. Using beach data provided by the Southern Water Authority covering a period from July 1976 to August 1985, the potential transport quantities were assessed using various sources.

SUMMARY

35. It is interesting to note that since 1973 the use and interpretation of physical models has not changed a great deal nor the modelling, as can be seen from comparison of Figures 8 and 9. What has changed is the instrumentation available, the use of video equipment and the time and cost of physical modelling. The costs of the modelling in 1975, were approximately £66,000. This included tests on the navigational locks. The cost of modelling the 1985 Scheme was approximately £55,000.

Fig. 8. Wave attack from easterly direction – 1985

Fig. 9. Wave attack from south south west direction – 1985

Due consideration must be taken of the use of the previous tests and information which cut costs considerably.

36. The use of models, videos and even visits by the public, or your clients, to testing laboratories has helped enormously in aiding the dissemination of results to action groups who are more geared up to fighting new installations along the coastline on environmental grounds than say a decade ago.

37. The increase in quality and use of computational modelling has changed the mode of approach used by the design engineer. He now has the use of tools at a reasonable cost to give him indications of the hydraulic performance of his scheme designs for harbour layouts before he commits larger sums to make physical models. The accuracy of predictions made by such models is also improving as comparison of Figures 5 and 6 testify. The cost of the first predictive study was £10,000 and the cost of the shingle study which came last of all was £3,000.

38. The mix of modelling techniques enabled the engineer to proceed with his design with confidence. The good agreement between the results also gives the engineer more confidence as there are occasions when time or economic situations preclude the use of physical models. It is comforting to know that computational modelling is now an acceptable tool to aid the engineer's design decisions when designing in that unreliable place - the sea.

ACKNOWLEDGEMENTS
The authors thank the Trustees of the Chatsworth Estate and Tarmac EZD for permission to give this paper and also Dr Steve Huntington and the staff of Hydraulics Research Ltd, Wallingford for their expertise and goodwill.

DESIGN OF BREAKWATERS

APPENDIX A
SPECIALISTS' REPORTS AND PAPERS

CIVIL ENGINEERING WORKS

01 Preliminary Report Regarding a Yacht Harbour
   (Ove Arup & Partners: October 1972)

   A preliminary investigation of the engineering problems
   associated with the design of a Yacht Harbour at the
   Crumbles.

02 Report Hydrographic Survey
   (Longdin & Browning: June 1973)

   The record of the hydrographic survey carried out between
   2 May and 6 July 1973. Includes level survey of seabed,
   of beaches and of coastal strip; current meter
   observations during spring tide; salinity observations and
   seabed soil samples.

   Also Appendix D/1 October 1973
   Appendix D/11   February 1974
   Appendix D/111   June 1974

   Record of seasonal variation in beach profiles.

03 Report on Coastal Changes in the Area
   (Ove Arup & Partners: September 1973)

   An investigation of littoral drift of beach material in
   the Eastbourne area.

04 Report Coastal Engineering Data
   (Ove Arup & Partners: September 1973)

   Data relevant to design of coastal works at the Crumbles;
   waves, winds, tides, tidal currents and geotechnical
   description of seabed samples.

05 Site Investigation Report Part 1 Site Work
   and
06 Site Investigation Report Part 2 Laboratory Testing
   (Ove Arup & Partners: October 1973)

   The record of a geotechnic land investigation covering the
   160 hectar site. Site and laboratory work carried out by
   Wimpey Laboratories Ltd, between 16 April and 26 July
   1973.

07 Report Wave Refraction Analysis
(Atkins Research and Development: October 1973)

A study of wave refraction to the proposed harbour
entrance.

08 Report on the Locks
(Ove Arup & Partners: October 1973)

A prediction of boat traffic generated by the locked
harbour and an outline specification of the necessary lock
facility.

09 Report Site Investigation for a Harbour at the Crumbles,
Eastbourne
(Soil Mechanics Ltd: November 1974)

The record of the geotechnic seabed investigation
providing data for the foundation design of the
breakwaters and for dredging. Site work carried out
between 4 June 1974 and 2 July 1974.

10 Report on Wave and Wind Conditions at the Crumbles,
Eastbourne
(Experimental and Electronic Laboratories, British
Hovercraft Corporation Ltd: December 1974)

The record of wind and wave observations throughout the
period 24 May to 1 July 1974. Also an analysis of the
wave data.

11 Geotechnical Report
(Ove Arup & Partners: January 1975)

Interpretive geotechnical report, applying the findings of
reports 5, 6 and 9 to the selected harbour design.

12 Report Hydraulic Model Study, Storm Wave Model of Tidal
Harbour
(Experimental and Electronic Laboratories, British
Hovercraft Corporation Ltd: May 1975)

A record of the 1:60 scale storm wave model study carried
out between 1 July 1974 and 7 March 1975.

13 Report Hydraulic Model Study, Lock Model Study
(Experimental and Electronic Laboratories, British
Hovercraft Corporation Ltd: May 1975)

A record of the 1:15 scale lock model study carried out
between 30 September 1974 and 28 March 1975.

DESIGN OF BREAKWATERS

14 Report Hydraulic Model Study, Breakwaters
   (Experimental and Electronic Laboratories, British
   Hovercraft Corporation Ltd: September 1975)

   A record of the 1:40 (approximately) scale model study of
   the proposed breakwater construction.

15 Report Coastal Engineering Data Report 2
   (Ove Arup & Partners: May 1975)

   Interpretive report, applying findings of reports 4 and 11
   to the selected harbour design.

16 Report Civil Engineering
   (Ove Arup & Partners: August 1975)

   A comprehensive report providing a factual record of all
   aspects of the harbour design produced in three parts:

   Civil Engineering Report: Text
   Civil Engineering Report: Drawings
   Civil Engineering Report: Construction
   Programmes and Cost Estimates

TRAFFIC GENERATION

17 Paper: Predicted Traffic Generation to and from the
   Crumbles
   (Ove Arup & Partners: February 1975)

BOATING FACILITIES

18. Paper: Feasibility, Size and Nature of the Proposed Yacht
    Harbour at the Crumbles, Eastbourne
    (Denys H Sessions: April 1979)

    Demand for boating facilities generally and at Eastbourne
    in particular; the facilities proposed: management policy.

HYDRAULIC TESTING

19. Report: Eastbourne Village Harbour, EX1417, HRL
    Wallingford
    (J V Smallman: February 1987)

    A preliminary study of wave effects.

20. Report. The Crumbles Harbour Village, Eastbourne, EX1544
    HRL, Wallingford
    (P J Beresford: March 1987)

    Random wave disturbance. Studies of a proposed marina to
    establish wave patterns inside harbour and assess beach
    erosion.

21.   Report.   The Crumbles Harbour Village, Eastbourne,
      EX1567, HRL, Wallingford
      (April 1987)

      Two dimensional random wave model tests of breakwater
      stability and Hydraulic performance on breakwater
      cross-sections.

22. Report. The Crumbles Harbour Village, Eastbourne, EX1607
     HRL, Wallingford
     (June 1987)

     Beach transport study.

APPENDIX B

Wind fetch lengths for the Eastbourne study (Wave prediction point near the site of the Royal Sovereign Light Vessel)

Mean wind direction (°N)	Fetch length (km)
5	13.5
15	14.6
25	16.6
35	19.3
45	34.3
55	38.9
65	214.2
75	81.0
85	72.0
95	73.8
105	75.6
115	75.6
125	86.4
135	86.4
145	86.4
155	86.4
165	84.6
175	85.5
185	93.6
195	145.8
205	151.2
215	160.2
225	165.2
235	176.4
245	over 500
255	over 500
265	111.6

Mean wind direction (°N)	Fetch length (km)
275	81.0
285	13.4
295	12.5
305	12.2
315	11.8
325	12.5
335	12.8
345	12.9
355	13.1

APPENDIX C

Predicted extreme offshore significant wave heights in metres for sea area near Eastbourne.

Storm duration is 3 hours

Direction (°N)	Return Period 1/1 year	1/10 year	1/50 year
45 - 75	2.5	2.9	3.1
75 - 105	2.6	3.1	3.5
105 - 135	2.1	2.8	3.2
135 - 165	2.5	3.1	3.4
165 - 195	3.7	4.5	4.9
195 - 225	4.9	5.8	6.3
225 - 255	5.4	6.2	6.7
255 - 285	4.1	5.3	6.0
All	5.6	6.6	7.3

8. The effect of the Lagos harbour moles (breakwaters) on the erosion of Victoria Beach

L. ONOLAJA, Federal Ministry of Works and Housing, Lagos, Nigeria

SYNOPSIS

   The Lagos area of the West African Coast is subject
to South Westerly waves giving rise to a littoral sand drift
from West to East which used to go through the Lagos Port
entrance channel. The natural coastline attained hydrodynamic equilibrium by maintaining an approximate orientation
of 45 degrees to the prevailing wave direction. The predominant coastal current is the Guinea Current which runs the
entire West African Coast from Sierra Leone to Cameroons.
Before the construction of the Lagos entrance moles between
1908 and 1912, the beaches on both sides of the channel were
stable. The construction of the moles locally interrupted
the littoral drift and caused a progradation of the Light
House Beach on the West side of the moles and a recession
of Victoria Beach on the east side of the moles.

   During the period 1912 - 1960 the recession of Victoria
Beach was spectacular. The erosion rate is recorded to have
been more than 1,500m near the east mole, decreasing to 800m
about 1.5km away from the mole.

   In 1958, the root of the east mole was endangered and
to avoid a break through, into the habour basin, a groyne
was built at the root of the mole.

   In 1960 articicial supply of sand to the beach was
implemented. However the erosion continued and progressed
up to 1970 when the shoreline was very close to the coastal
road (60m in some places).

   From 1970 to 1981, the shoreline was subject to seasonal
and yearly fluctuations within a range of about 100 metres.

   Various studies have been carried out from 1948 to 1986
and many solution proposed. The hydraulic model studies
was completed early in 1986 and the final recommended solution to the erosion will also be presented.

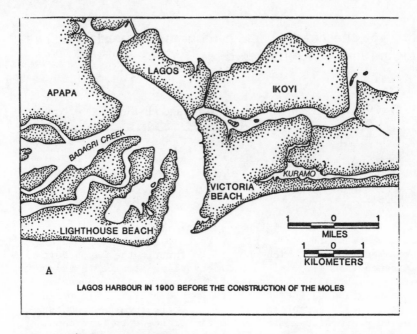

A

LAGOS HARBOUR IN 1900 BEFORE THE CONSTRUCTION OF THE MOLES

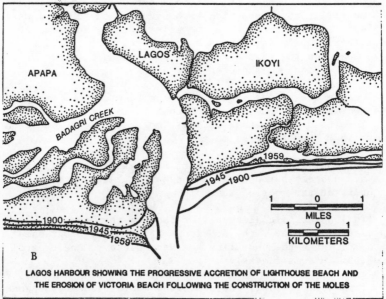

B

LAGOS HARBOUR SHOWING THE PROGRESSIVE ACCRETION OF LIGHTHOUSE BEACH AND
THE EROSION OF VICTORIA BEACH FOLLOWING THE CONSTRUCTION OF THE MOLES

Fig 1 – The Lagos coastline before (A) and after (B)
the construction of the West and East moles (Breakwaters)

INTRODUCTION

1.   Prior to 1908, erosion was not a worrying problem
in the coastal areas of Lagos. The year 1908 is particularly
significant when discussing the Victoria Beach erosion
because this was when works on the Lagos Harbour Moles(Breakwaters) started. The works were completed in 1912. The
moles were constructed at more on less right angle to
the coastline to stabilise the (dredged) entrance into
the Lagos habour. Although this objective was achieved,
the moles triggered off a phenomenal beach erosion problem
east of the inlet at Victoria Island. Since then, shoreline
erosion is recorded to have been more than 1,500m near
the east mole, decreasing to 800m about 1.5km away from
the mole.
2.   The problem associated with Victoria Beach erosion
has been so enormous that the Federal Government had
over the years commissioned various types of studies,
both short and long term, to investigate the problem with
a view to finding a long term solution to it. Some of these
include investigations by Delft Hydraulic Laboratory, 1951,
1964 and 1975; Stanley Consultants 1968;Fowora Renardat
Associates, 1981; Webb 1960, Osoba 1971; Usoroh 1977, Ibe
et al, 1984; Oyegoke et al 1984, 1985; Ibe et al 1986.

VICTORIA BEACH

3.   Victoria Beach is part of the beadridge - barrier -
lagoon system which is located to the western extremity
of Nigeria, along the coast of West Africa. It is located
immediately east of the eastern breakwater on the down drift
side of the natural inlet into the Lagos Habour(Fig. 2).
It is also the seaward area of the piece of land known as
Victoria Island which in the past, has undergone and is
at present, undergoing active development by both private
individuals and municipal authorities. The active beach
is an average 30m. wide and slopes at 5° from the approximately 1m erosion scarp offshorewards.
However, much of this beach is being eroded away so fast
that valuable structures on Victoria Island are being threat-
ened by the occasional storm surge.
4.   Before the construction of the breakwaters in 1908, the
active beach at Victoria Island was almost in line with the
lighthouse beach to the west of the inlet, forming a contemporary barrier bar system and having a consistent west-east
transport of sand along the coast forming an underwater
sand bar across a tidal inlet. This was the state of the
beach/harbour prior to 1908 when construction of three moles
(breakwaters) commenced. (Fig. 1a). The construction,
however, upset the state of equilibrium of the beach which
existed before the moles were constructed. This has resulted
in the beach to the west of the harbour growing seaward
with the accretion of sand there, while the beach to the
east, Victoria Beach, is actively eroding. The erosion

Fig 2 – Aerial photo of Lagos Harbour and Victoria Beach, showing study stations

on Victoria Beach has been of major concern to the government and attempts have been made to stabilize the beach.
Regular pumping of sand into the eroding beach is presently
being used to stabilize the beach. A permanent solution
to the problem is now being looked into.

THE EROSION PHENOMENON
5.   The construction of the breakwaters produced the
following effects viz:

(a)   Altered the direction and mode of the littoral
      transport
(b)   Created an imbalance in the rate of supply
      and loss of materials in the coastal region
      of the area which resulted in Victoria beach
      eroding while the light house west of the mole
      is accreting.

It has been estimated that between 1912 to 1957 the immediate area west of the light house mole(west mole)
prograded about 600m seaward while Victoria Beach, adjacent
to the east mole, retreated about an average 1200m inland.
From these figures, it is clear that rate of erosion on
Victoria Beach far outstrips the rate of accretion of
the light house beach.
6.   By 1958, a breakthrough of the sea into the harbour
basin was imminent. To forestall this, the eastern breakwater was reinforced by a groyne. In the same year, 1958,
periodic sand replenishment measures to check the rapid
retreat of the coastline commenced (Table 1). Today,
the impact of this remedial measures – sand replenishment
– is yet to be felt as the shoreline continues to retreat
at increasingly alarming rates. For instance, a tenta-
tive super-imposition of a 1983 serial photography on
that of 1949 (Fig 3a and 3b) show that the shoreline at
Victoria Beach has retreated to a total of close to 2km.
in the post breakwater construction period (Ibe, 1986)
Erosion studies by the Nigerian Institute for Oceanography
and Marine Research Lagos show that between 1986 to present
day, more than 80m of beach had been lost at the most
severe sites along the beach.

OCEANOGRAPHIC PARAMETERS INFLUENCING EROSION AT VICTORIA
BEACH
7.   The identified parmeters influencing erosion on
Victoria Beach are as follows:

(a)   Tidal Influences
(b)   Wave Characteristics
(c)   Ocean Currents
(d)   Sediment Size
(e)   Coastal Morphology
(f)   Wind

TABLE 1. Erosion control measures applied from 1958 to present at Victoria Island

1958	Construction of a groyne at the foot of the eastern breakwater to avoid the undermining of the breakwater by erosion
1958-1960	Dumping of sand dredged from the commodore channel at the extremity of the eastern breakwater for dispersal along the beach by waves
1960-1968	Permanent pumping station built on the eastern breakwater supplied an average of 0.66 M m^3 pa of sand from the commodore channel to the beach; in between, in 1964, a zig-zag timber groyne (palisade) running parallel to the coastline was driven in some 26 m from the shoreline
1969-1974	Some artificial sand replenishment was carried out; however, no reliable records of quantities or frequency are available
1974-1975	3 M m^3 of sand dumped and spread on the beach
1981	2 M m^3 of sand dumped and spread on the beach
1985	3 M m^3 of sand dumped and spread on the beach

8. Tidal Influences:
Tides play a very important part in sediment transport especially the Onshore - Offshore Sediment motion. The tides
at Victoria Island are semi diurnal in nature with significant diurnal inequalities. The mean tidal range is approximately 1m although during Spring tides, the range is up
to 1.5m. The tides approach the coast from a South westerly
direction and create strong semi-permanent currents which
occur at high angles to the coast carrying sediments away
from the beach out to the sea.

Wave Characteristic:
9. The breaking waves which are predominantly plunging
are on the average 1.60m high. The waves come from a south-
west direction and arrive at an oblique angle of 7-12°
generating longshore currents less than 1m/s. The wave
climate is more intense during the meteorologically harsher
wet season.

Ocean Currents:
10. The ocean current present here is the Guinea Current
which flows seasonally and in the direction of longshore
transport i.e. west to east. Because of the sheltering
influence of the east mole, the effect of this current
is not expected to be a decisive factor in sediment transport
on Victoria Beach. The velocity of the Ocean current is
between 0.5.14m/sec. to 1.03m/sec. on the Victoria Beach.
The breaking waves at Victoria Island also set up localised
rip currents with water flowing back through the breaker
line in sectors a few tens of meters wide.

Sediment Size:
11. The sediment size of beach material greatly influences
the rate of sediment transport along any coast. Fig 5
shows analysis of sand samples obtained from Victoria beach.
Sediments on the Island are medium to coarse sand with
a mixture of broken shells. Sediments on the beach ridges
are better sorted than those from the active beach.

Wind:
12. Analysis of wind records over a period of two years
shows that the southwesterly winds are the prevailing (more
frequent) and predominant(stronger) winds at Victoria Island.
The winds are largely confined to azimuth 215° - 270° and
have velocities generally exceeding 12km. per hr. The
winds are more consistent and stronger during the rainy
season when near storm conditions in the form of thunder-
storms and squall lines are more frequent and wind speeds
on such occasions exceed 20km. per hr.

VICTORIA BEACH PROFILES
13. Ten profiling stations established on the beach were
resurveyed approximately bi-monthly. This profile survey
has been going on since April 1986 immediately after the
sand replenishment scheme in which about 3 million m³ of
sand was pumped on the beach and results obtained show

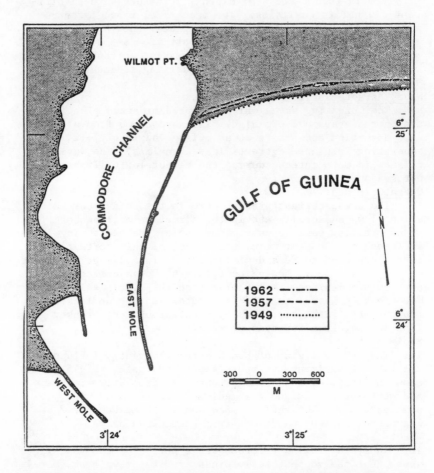

Fig 3A - Sequential aerial photo analysis of shoreline
migration at Victoria Beach between 1949 and 1962

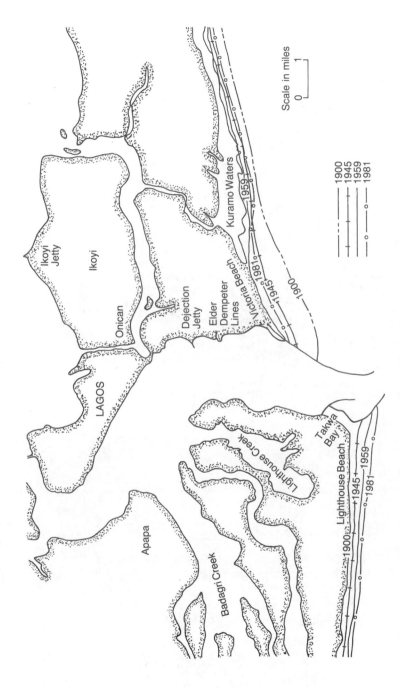

Fig 3B – Lagos Harbour showing the progressive accretion of Lighthouse Beach and the erosion of Victoria Beach following the construction of the moles

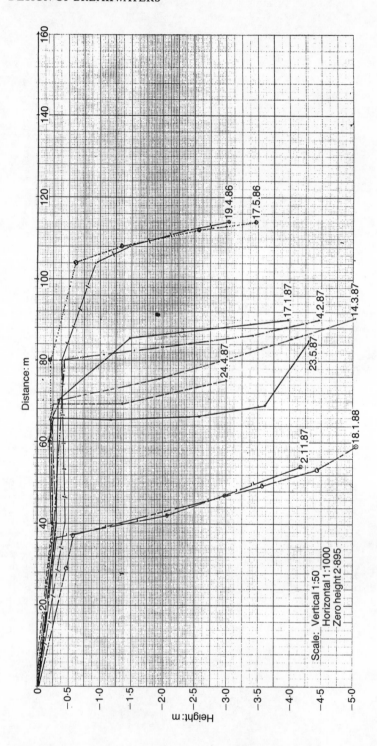

Fig 4 – Profile 2 PTB

a startling continuous erosion rate (Table 2).
14. Results of two years Victoria beach profiling,
(Table 2),show a predominantly eroding beach with high eroding
intensity between profiles 1 and 4 i.e. areas close to the
eastern breakwater. Conclusively, the results show that:

(a) Erosion was a continuous process on the Victoria
beach during the study period

(b) The most pronounced erosion occurred in areas
immediately to the east mole (about 700m - 800m
to the east mole i.e. stations 1 to 4). The
erosion rate decreased eastwards.

It is apparent, that the erosion menace, due largely to
the presence of the harbour moles has defied all previous
solutions. It is in the light of this that a detailed study
of the major proposed solutions using a Hydraulic Model
of Victoria Beach was carried out, at the Hydraulic Research
Unit of the University of Lagos.

REVIEW OF MAJOR PROPOSED SOLUTIONS
15. Efforts in the past aimed at abating the the Victoria
beach erosion problem consisted mainly of sand replenishment
programmes. The failure of these programmes to stem the
continued retreat of the coastline has led to calls for
the implementation of certain proposed permanent solutions.
Seven possible solutions to the problemwere identified as
follows:

(a) Initial reclamation coupled with regular sand
nourishment,

(b) Building or groynes normal to the shoreline
with stones, tetrapods, etc.

(c) Re-alignment of the present three breakwaters
(moles) i.e. harbour entrance

(d) Construction of Off-shore breakwaters

(e) Construction of a Seawall

(f) Use of Artificial Sea-weeds; and

(g) Combination of any of the above solutions.

The outcome of the comparative assessment of the proposed
solutions with regards to their technical and cost effectiveness is as follows:

(a) Initial reclamation coupled with regular sand
nourishment: This proposal involves the supply
of 1,000,000 cubic metres of sand every year and
appears to be the most economical solution to
the problem. However the idea of having to reclaim
the beach every year, as a result of continous
erosion, does not commend itself as an economical
solution .

(b) Building groynes: The action of erosion on the
beach would begin where ever the last groyne ends.
This cannot therefore by itself be an appropriate
solution.

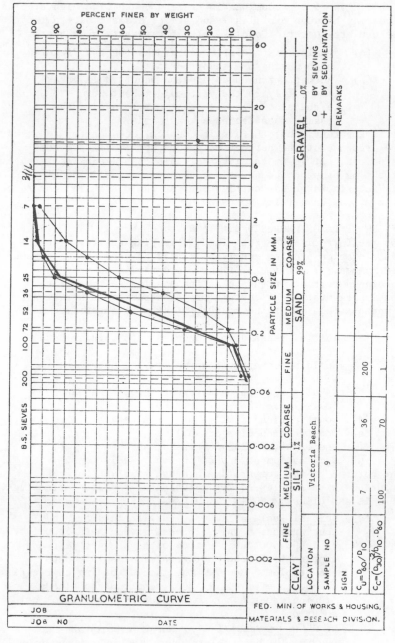

Fig 5A – Analysis of sand sample obtained from Victoria Beach during the 1985 replenishment

FEDERAL MINISTRY OF WORKS

MATERIALS BRANCH

SIEVE ANALYSIS

SOIL SAMPLE ....

.................TOTAL WEIGHT OF DRY SOIL IN GM.. 603gm .... LAB. REG. NO.........

LOCATION .. .Victoria Beach ....WEIGHT OF DRY SOIL

AFTER WA SHING IN GM................. DATE 3/7/85

BORING NO........SAMPLE DEPTH.........

SAMPLE NO........ 9 ....WASHED THROUGH SIEVE

NO. 200: 9 ....OPERATOR......

SPECIFIC GRAVITY.....

SIEVE NO.	SIEVE OPENING IN MM	WT SIEVE + SOIL IN GM	WT. SOIL RETAINED IN GM	PERCENT-RETAINED	CUMULATIVE PERCENT RETAINED	PERCENT
7			2	0.33	0.33	99.67
14			7	1.16	1.49	98.51
25			68	11.28	12.77	87.23
36			107	17.74	30.51	69.49
52			147	24.38	54.89	45.11
100			229	37.98	92.87	7.13
200			40	6.63	99.50	0.50
Pan			3	0.50	100.00	0.00

REMARKS ....

Fig 5B - Sieve analysis

TABLE 2. Erosional trend of active beach surface and berm at Victoria Beach over two-year period

PROFILES		Active Beach Surface			Actual Berm		PERIOD
		Eroding(M)	Rate (M/Month)		Eroding (M)	Rate (M/Month)	
1	-	35.0m	2.9	-	21	1.75	APRIL 86 to April 1987
	-	-	-	-	-	-	APRIL 1987 to JAN. 1988
2	-	96.0	8.0		87.0	7.25	APRIL 86 to APRIL 1987
		87.0	7.25		81.0	6.75	APRIL 87 to JAN 1988
3		60.0	5.0		45.0	3.85	APRIL 86 to APRIL 1987
		20.0	1.67		30.0	2.50	APRIL 87 to JAN 1988
4		37.0	3.00		28.0	2.33	APRIL 86 to APRIL 1987
		19.0	1.58		18.0	1.50	APRIL 87 to JAN. 1988
5		43.0	3.58		18.0	1.50	APRIL 86 to APRIL 87
		NO SIGNIFICANT CHANGE			2.0	0.67	APRIL 87 to JAN. 88
6		41.0	3.42		24.0	2.0	APRIL 86 to APRIL 87
		10.0	0.83		7.0	0.58	APRIL 87 to JAN. 88
7		31.50 (in 15 months	2.10		10.0	0.67	SEPT. 1986 to DEC. 1987
8	-	-	-		-	-	-
9		34.0	2.40		15.0	1.00	SEPT. 86 to DEC. 1987
10	-	18.0	1.40		2.0	0.13m	SEPT. 86 to DEC. 1987

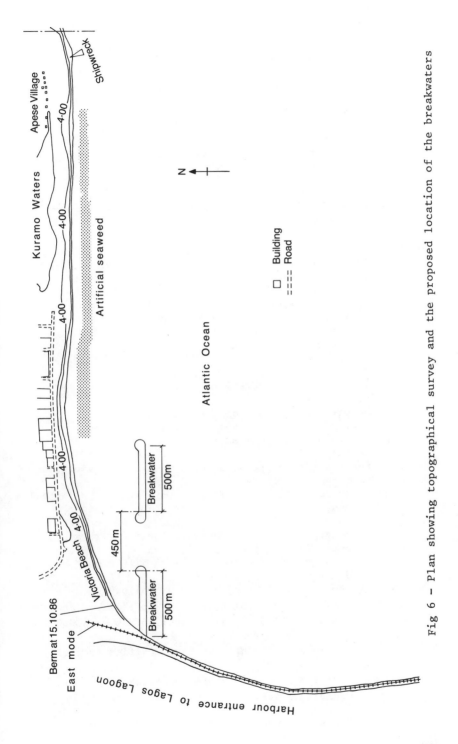

Fig 6 – Plan showing topographical survey and the proposed location of the breakwaters

143

    (c) <u>Re-alignment of the existing three breakwater</u>
<u>(harbour entrance)</u>: This was found to be the most
uneconomical solution as it would cost not less
than ₦ 200 million to excute.

    (d) <u>Construction of Seawall</u>: The nature of the conti-
nental shelf as well as the problems of strong
lee-erosion and vortices created at the tip of the
seawall would not make it technically advisable.
However, the cost of constructing a seawall 5km.
long would be astronomical while the wall will
prevent the use of the beach, for recreational
facilities such as picnicking and bathing

HYDRAULIC MODEL STUDIES

16. The hydraulic Model Studies of Victoria Beach sea defence system was carried out at the Hydraulic Research Unit
of the University of Lagos. The studies include the
following:

    (a) The gathering and processing of eseential
coastal hydrodynamic and oceanographic data for
the purpose of the model investigation and other
engineering works relating to the Victoria Beach.

    (b) An exhaustive review of all major proposals and
studies relating to erosion and flooding of
the Victoria beach since the harbour moles
were introduced. This includes the examination
of solutions adopted for erosion on beaches
with similar characteristics to the Victoria
Beach.

    (c) The preparation, building and calibration of a
three dimentional movable bed coastal model
encompassing approximately 30 sq km. of the Lagos
coastal region.

    (d) Investigation of proposed solutions of the Victoria
Beach Sea Defence System.

The hydraulic model was built in a wave basin and initially
tested to obtain the desired scales. The modelling
facilities used include a wave basin, a wave generator,
a large underground water reservoir equiped with a pump
cellar containing a battery of 11No. 4h/p. Centrifugal
pumps and an over head reservoir along with a host of
other hydraulic laboratory equipments.

Some of the results arrived at are as follows:

    (a) Time scale for bed evolution = 53 min : 1 year or 1:9924

    (b) The sediment transport scale = 0.013957cub.m :380,000cub.m
or 1 : 27,226481.

The possible solutions identified were modified to give
the following four modified alternatives.

    (a) Modified alternative A:- Breakwaters only

    (b) Modified alternative B:- Breakwaters and Groyne
System

    (c) Synthetic Seaweeds without sand Nourishment

(d) Modified alternative D:- Synthetic Seaweeds with sand Nourishment.

The alternatives were tested for a total period equivalent to 20 years in real time and the results of each solution is presented below:

17. Modified alternative A - Offshore Breakwaters Only: Observations made were:

(a) No form of erosion was recorded throughout period of the experiment in areas immediately east of the east mole. Similarly, areas protected by other breakwaters were found to be stabilised. Also, it was observed that the water in the sheltered zones were relatively calm.

(b) Severe erosion was observed in the unprotected areas between two offshore breakwaters.

(c) Immediately after the most easterly breakwater which is about 3 km. east of east mole, severe erosion of about 0.75 km. eastwards was recorded. Hence the erosion continued to the east in a gradual but constant manner.

18. From the above studies, it is clear that this alternative consisting of breakwaters only will stabilize, in part, the 6 km. stretch of the protected Victoria beach only. It will however spark off new erosional trends east of the breakwaters.

19. Modified alternative B:- Breakwaters and Groyne System System: Observations made were:-

(a) The area immediately to the east of the first east mole was adequately stabilized just like in alternative A.

(b) Areas protected by the other breakwater were equally stabilized.

(c) Sheltered areas of other breakwaters - apart from the first one nearest to the east mole - indicate substantial build up of materials.

(d) The groyne protected the areas of the beach in which they are located whereas in the unprotected areas between the detached breakwaters, the coast continued to erode.

(e) Just like the case of alternative A, the erosion continued unabated, after the east most breakwater.

20. From above, it can be seen that this proposal just like alternative A, stabilized the beach to some extent, however, erosion is shifted further east to the unprotected coast.

21. Modified alternative C: Synthetic Seaweed without sand Nourishment:

During and after the experiment, the following observations were made:

(a) Between the root of the east mole and the beginning of the weeds, the coastline showed severe erosion as before.

(b) Immediately after the weed, going eastwards, the area recorded a high degree of erosion.

(c) The entire zone covered by the Synthetic seaweed was actively stabilized coupled with substantial buildup of sand and subsequent shoreline gain.

22. It is therefore obvious, that the synthetic seaweeds cannot be used to stabilize the Victoria beach, unless the entire shoreline or the area of interest is covered completely with the weeds. Though the degree of erosion is reduced slightly immediately after the seaweeds, it continues vigorously further east, hence it only shifts the problem eastwards.

23. Modified alternative D: Synthetic weeds with sand Nourishment:

All the features present in alternative C appeared except that no erosion was recorded in the diffraction zone of the eastmole. Also the severity of erosion immediately after the eastern end of the weeds was reduced. An effective means of checking erosion within the diffraction zone of the eastmole will be a periodic sand replenishment within this area. Erosion is however anticipated further east after the area of influence of the weeds.

SUMMARY AND CONCLUSION

24. The recommended solution to the flooding and erosion problems of Victoria Beach, arrived at from the results of the Hydraulic Model Studies, consists of the construction of two offshore breakwaters in combination with the artificial seaweed, after an initial reclamation of the beach with 5 million cubic meters of sand (Fig. 6). The final design of the scheme is in progress. The offshore breakwaters would form a defensive wall against possible future ocean surge, provide a calm beach for bathing and form a protective device for the reclaimed beach, while the artificial seaweed would reduce the rate of erosion and promote accretion at an economical cost.

TABLE 3 - Profile 2 PTB

19th April, 1986			17th May 1985		
Distance(M)		Profile Height(M)	Distance		Profile Height(M)
0.00	2.895	0.000	0.00	2.895	0.000
40.00	2.490	-0.405	40.00	2.670	-0.225
80.00	2.500	-0.395	80.00	2.685	-0.210
110.00	2.020	-0.875	110.00	2.255	0.640
120.00	1.450	-1.445	120.00	1.550	-1.345
130.00	0.565	-2.330	130.00	0.275	-2.82
135.00	-0.210	-3.105	135.00	-0.580	-3.475
17th January 1987			14th February 1987		
0.00	4.789	0.000	0.00	4.789	0.000
40.00	4.524	-0.265	40.00	4.535	-0.254
75.00	4.430	-0.359	75.00	4.422	-0.367
85.00	3.301	-1.488	79.00	4.398	-0.391
90.00	0.851	-3.938	85.00	1.362	-3.427
			90.00	0.329	-4.460
14th March 1987			24th April 1987		
0.00	4.789	0.000	0.00	4.789	0.000
40.00	4.526	-0.263	40.00	4.539	-0.250
65.00	4.594	-0.195	65.00	4.594	-0.195
68.00	4.516	-0.273 Berm	67.00	4.539	-0.250 Berm
75.00	2.929	-1.860	67.00	3.449	-1.340 Base
85.00	0.698	-4.091	75.00	1.884	-2.905
90.00	-0.244	-5.033	80.00	0.878	-3.910
			85.00	0.104	-4.685
23rd May 1987			2nd November, 1987		
0.00	4.789	0.000	0.00	4.789	0.000
40.00	4.539	-0.250	36.50	4.449	-0.340 Berm
60.00	4.594	-0.195	42.00	2.714	-2.075
64.00	4.582	-0.207 Berm	47.00	1.809	-2.980
64.00	3.566	-1.223 Base	54.00	0.594	-4.195
70.00	2.192	-2.597			
75.00	1.167	-3.622			
80.00	0.262	-4.527			
18th January, 1988					
0.00	4.789	0.000			
28.00	4.383	-0.406			
36.05	4.243	-0.546			
39.05	3.327	-1.462			
48.20	1.239	-3.55			
53.20	0.368	-4.421			
58.00	-0.363	-5.152			

TABLE 4 - Profile 3 PGD 86/6

19th April 1986			17th May 1986		
Distance(M)		Profile Height(M)	Distance(M)		Profile Height(M)
0.00	1.093	0.000	0.00	1.093	0.000
40.00	1.083	−0.010	40.00	1.013	−0.080
84.00	1.038	−0.055	84.00	1.033	−0.060
124.00	2.083	+0.990	124.00	2.078	+0.985
164.00	2.823	+1.730	164.00	2.848	+1.755
204.00	2.483	+1.390	204.00	2.448	1.355
245.00	2.633	+1.540	245.00	2.583	1.490
255.00	2.303	+1.210	255.00	1.093	0.000
260.00	1.198	+0.105	260.00	0.023	−1.070
265.00	0.323	−0.770	265.00	−0.937	−2.030
18th Sept. 1986			16th October 1986		
0.00	3.218	−0.000	0.00	3.218	0.000
20.00	2.825	−0.390	20.00	2.828	−0.390
40.00	3.148	−0.070	40.00	3.148	−0.070
60.00	3.188	−0.030	60.00	3.188	−0.030
80.00	3.208	−0.010	80.00	3.208	−0.010
100.00	3.378	+0.160	100.00	3.378	+0.160
120.00	4.308	1.090	120.00	3.538	+0.320
140.00	4.963	1.745	140.00	4.318	+1.100
160.00	5.103	1.885	160.00	5.068	1.850
180.00	4.983	1.765	180.00	5.213	1.995
200.00	4.783	1.565	200.00	4.983	1.765
220.00	4.263	1.545	220.00	4.828	1.610
230.00	3.438	0.220	229.80	4.848	1.630
235.00	2.518	−0.700	234.80	3.528	0.310
240.00	1.768	−1.450	239.80	2.548	−0.670
17th January 1987			14th February 1987		
0.00	2.978	0.000	0.00	2.978	0.000
140.00	4.665	1.687	140.00	4.652	1.674
180.00	4.708	1.730	180.00	4.700	1.722
205.00	4.557	1.579	205.00	4.557	1.679
215.00	2.830	−0.148	210.00	4.008	1.030
220.00	0.839	−2.139	215.00	2.269	−0.709
			220.00	1.996	−0.982

TABLE 4 - continued

14th March 1987			24th April, 1987		
Distance(M)		Profile Height(M)	Distance(M)		Profile Height(M)
0.00	2.978	0.000	0.00	2.978	0.000
140.00	4.642	1.664	140.00	4.657	1.679
180.00	4.704	1.726	180.00	4.716	1.738
195.00	4.597	1.609	195.00	4.588	1.610
201.40	4.570	1.592 Berm	200.00	4.558	1.580 Berm
205.00	3.610	0.632	200.00	3.903	0.925 Base
215.00	1.797	−1.181	215.00	1.118	−1.860
220.00	0.893	−2.085	220.00	0.273	−2.705
225.00	−0.820	−3.798			
23rd May 1987			2nd November 1987		
0.00	2.978	0.000	0.00	2.978	0.000
140.00	4.657	1.679	94.95	3.243	0.265
180.00	4.716	1.738	153.15	4.683	1.705
195.00	4.619	1.641	168.88	4.678	1.700 Berm
196.30	4.645	1.667 Berm	179.23	2.543	0.435
196.30	4.023	1.045 Base	184.23	1.748	−1.23
205.00	2.462	−0.516	189.23	0.968	−2.01
210.00	1.540	−1.438			
215.00	0.660	−2.318			
18th January 1988					
0.00	2.978	0.000			
97.50	3.014	0.036			
124.20	3.947	0.969			
177.65	4.486	1.508			
188.95	2.383	−0.595			
196.95	0.732	−2.246			
201.95	−0.108	−3.086			

TABLE 5 - Profile 10: After shipwreck (Last station)PGD 86/17

19th April 1986			10th May 1986		
Distance(M)		Profile Height(M)	Distance (M)		Profile Height(M)
0.00	0.482	0.00	0.00	0.482	0.00
40.00	0.502	+0.020	40.00	0.502	+0.020
79.00	1.002	+0.520	79.00	1.472	0.990
89.00	2.442	1.960	89.00	2.482	2.000
109.00	1.782	1.300	109.00	1.962	1.480
119.00	1.207	0.725	119.00	0.967	0.485
124.00	0.557	0.357	124.00	0.367	-0.115
18th September,1986			15th October 1986		
0.00	3.769	0.000	0.00	3.769	0.000
20.00	3.499	-0.270	20.00	3.449	-0.320
40.00	4.019	0.250	40.00	3.854	0.085
60.00	4.299	0.530	60.00	4.204	0.435
80.00	4.339	0.570	80.00	4.479	0.710
90.00	6.069	2.300	90.00	5.580	1.811
100.00	5.769	2.000	100.00	6.134	2.365
110.00	5.719	1.950	110.00	5.874	2.105
120.00	5.057	1.288	120.00	5.454	1.685
130.00	4.389	0.620	130.00	4.844	1.075
			135.00	4.354	0.585
14th December, 1986					
0.00	1.842	0.000			
50.56	2.182	0.340			
81.66	2.852	1.010			
88.93	3.607	1.763			
98.93	4.287	2.445			
108.93	4.302	2.460			
113.93	3.890	2.048			
118.93	3.327	1.485			
123.93	2.932	1.090			
128.93	2.152	0.310			

TABLE 5 - continued

17th January, 1987			27th February, 1987		
Distance(M)		Profile Height(M)	Distance(M)		Profile Height
0.00	1.842	0.00	0.00	1.842	0.00
70.00	2.401	0.559	70.00	2.355	+0.513
120.00	3.714	1.872	110.00	3.920	+2.078
130.00	3.019	1.177	120.00	3.688	+1.846
140.00	1.193	0.649	130.00	1.384	−0.458
27th March 1987			24th April, 1987		
0.00	1.842	0.00	0.00	1.842	0.00
70.00	2.350	0.508	70.00	2.367	+0.525
110.00	4.062	2.220	110.00	4.073	+2.231
120.00	3.708	1.966	117.10	4.058	+2.216
125.00	2.314	0.472	117.10	3.292	+1.450
130.00	0.827	1.015	130.65	2.130	+0.288
			135.65	1.118	−0.724
			140.65	0.173	+1.669
23rd May 1987					
0.00	1.842	0.00			
70.00	2.367	0.525			
110.00	4.117	2.275			
115.25	4.139	2.269			
120.00	3.111	1.260			
130.00	1.871	0.029			
140.00	0.976	0.866			
145.00	0.591	1.251			

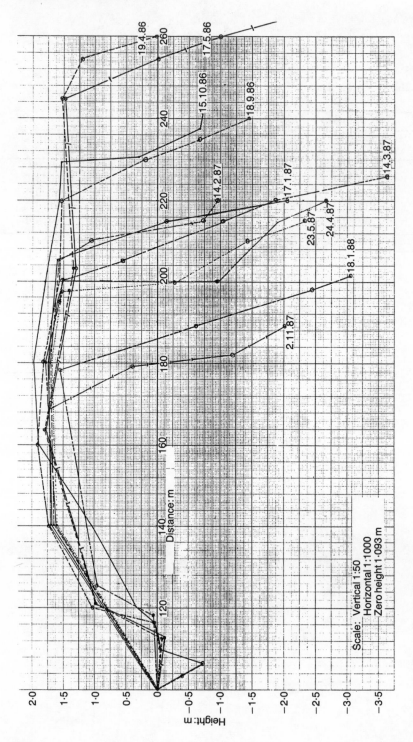

Fig 7 – Profile 3 PGD 86/6

Fig 8 - Hydraulic model studies (modified alternative B)
- breakwaters and groyne system

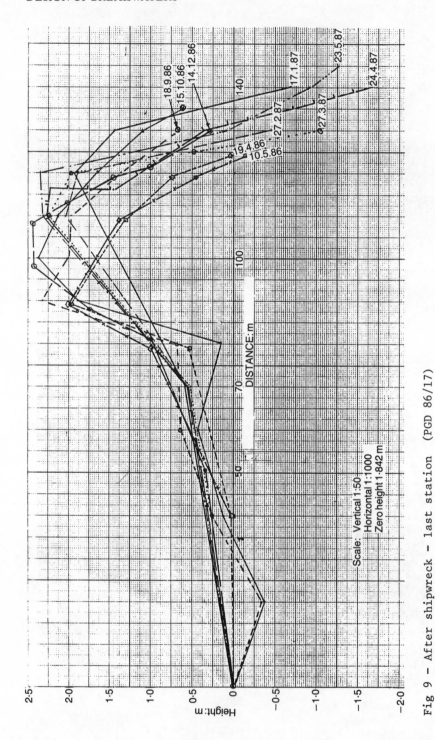

Fig 9 – After shipwreck – last station   (PGD 86/17)

REFERENCES
1.   FISHER B. A. O. et al.   Victoria beach erosion:
Seminar by the Federation of Building and Civil Engineering
Contractors of Nigeria. 1986.
2.   IBE A. C. et al.   Harbour development related erosion
at Victoria Island, Lagos, 1986.
3.   IBE A. C. et al.   Protecting Victoria Island against
sea incursion: A position paper submitted to the Ad-Hoc
Committee on Victoria Island erosion problem, 1984
4.   OYEGOKE E. S. et al.   Hydraulic model studies of
Bar Beach sea defence system - Progress report I and
II., 1986
5.   OYEGOKE E. S. et al.   The importance of hydraulic
model studies for finding scientifically economic solutions to the Lagos Bar Beach surge, 1986.
6.   OYEGOKE E. S. et al.   On Re-examination of Modified
Alternative to the Bar Beach Sea Defence system. 1986
7.   SILVESTER R. et al.   Developments in geotechnical
Engineering, 1974.
8.   ONOLAJA L.   "A review of Coastal Process, Problems,
Defences and Federal Government Efforts in Combating
the Problems of Flood and Erosion along the Nigerian
Coastline"
Proceeding of the National Seminar on Flood and Erosion
Along the Nigerian and Similar Coastline, 1986.

8a. Blanket theory revisited or More than a decade down under

C. T. BROWN, Seabee Developments (UK), Bedford

SYNOPSIS. For nearly twenty years the Author has lived and practised outside the United Kingdom in the fields of Coastal and Maritime engineering. In the early 70s as a result of a exposure to Pattern Placed Tribars and the requirements of a particular job and the early papers on Cobs, he developed an interest in the linear form of Hudson's Equation and the implications of this for the design of armour units. On this first attendance at a British Breakwater conference, an attempt is made to summarise theoretical, experimental and practical results of over a decade of investigation into design using Blanket Theory, being a linear theory which describes the behaviour of coherent or discrete armour layers having negligible stiffness and characterised by independent design of layer thickness, porosity, roughness, element mass and strength and the adoption of suitable factors of safety in the design process. The paper compares the Author's original concepts with more recent variations on the theme and rehearses the differences of approach between them.

THE ORIGINAL CONCEPT.

1. Historical design of armour for breakwaters and revetments was based on the weight of the units (ref 1,2). Wave period effects, if considered at all assumed some linearity between wave height and period at the wave break condition (ref 3). Thus most formulae could be characterised by a relationship of the form:

$$W = \frac{K. H^3 y}{(Sr-1)^3.\emptyset a} \qquad (1)$$

2. Hudson's conceptual model for the theoretical derivation for the stability criteria was a fixed jet of water perpendicular to the slope and related the destabilising forces to the momentum of the jet and the stabilising forces to gravitational (and friction) forces of the unit mass on the slope. He derived a linear stability relationship between the incident wave height and a "characteristic armour dimension" dependant upon the cosine of the slope angle. This Stability Number Ns was described as the relationship between the apparent diameter or equivalent armour cube size and the incident design wave. The presenr author believes that a more rigorous treatment of the equations shows

that it is the slope normal dimension.

$$Ns \quad = \quad \frac{\{y\}^{1/3} . H/(Sr-1)}{\{W\}} \qquad (2)$$

$$= \quad k.kv_{2/3}/Cq \qquad \text{from theory}$$

$$= \quad a \ (cot \ a)^{1/3} \qquad \text{from experiments}$$

where   k   =   f(H/L) such that Hb = k. db

            kv   =   solidity factor (1-p)

and   Cq   =   Drag and effective inertia coefficent

This equation transforms to the well known cubical "Hudson Equation" to produce a design formula for rocks in terms of their mass or weight when no account is taken of the orientation or shape of the rocks or of their grading curve.

3. In reviewing Hudson's 1959 paper apart from discovering that he had said it all first, it will be noticed that he had ascribed a fixed direction to the wave jet. This, together with the lack of variation in velocity, appeared to be one of the possible reasons for the absence of wave period effects in the formula, so the analysis was reworked using a rotating stream-tube whose velocity was related to that of the impacting wave-jet (ie- breaker type) and the stability for variable water and particle directions of motion examined .

4. The resulting curves show the deleterious effects of a wide departure angle (ie rocks etc) and the rapidly diminishing stability as outwash approaches the minimum departure angle with the same locus of minimum stability the same as Hudsons theoretical value and close to the empirical cot angle function.

5. An empirical constant relating to the efficiency and/or drag factors of a particular unit is required to be determined experimentally and mathematically the stability can be shown to be dependant on the direction (angles) of water flow and particle movement (i.e. wave breaker type and particle shape and interaction). The relevant angle functions are:

General Equation:

$$R = H/C^B \ x \ (1-p) \ x \ (Sr - 1) \ x \ \emptyset a \qquad (3)$$

$\emptyset a$: This is the function relating water and particle movement directions and the force vectors. By considering various limit state conditions the following simpler relationships are derived:

uplift: coherent system $cos \ a \ /sin^2(a+b)$ } 3 /

          unitary system $sin \ d \ /cos^2(d-b)$ } $\sqrt{}$ cot a

sliding: instability $tan \ a > u$

          downslope $\dfrac{(ucos \ a - sin \ a)}{(k1.sin(2a + 2b)+k2usin^2(a = b)}$

upslope $\qquad$ $\dfrac{(u\cos\ a\ \pm\ \sin\ a)}{(k1.\sin(2a\ +\ 2b)+k2u\sin^2(a\ =\ b)}$

where   a   =   angle of slope
           b   =   direction of flow
           d   =   direction of incipient motion of particle
           u   =   coefficient of friction

6.   For the minimum stability cases, the angle functions may be simplified without a great loss of accuracy (and with empirically determined coefficients to represent the drag factors etc carefully determined to be the minimum factors) to:

Uplift $(\cot\ a)^{\wedge}1/3$

Sliding $\cot\ a$

7.   Practical results for discrete and coherent armour systems demonstrate a very similar curve for spilling and plunging waves, but for the outwash case the stability function remains low where the maximum outflow velocity occurs before the arrival of the next wave (Xsi > 3). This is also the value of Xsi at which its relationship to the level of runup ceases to be linear and becomes either constant or tends to diminish. This point was missed in the above exposition and its incorporation, whereby at Xsi > 3 runup is sensibly constant so too is the outflow velocity and the angle of outflow at peak velocity is constant as this now occurs before the arrival of the next wave. This substitution leads to a set of predicted stability curves with a quite remarkable degree of similarity to the experimental curves which suggest that relative stability under varying conditions is inversely proportional to the square root of run-up.

8.   The gabion results showed also the Xsi curve effect reported by Gunbak and Bruun. The effect of directional attack on the value of Xsi must be born in mind if one becomes tempted to take advantage of the high stability at Xsi less than 3.

9.   Subsequent analysis of experimental work (ref 17) has lead to the development of a similar relationship using the Irribarren No Xsi, which is the ratio of wave steepness to slope steepness, i.e. it describes the water trajectories implicitly.

$$\text{Ns} \quad = \quad C \ . \ \{\text{COS} \ a \ / \ \text{Xsi}^{0.5} \ \} \qquad\qquad (4)$$

$$\text{or} \quad D \quad = \quad \text{Hs} \ / \ \emptyset \ x \ (Sr - 1) \ x \ \cos\ a \ /\text{Xsi}_{0.5} \qquad (5)$$

$$\text{cf} \ R \ x \ (1-p) = \quad H \ / \ C_B 9 \ x \ (Sr - 1) \ x \ \cos\ a \ /\sin^{\wedge}2(a+b) \ (6)$$

R is defined as the slope normal layer thickness of the armour system whilst D is a notional thickness of an equivalent cube of the same volume as the armour element.

10. By inspection the minimum stability occurs occurs when the

water flow is normal to the slope or parallel to the minimum departure angle (sin(a+b)=1). The stability coefficent in the Blanket Theory form is defined at this minimum stability. The PIANC method defines the stability coefficient at Xsi=1, but this is not at the point of minimum stability and is thus a higher value. For slopes steeper than about 20 degrees values of Xsi=1 cannot be reached because of limiting wave steepness. The ratio between the angle functions is about 1.6 for minimum values determined at steep slopes.

11. The use of stability coefficients defined at other than the minimum value in a formula approach is a new development in revetment design and one which may be liable to mis-interpretation. As the Xsi value at minimum stability must be input for a broad spectrum design the minimum value must still be sought especially for any new variation in construction or design. For a formula approach the use of the minimum established stability ensures that minimum stability case is covered. The best approach is probably the chart approach.

12. The development of a predictive conceptual model which agrees well with experimental results means that the model can be used as a tool either to design better armour units or to use what is available to the best advantage. These "Blanket Theory" arguments were rehearsed in the early 1978 papers at Antwerp, Hamburg and Adelaide and the available references scoured for suitable units to compare. The theoretical expression in its fullest extent predicts both sliding and uplift failure modes, as experienced during the testing for Gabion revetments and in early Seabee trials with insufficient berms.

13. It is emphasised that the theory relates the slope normal surcharge i.e. the net normal height of the element to the disturbing effects of the incident wave. This is a crucial factor as it teaches that intelligent placement of stones with their long axes normal to the slope should be much more stable. It also suggests how to make armour units suit your own construction requirements and explains why Seabees work. Thus the dimension that should be taken for solid units is the product of the slope normal real height of the armour element (R) and the armour solidity (1 - porosity or voids ratio), i.e. the net normal height of the element.

14. Whilst Hudson was unable to derive any more detailed relationship from his data than the simple cot function, he did suggest that a more complete formula might contain the following factors:

for random rocks: $\qquad Ns = f(a, H/L, d/L, Damage)$ (7)

for concrete armour $\qquad Ns = f(a, H/L, d/L, R)$ (8)

and more generally

$$Ns = f(a,Cd,Cm,ka,kv,1/v\char`^2(dv/dt),H/L,d/L,H/d,P,r,h,m) \text{ etc} \quad (9)$$

Now that we have defined $Xsi= \{tan\ a/(H/Lo)^{0.5}\}$ and given Gordon's classic description of time/damage effects (confirmed independently by Broderick) (refs 4,19)

$$Ns = a\ D\char`^n \quad\quad\quad\quad (10)$$

it can be seen that the major items still missing are d/L and or H/d and the relationship between underlayer voids, grading etc.

15. It could be said that the problems that have occurred in the use of this formula have been because of a failure to consider the original waivers and a lack of updating. Whilst the simple Hudson formula may be wanting in essence the minor refinements made by the author to his conceptual model demonstrate a striking resemblance to the actual results.

16. Thus when the armour layer has been so improved that the units are not noticeably the weakest intrinsic link, failure modes may include sliding, wholesale uplift (profile modification, flapping and understreaming) and structural failure (breakage, shear, crushing etc). Gabions were an obvious coherent form and should obey the same rules. However, on steep slopes they fail by sliding unless adequately restrained. This failure mode is also present with blocks and must be restrained by an adequately surcharged toe or toe berm.

17. Of the many single layer units that appeared to be practical examples of Blanket Theory, Tribars, Svee Blocks, Gobi Blocks (now known as Antifers) Stolk Cubes and Cobs appealed as they all were used in a way to define the porosity. However Svee Blocks appeared to require a very accuratly prepared substrate whilst Tribars were difficult to lay and the Cubical shapes of Cobs and Gobis also required accurate preparation is shear was not to be an important factor in large units. Reports on stability of Cobs suggested also that when placed in columnar array a single unit loss could result in the loss of the whole upper part of the column. (Ref) The hydraulics of a degree of lateral porosity also appealed more than the Gobis single porosity, but why symmetrical porosity? The Cob and the Shed, whilst appearing to be variable in practice appear to be unique forms and thus belong in the same camp as other fixed geometry units as regards variablity but quite obviously in their own league as regards to stability, except for the Pattern Placed Tribar which achieved a notable Kd in tests of 125.

18. Bearing all this in mind, a shape was sought which would allow simple manufacture of variable sizes and accept dislocations in array without induced nasty stresses. The hexagon was found to comply and allows a unit with remarkably constant section thicknesses so minimising temperature stresses. So the Seabee was born, with optional slots in (some) sides legs, bumps etc etc. Central to the Seabee system is the idea that the

engineer, be he designer or contractor, should be in control of all aspects of sizing the product to suit the need.

19. To begin testing however aluminium and ceramic extrusions were used and these had no lateral porosity. Initial testing was directed towards demonstrating the validity of the surcharge concept by varying both height and porosity. Initial service values were close to those reported for the reported multiaxis porosity units (Cobs, Cobwebs, Pattern Tribars) and the product first saw the light of day as an hexagonal vitrified clay extrusion. Completion of the prototype structure lead to the development of the multicell ceramic unit which has now been in production for ten years.

20. During early project testing many aspects were instigated, including:

- stability of armour layer outside the impact zone
- behaviour of broken units
- effect of missing units, extraction of underlayer
- construction stability of working edge in construction wave climate
- ultimate behaviour design
- run-up control
- toe-scour and slippage control
- manufacturing and casting techniques

21. Roughnesses were modelled on the underside and top-side of the units, mainly in the form of splitter walls to divert flow across the pores. No significant benefit was found and unit costs were increased due to more complex forming. We also experimented with a large voided crest caisson unit with an entry slot at armour surface level to absorb run-up. It worked very effectively but was quite expensive. It was found that the most economical way to control run-up was to eliminate the armour layer as soon as possible and allow direct access of the run-up to the secondary zone armour as well as providing a thicker zone than normal. Elongation of units at toe and crest allowed superior behaviour, and the use of loose shear-rocks or "keys" in every other toe unit provided a caisson toe unit that could allow up to 2.5m local scour without inducing slippage of the toe.

22. Construction tolerances were carefully modelled, with inter-unit gaps of up to 0.5 m and severe surface irregularity of the sub-layer. Failure modes were investigated and for low aspect ratio units , found to be progressive and quite well-behaved. In one case an armour unit was ejected from the slope by an oversize underlayer stone so thet the stone took the unit's place in the array and the unit sat on the surface of adjacent units. Where it remained for a further 30 mins!

23. By mistake, some of the fancier geometry units were assembled using PVA glue, which is water soluble. This was not really noticed until we tried to remove them when they came to

bits in our hands. But they survived the waves. Similar tests were run with split and broken units comprising from 2 to 6 pieces. Patches of two or three units were quite stable, but areas of over ten units, if placed to construction tolerances, gradually moved around and lost the appearance of normal units.

24. The effect of missing units was tested and led to the adoption of a maximum diameter of 2.5R. Above this, substrate material can be lost during the design storm by continuous beaching.

25. The first large concrete units (4 tonnes) are almost exactly the same proportions as the early squat ceramic test units because they were so cheap to make. The simple unit rolled out of the factory at a rate of 18 per mould per day with a production cycle time of about 15 minutes per unit.

26. To date some twenty installations have been completed for wave climates from 0.5 to over 5m and in water depths from tidal to over 5m. In addition a large number of Contractor supported alternate bids have been made for Breakwaters for conditions up to and including 12m overtopping waves in Australia, Oceania, S.E Asia, Pacific, Caribbean, USA and Europe.

27. In something approaching 90% of these cases the Seabee design has been significantle cheaper whilst being designed with a considerable factor of safety against damage and bearing an increased profit margin for the Contractor. In the majority of cases the alternative design is not received with any great sign of relish and the conforming design has on many occasions proceeded although apparently at a lost saving.

28. Of 15 such bids, some supported by more than one contractor, two have failed to proceed, six were awarde to rock, one to Dolosse, one to Hanbars and five to Seabees. Of the Rock projects, two apparently proceeded OK (no more heard) and four ran into major supply problems. One required a major redesign and over-ran by 150%, largely at the contractors expense, one required removal of approx 100% additional waste material rfom the quarry and was redesigned during construction, one had severe supply probles but completed and in the fourth the quarry did not work as forecast and the contract was determined pending a redesign. The dolosse project completed on time but was a hassle to measure whilst the Seabee and hanbar jobs were all apparently completed on time and at a profit.

29. It might be surmised that whilst the risk of the apparently new may be too much to chance, the age-old gamble with the quarry is always worth another throw. But at whose expense?

30. In preparing a detailed design, the usage of armour material over the armoured area is modified and refined so that the factor of safety is reasonably uniform over the whole of the

structure. Thus varying amounts of armour surcharge are provided to various areas in accordance with incident wave conditions and location on the slope.

31. The following order or algorithm is recommended for the detailed design study.

.1 Choose crest elevation to suit wave transmission criteria.

.2 Check for adequate height above extreme water levels.

.3 Design storm zone armour layer for design wave spectra

.4 Optimise armour unit sizes and distribution

.5 Determine underlayer sizes from quarry grading curves and compare with chosen unit sizes

.6 Determine core material grading(s)

.7 Detail toe berms

.8 Compare required grading with projected quarry yields and revise and optimise as required

.9 Conduct model study; iterate.

32. The crest elevation is determined by establishing the allowable transmitted wave heights and comparing with the known wave climate to give required wave or water transmission criteria. These criteria can be matched against available test data to give an estimate of required minimum crest level. Testing carried out in a random wave flume for two projects in Australia has given the indicative data for wave transmission shown in fig Minimum recommended crest super-elevation (Zc) is one third significant wave height so that the crest level may be established from the point of view of ultimate stability rather than minimum service conditions.

33. The Storm Armour Zone is considered to lie between the levels of one half of wave height below low water mark and one quarter the wave height above storm water level. If the variation in water levels has a significant effect on wave height at the structure, an envelope surcharge diagram can be prepared for the structure, showing the variation of required surcharge with water level. Outside the wave rotor zone the armour layer is indirectly affected by the wave jet and armour surcharge may be progressively reduced to about 60% of this value above and below these limits and on the crest and lee of breakwaters meeting the above crest level criteria.

34. The maximum irregularity in the placement of units is a 30% height projection. This also applies to armour zones, so that three zones of armour may be used, being 62% 78% and 100%

164

of storm armour requirement. Maximum squatness of units is an aspect ratio (R/D) of 0.4, as described above. Optimisation will frequently be found to allow the use of much greater radii with available plant.

35. The Armour layer (or carapace) can also be optimised along the length of the breakwater or seawall in conformity with incident wave heights, given due consideration to any future changes in bathymetry.

36. The most convenient method of presenting this information is in the form of a side elevation of the structure,with the required zone surcharge requirements drawn to a convenient vertical scale. Given a maximum number of unit size choices of say,between 3 and 5 (to avoid logistical problems), the optimum choice of sizes can be made. (An example is shown in Fig. 5.)

37. Calculations should be carried out to check the strength attributes of the units for larger waves. Important values are the bridging strength of the unit under impact load and the effect of internal impact pressure on the hoop strength of the unit. Calculations are made using the bulk modulus for concrete and the net tensile stress as calculated is compared with allowable tensile stresses in concrete. For design significant waves of the order of 12m (40 ft) consideration may be given to reducing the porosity in the wave impact zone to reduce these stresses or providing radial shock relief vents.

38. Minimum underlayer thickness is 0.65 times the maximum armour layer thickness. Thicker layers may be used to advantage especially when they are constructed from porous large widely graded material. The effectiveness of this increased porous volume on service functions such as run-up and reflection diminishes with thickness and it is probably uneconomic to exceed 1.5 R in thickness.

39. The maximum size of underlayer should be such that the rocks do not act as plugs in the unit voids. This requires that equivalent diameter of underlayer grading should not exceed 1.2 times void diameter. On the other hand underlayer material may be as fine as one sixth unit base diameter. Material finer than this may be washed out of the hole left by a missing unit.

40. Optimum size of underlayer is determined also from considerations of construction damage and material availability as well as ultimate behaviour and thus the final choice of armour size may be affected.

41. To provide a damage resistant structure against depth limited waves instead designing for a ' statistical storm ' may require only an additional 15% concrete in the total armour layer, or a 3% increase in total project estimate. This is probably less than the difference between the two lowest bidders.

42. The inverting form system used at Abbot Point allowed up to eighteen 4 tonne units to be cast per day per form, using a zero slump low water/cement ratio concrete, with a six man crew.

43. A lower cost alternative is to use gang-forms so that each setup and strike sequence produces 4 or 6 units at a time. Depending on site scheduling and daily work rate, two or three pours per day are possible, so that 32 or 48 units may be produced per day with only four sets of 4-unit gang-forms.

44. Placement tolerances allowed by the system are much greater than those utilised on jobs constructed to date. Hydraulic model testing is usually carried out with surface irregularity of 20 to 30 per cent and with packing factors of between 90 and 95 per cent and also use broken and missing units.

45 The Seabee is intended to be self-locating with diving inspection only to establish the adequacy of placement. This is one of the reasons for the hexagonal shape, in that it provides a natural notch into which a unit is lowered.

46. All techniques are much enhanced if rigid boom equipment is used for placement. Such equipment has a much faster cycle time and can allow remote release of the unit as well as digitised location techniques.

47. Placement underwater can make use of both the tolerance and the reduced surcharge requirements to produce an acceptable result. It should be remembered that a planar, tight surface is not a requirement for success, least of all underwater. Such techniques should be reserved only for those areas where appearance has a very high priority (i.e where rock has already be discounted).

48. Some research has been undertaken into the behaviour of randomly placed unit, which appear to shakedown into a satisfactory layer within a few hundred waves. However, an extended programme of research and investigation into both the effect of wave time-history and material properties would be required to pursue this further.

49. The most detailed tailored design, which was subjected to several series of tests, was the alternative design proposed for the overtopping breakwater at Eden. A minimum of three different sizes were recommended, with further benefits resulting from the use of a total of five different sizes. Abbot Point uses two sizes and Wewak three.

50. Proposals for an extension to the Eden Breakwater with a requirement for 60,000 tonnes of 15-25t armour rock or 30,000 tonnes of Hanbars, developed for somewhat less than the 1 in 50 yr exposure (with 5-10% damage and overtopping) were advertised for tender. An alternative design was developed and tested which

offered zero damage behaviour in the 1 in 250 year event at a suitable water level, whilst reducing armour mass to 12,500 tonnes and unit size to a range of 6 to 12 tonnes (Contractors choice). 4 out of 7 contractors bid the design as the lowest bid including the extra risk of it being their own alternative, showing savings of the order of 25% less than the Hanbars and rock designs.

CONCLUSIONS

51. The most satisfactory design tools are those where empirical results and conceptual model predictions are in close agreement. The simple amendment of allowing the jet to rotate in Hudson's conceptual momentum model can be seen to give theoretical stability curves very similar to the general run of empirical results which suggests that the model is a good tool for analysing the behaviour of armour and its desirable attributes.

52. By virtue of design flexibility offered by variable geometry arrayed armour units implicit in the linear Blanket form of the theory, it is possible to design for a factor of safety against damage and still proportion a unit that makes the optimum use of resources. Breakwaters and Seawalls are compromise structures and a balance must be struck between hydraulics, durability and economy. It is rare for the balance at one site to be the same as at another and so the ability to reflect these variations iis important. Single layer variable geometry armour units offer realisable economies, increased predictability and usually reduced or zero maintenance.

53. The control of run-up and reflection control are important but perhaps even more important is the maintenance of an adequate integrity of the unit and the armoured slope even after minor damage. The hexagonal shape of the Seabee allows a very lightly stressed unit to be placed in quite poor arrays. If the Seabee units break the pieces still do their part of the work. It also takes a great deal to break them.

54. The unit has shown a repeated ability to compete on economic terms with rock from very short hauls and is recommended for consideration. An engineer can however make many more meaningful decisions with Seabees than he can with fixed geometry units.

55. The contemporary Australian practice of putting up a number of armour designs for tender, although not discussed above, is strongly recommended as it allows the tenderers to reflect market place economies not usually visible to the designer.

56. The problem of risk in supply from quarries is one that is poorly allowed for by our profession at present. We seem to be eternal optimists.

57. Despite his obvious interest, the Author is concerned at the

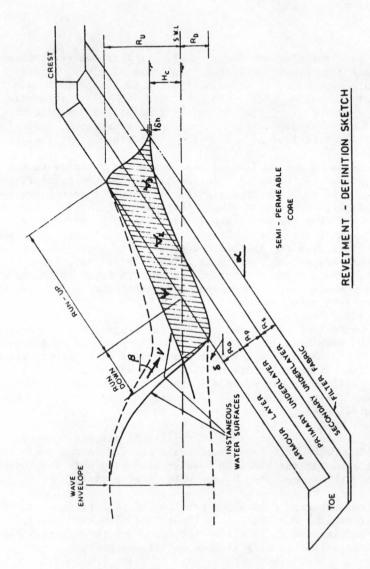

Fig. 1. Elements of the slope

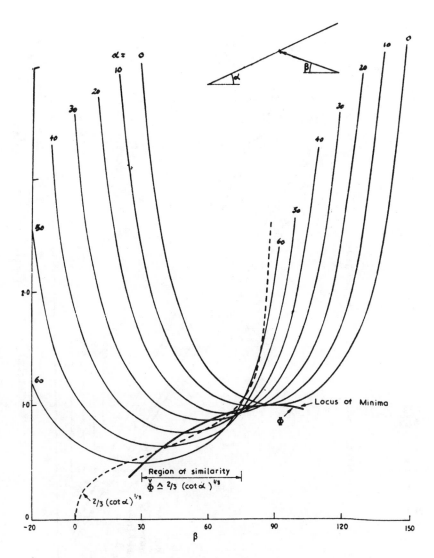

Fig. 2.   Linear theory stability curves

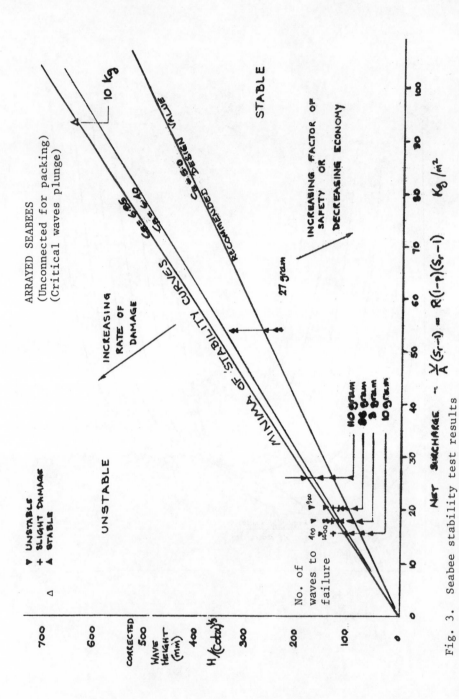

Fig. 3. Seabee stability test results

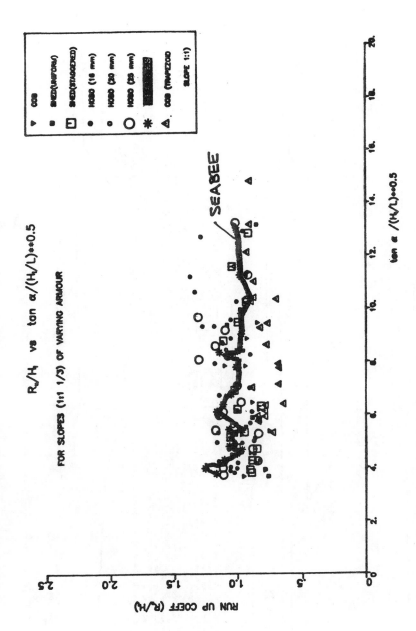

Fig. 4. Seabee and hollow block service behaviour

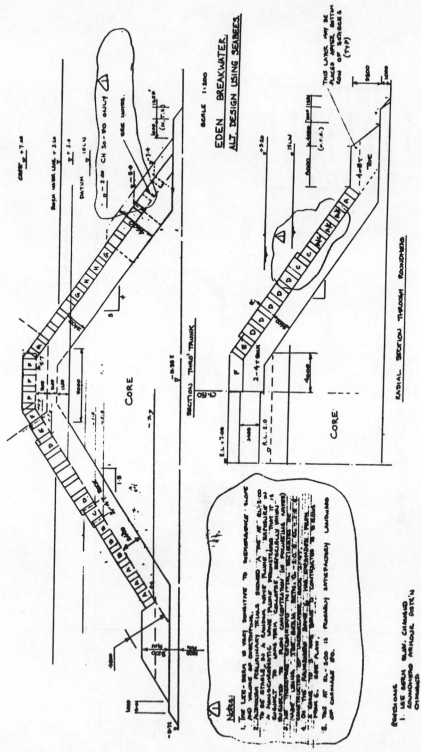

Fig. 5. Typical breakwater profile with variable geometry

172

Fig. 6. Typical gang formwork

longevity of any armour system, rock or concrete that is designed on the basis of substantial and repeated movements. Most rock has some weaknesses and concrete is not the world's strongest or most crush resistant material. More in situ measurements are require of prototype wear rate.

58. Slope stability and improved knowledge of substrate behaviour may well require different or more complex profiles, but I believe that Seabee type of armouring will eventually be seen on the best breakwaters, because they will offer the best economies and the most predictability. It is encouraging to see the increasing use of the linear representation of armour stability as it implies an increasing willingness to consider single layer armour systems.

REFERENCES

1. CASTRO, Eduardo, Disques de escolleru, Revista de Obras Publicas, Lisbon 1933

2. Epstein Tyrrel, Design of Rubble Mound Breakwaters, Proc 17th PIANC Congress July 1949.

3. Hudson R.Y. Laboratory Investigations of Rubble Mound Breakwaters, Journal Waterways & Harbours Division Proc A.C.E. No WW3 Sept 1959 pp 93-121 and discussion.

4 Gordon A.D. The Stability of Armour Units against Breaking Waves, M.Eng Sc Thesis, University of NSW, March 1973.

5 Per Bruun & Ali Riza Gunbak, Hydraulic and Friction Parameters affecting Stability of Rubble Mounds. PIANC bulletin No 24 1976.

6 Anon. Artificial armouring of Marine Structures, Dock and Harbour Authority, Vol LI No 601, November 1970.

7 Anon. Tests of Modified Cob Block, HRS Wallingford Report No Ex 622, April 1973

8 Report on Breakwater Design for Stage 1, Port Kembla Coal Loader Reclamation PPSK August 1974

9 W.Allsop, Pers Comm Nov 1977.

10 C.T.Brown Seabees, A Third Generation Armour Unit. 7th Int'l Harbour Congress, Antwerp.1978

11. C.T.Brown Blanket Theory & Low Cost Revetments 16th ICCE, Hamburg August 1978

12. C.T.Brown Armour Units - Random Mass or Disciplined Array?

Coastal Structures, ASCE Specialty Conference,Washington.
March 1979

13   C.T.Brown Gabion Report. W.R.L.Research Report No 156.
      (University of N.S.W.)October   1979

14.   C.T.Brown Seabees in Service.Coastal Structures A.S.C.E.
      Specialty conference, Washington D.C. March 1983

15   P.C. Barber & T. C. Lloyd "The Diode wave dissipation
      block" Proc ICE Paper 8826. Nov 1984 And discussion,
      1986.

16.   C.T.Brown Seabee Users Manual, 2nd Edition February 1985

17.   PIANC Guidelines for the design and construction of
      Flexible Revetments etc, Report of WG4, PT1 May 1986

18.   N.W.H Allsop Sea Walls a Literature Review. Report No
      Ex 1490 September 1986

19.   L.Broderick Riprap Stability: a Progress Report. Coastal
      Structures,ASCE Specialty Conference,Washington. March 1979

9. Single layer armour units

J. A. DUNSTER, Shephard Hill & Co. Ltd, A. R. WILKINSON,
Coode Blizard, and N. W. H. ALLSOP, Hydraulics Research,
Wallingford

SYNOPSIS. This paper reviews the use to-date of hollow
single layer armour units with specific reference to 'COBs'
and 'SHEDs' including details of hydraulic performance,
project designs, manufacturing and construction methods.
A description is given of present research to describe
more fully their performance and to identify limiting
conditions for their use.

INTRODUCTION

1. Hollow block armour units have now been in use since
the early 1970's (Ref 1). Several different types of unit
have been developed such as the 'COB', the 'SHED', the
'DIODE' (Ref 2) and the 'SEABEE' (Ref 3). General details
of these units are shown in figure 1. As the 'SEABEE'
has porosity in only one plane and the 'DIODE' has been
installed in only one location the rest of the paper will
concentrate on the COB and the SHED.

2. Hollow block armour differs from conventional
breakwater armour in two fundamental ways. Firstly the
wave energy is destroyed within the large void of the unit
and not in voids between adjacent units. Secondly the
units are placed in close contact forming a stable
structural surface layer of fixed high porosity. The
armour is not forced apart by the waves and therefore
does not move like random rock and concrete armour. This
additional stability allows the use of much smaller
individual units in single instead of double layers
resulting in greatly reduced quantities of armour. The
quantity of embankment fill can also be reduced by the use
of steeper slopes than can be used with random armour.
These reductions result in major cost savings.

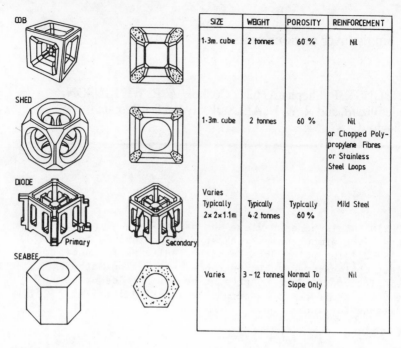

	SIZE	WEIGHT	POROSITY	REINFORCEMENT
COB	1·3m. cube	2 tonnes	60 %	Nil
SHED	1·3m. cube	2 tonnes	60 %	Nil or Chopped Polypropylene Fibres or Stainless Steel Loops
DIODE Primary / Secondary	Varies Typically 2×2×1.1m	Typically 4·2 tonnes	Typically 60 %	Mild Steel
SEABEE	Varies	3 – 12 tonnes	Normal To Slope Only	Nil

Fig. 1

<u>**HYDRAULIC PERFORMANCE**</u> (Notation listed at end of paper)

3.   The main design parameters required in the design of conventional rubble mound armour layers are:

a)   the degree of armour movement under wave action;
b)   wave run-up, and run-down, levels;
c)   the severity of wave reflections.

4.   Hydraulic model tests have been conducted to measure the resistance to armour unit movement for COBs or SHEDs at Wimpey Laboratories; Queen's University, Belfast; Imperial College, London; and at Hydraulics Research, Wallingford. In each instance it has proved extremely difficult to fail the armour layer by displacing individual armour units. This may be illustrated by comparing the values for a design dimensionless wave height, $H_s/\Delta D_n$, where $D_n = (W/\rho_c)^{1/3}$, and $\Delta = (\rho_c/\rho_w)-1$, with values for the same parameter for other armour units:

$$H_s/\Delta D_n$$

	$H_s/\Delta D_n$
Rock	1.5-2.0
Cube	2.4
Dolos	2.7
Accropode	2.5-3.7

5. In random wave tests of 2 tonne SHED units Allsop
(Ref 5) noted that minimal armour movement occurred for
waves heights up to $H_s/\Delta D_n = 5.7$.

6. Wave run-up levels on SHED units have been presented
previously (Refs 4,5), and compared with run-up levels on
smooth slopes. For values of the dimensionless wave
height $H_s/\Delta D_n$ between 3.0 and 5.5, and for values of the
Iribarren number, $Ir = \tan\alpha/(S_m)^{\frac{1}{2}}$ between 2.5 and 4.0, the
significant run-up level R_s was consistently around 62% of
that on the equivalent smooth slope. The 2% run-up level,
R_2, was generally around $1.5\ R_s$.

7. Wave reflections on armoured rubble slopes have
recently been considered in detail by Allsop & Hettiarachchi
(Ref 6). They have re-analysed measurements of regular
wave reflections from COB armoured slopes, and reflections
of random waves from SHED slopes. Using the surf
similarity parameter or Iribarren number, $Ir = \tan\alpha\,(H/L)^{\frac{1}{2}}$
for regular waves, and the modified Iribarren number,
$Ir' = \tan\alpha/(H_s/L_p)^{\frac{1}{2}}$, the coefficient of wave reflections,
C_r, were defined:

COBs, regular waves,
$$C_r = \frac{0.5\ Ir^2}{Ir^2 + 6.54}$$

SHEDs, random waves,
$$C_r = \frac{0.49\ Ir'^2}{Ir'^2 + 7.94}$$

8. An interesting series of experiments have been
conducted by Hettiarachchi (Ref 7) who measured wave forces
on a single SHED armour unit. The results are presented as
force parallel to the slope, + upward, - downward; and as
+ ve and - ve lift forces normal to the slope. In the
preliminary analysis the results of force measurements are
presented scaled by corresponding component of the
submerged weight of the unit. The maximum scaled positive
lift force on the single unit only exceeded unity for
steep waves, $H/L_o > 0.06$. The wave force up the slope,
scaled by the slope-parallel component of the submerged
weight, reached maximum values around 4-5 for
$0.055 < H/L_o < 0.070$. It is expected that further analysis
of these results will be presented at this conference
(Ref 8).

PERFORMANCE IN SERVICE

9. COBs and SHEDs have now been used effectively on
various projects in different parts of the world. Here we
briefly describe their use to-date.

Jersey, Channel Islands - La Collette Harbour (fig. 2).

10. COBs were placed on the La Collette Breakwater during
1973 and 1974. This was the first ever use of hollow
block units and was preceded by extensive model tests.

179

The breakwater is 770 metres long, faces the prevailing
storms from the S.W. and is armoured with about 8700
two tonne COBs. The approaching waves were recorded on a
wave recorder about two miles away to the S.W. The
estimated height of the maximum waves at the breakwater is
over 7 metres. A pressure wave gauge has recently been
installed close to the breakwater to obtain more exact
information. The slope of 1 in 1.33 was selected by the
stability of the stone before placing the COBs. This steep
slope greatly reduced the quantity of stone and the cost of
the breakwater. The concrete toe wall and the size and
weight of the COBs are, with hindsight, probably hydraulically overdesigned, but the 2 tonne units are the most
economical to handle and place. The COBs have been in
place at La Collette for 14 years and no sign of movement
of units has been observed. Some of the underlying stone
is considerably smaller than the apertures in the COBs, but
it is not dragged out by the waves. Although some units
have exhibited cracks since the time of construction no
failure of units has occurred and stability does not appear
to have been affected.

Das Island, Arabian Gulf (fig. 3).

11. COBs were used in coastal defences on Das Island in
1977. Sand filling had to be protected and this was done
by placing "Fabriform" on the sand slope at an angle of
1 in 1.33. COBs were then placed on top of the "Fabriform".
The COBs destroy the wave energy and reduce the run-up,
while the "Fabriform" contains the sand but allows escape
of water pressure from behind. Maximum wave heights were
only about 1.5 metres here, but 2 tonne COBs were used for
convenience of placing. The concrete filled nylon sack
of the "Fabriform" destroys less wave energy than rubble
below the units, but, shows many possible uses where stone
is not available.

Jersey Channel Islands - Sea Wall, St.Helier (fig. 4).

12. SHEDs were first used to provide the primary
armouring layer of a new sea wall at St.Helier. The wall
was completed in June 1983 at a cost of £2¼ million and
incorporates some 4200 two tonne SHED Armour Blocks on a
slope of 1 in 1.5, designed to withstand a significant
wave height of 3.5m in a maximum tidal range of 12m. About
25% of the cement in the concrete mix was replaced by PFA
to assist workability and increase resistance to sulphate
attack, and also to eliminate the dangers of alkali
aggregate reaction, which has been a problem in Jersey in
the past. Chopped polypropylene fibre reinforcement was
also incorporated in the mix, in order to increase the
impact resistance of the concrete. A small number of
units cracked due to toe and bank settlement during
construction but no further cracking or movements have

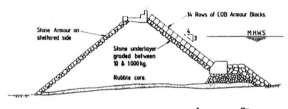

Fig. 2. La Collette Harbour, Jersey

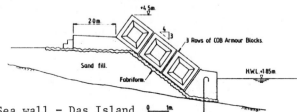

Fig. 3. Sea wall – Das Island

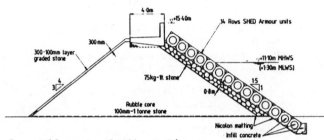

Fig. 4. Sea wall west of Albert Pier,
Jersey

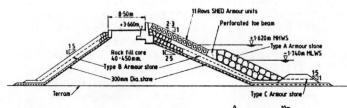

Fig. 5. North breakwater – Bangor,
Northern Ireland

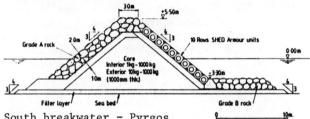

Fig. 6. South breakwater – Pyrgos
Marina, Limmasol, Cyprus

181

been recorded. Random armour breakwaters are dangerous
for the public and access to them needs to be limited
whereas hollow block armour is relatively safe and this
sea wall provides a safe play area for children.

Bangor, Northern Ireland - North Breakwater (fig. 5).
13.   Some 2700 two tonne SHEDs were introduced during the
construction of this breakwater in 1982/83 when local
quarries were unable to supply the specified 6 to 8 tonne
armouring rock in sufficient quantities within the
allowable time scale. The approximate cost of the scheme
was £2.3 million. The use of SHEDs on the upper part of
the structure above low tide level arose from the change
of design during construction.

Limassol, Cyprus - Pyrgos Marina (fig. 6).
14.   About 4000 SHEDs were placed on a breakwater about
450 metres long. About 2000 of these units were
permanently under water. The Contractor was able to
place an average of 40 SHEDs a day working at depths of
up to 6 metres below water. On most previous projects high
tidal movements allowed most units to be placed above the
water line. The toe wall was formed of precast concrete
units supported by a rubble apron. The maximum waves
expected on this breakwater are about 5 metres in height.
The breakwater was completed in 1984.

Fort Jalali, Oman - Sea Defence Revetment (fig. 7).
15.   A 250 metre long revetment formed with 5 tonne
DOLOS units at Fort Jalali failed in service and was
reconstructed using about 3500 2 tonne SHEDs more than
half of these being placed under water in depths of up
to 6.5 metres. The contract was started in April 1985
and completed in November of the same year. Some 65 units
exhibited cracking on completion and some settlement of
the revetment has occurred since construction. Cracking
is mainly due to point loading between units arising
from the settlement and is not considered a problem for
future stability. Considerable crustaceous accretion
has occurred on and between units since construction
adding to the fixity of the SHEDs in the armour layer.

Muscat, Oman - Al Inshirah (fig. 8).
16.   Two breakwaters were constructed in 1986 to protect
a leisure development between Muscat and Muttrah. About
1300 two tonne COBs were used to armour the breakwaters
against maximum waves of about 3.4 metres. An underwater
in situ concrete toe wall was formed and COBs were placed
above water level and also below water level to a depth
of about 10 metres. The work was completed in less than
5 months.

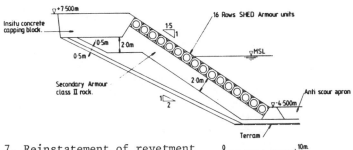

Fig. 7. Reinstatement of revetment
- Fort Jalali

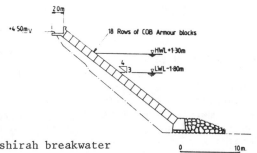

Fig. 8. Al Inshirah breakwater
- Muscat, Oman

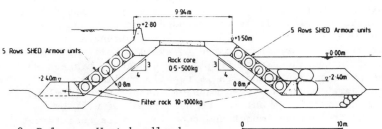

Fig. 9. Dalyan - West headland,
Istanbul, Turkey

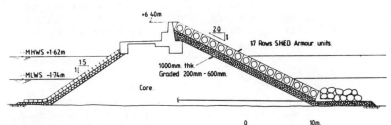

Fig. 10. Pickie breakwater - Bangor,
Northern Ireland

Istanbul, Turkey - Kadikoy Breakwaters (fig. 9).
17. A series of ten breakwaters have been constructed
to protect five tourist beaches on the Marmara sea near
Istanbul. Approximately 9000 2 tonne SHEDs have been
placed mainly underwater. The contract was started in
July 1987 and completed in February 1988.

Bangor, Northern Ireland - Pickie Breakwater (fig. 10).
18. This second breakwater at Bangor is under
construction at present. It will incorporate approximately
3500 2 tonne SHEDs and is the first structure making
use of special spacer rings to correct the alignment of
units arising from tolerances in the placing of units
under water. It also makes use of stainless steel ring
reinforcement in the units to limit damage in the event
of ship collision during a storm. The 1 in 2 slope has
been used partly due to foundation stability but also
to minimise wave reflection and run-up.

PRODUCTION OF ARMOUR UNITS
19. Both COBs and SHEDs are manufactured either on site
or in a factory using special steel external moulds with
collapsible internal formers. In the case of the COB
the internal former is made of steel plates whereas the
SHED uses an inflated spherical bladder which prevents
shrinkage cracking of the block prior to stripping the
mould. One or more casts can be achieved daily and in
the case of the SHED the inner bladder can be removed
within a few hours if required.Concrete is placed using
internal vibrators or by placing the complete mould on a
vibrating table. It is recommended that high quality
concrete is used to provide high early strength and
durability. Care is required to avoid alkali aggregate
reaction and the use of PFA is advantaegous in areas where
the problem exists and PFA is readily available.
Considerable use has been made of chopped polypropylene
fibre as reinforcement in the proportion of 0.2% by
weight to improve impact and handling stress resistance.

PLACING OF ARMOUR UNITS
20. Above water the placing presents no problems and can
be carried out using conventional hydraulic excavators to
prepare the bed and cranes for placing the units. Outputs
of 50 or more units per day can easily be achieved under
these conditions with one gang.
21. For underwater operation the same equipment can be
used with divers carrying out the final placing and release
of units at about half the output achieved above water.
Where continuous swell or wave action exists and underwater
visibility is poor special equipment needs to be developed

to place units underwater without the use of divers. No project has required this to date but schemes have been prepared in outline to deal with this eventuality.

22. Accuracy of placing underwater is more difficult than above water and in the case of the Pickie breakwater at Bangor special spacer units are being incorporated to achieve precise lines and levels above low water level. Similar spacers can be used to construct round heads with greater precision and interlock if required.

23. The only time the breakwater is vulnerable to failure is during construction and to minimise this risk the following actions can be taken :-

a) All operations of placing core, bedstone and armour should be carried out closely together at the same rate of linear production to avoid long lengths of exposed embankment.

b) In the event of storm warnings the leading face of armour units can be reinforced temporarily by a row of interlocking double units made by glueing standard units together with epoxy resin adhesives. The exposed core and bed material can be temporarily protected by special gabion mattresses kept ready for use when required.

24. Placing patterns to date have allowed variation from full brick type bonding to straight joint lines and no special units have been used to maintain bonding patterns around curves. It is debatable whether special bonding patterns are necessary but the following points have arisen from experience to date :-

a) Bonding in model tests increases the strength of the armour layer when one or more units are removed.

b) In full scale structures bonding causes cracking of units as point loads occur on units when embankment or toe settlement occurs.

25. On none of the breakwaters constructed so far has any unit been plucked out and in no case has the cracking of units caused further disintegration. Further study of this aspect is required to determine the most satisfactory solution.

PRESENT RESEARCH

26. Any assessment of the performance of concrete armour units must include details of both the hydraulic and structural performance of the armour. A recent review of conventional concrete armour units suggests that many of the instances of armour layer failure have been precipitated, or accelerated by structural failure of individual armour units (Ref 4). Severe damage to breakwaters armoured with Dolosse, Tetrapods and Tribars

has highlighted major gaps in understanding of armour layer performance. It has been noted that the assessment of wave and settlement loads acting on an individual unit, and of the strength of the unit to resist these loads, is very difficult and the results are presently uncertain. This is aggravated by the random nature of interlock of conventional armour layers.

27. Regularly placed single layer units offer considerable hydraulic stability for a given armour unit size, often yielding considerable savings over conventional armour. However, it is often not possible to exploit fully these advantages as the designer may be uncertain of the imposed loads, or of the capacity of the unit to resist those loads.

28. A collaborative research project has been set up to study both the hydraulic and structural performance of single layer armour units, in the laboratory and in the field. The work is supported by a research club funded by its industrial and reseach members, the Science and Engineering Research Council, and the Department of the Environment. The project is intended to provide the designer with the means to determine the hydraulic performance and the armour unit stresses for a wide range of structural configurations and wave conditions. It is expected that this will require the derivation of a range of advanced mathematical models for hydraulic and stress analysis, as well as data from measurements in structures and hydraulics laboratories, and in the field.

29. To date the research work has been conducted principally by Bristol University, Plymouth Polytechnic and Hydraulics Research. Active participation from the other members of the research club has been essential in field and laboratory work, in the provision of armour units for testing, and in deploying and operating field instruments. Already 2 full size SHED units have been tested to destruction in the structures laboratory at Bristol. A further 8 SHED units have each been equipped with 4 vibrating wire strain gauges, and have been deployed on Pickie Breakwater, Bangor, Co. Down. Field measurements of wave run-up and impact pressures have been made during winter 1987/88 on La Collette Breakwater, St.Helier, Jersey. It is intended that similar measurements will be made at Bangor in winter 1988/89, together with armour unit strains and phreatic water levels within the breakwater. Preliminary mathematical models of wave action on steep slopes, wave force and armour unit stresses are under development.

30. In parallel with this research, a separate project has also been started to examine the geotechnical stability of rubble mounds under wave action. This is of particular importance for steep slopes armoured with relatively small

units. This research is led by Hydraulics Research,
supported by mathematicians at Bristol University and
geotechnical engineers at Sheffield University.

CONCLUSIONS

1. The use of hollow single layer armour units overcomes
the inherent design and stability problems associated with
random unit breakwaters.

2. Failure of breakwaters faced with single layer hollow
armour is likely to arise from embankment collapse rather
than surface erosion.

3. Units of 2 tonne weight are adequate for waves up to
or exceeding 7 metres.

4. Further research is required to describe more fully
the performance of these types of unit and to identify
limiting conditions for their use.

5. Correct design of layer boundaries at the toe, top
and ends is vital to maintain structural strength of the
armour layer.

NOTATION

H = Wave height

H_s = Significant wave height

D_n = Nominal unit diameter

W = Characteristic unit weight

ρ_c = Unit density

ρ_w = Water density

α = Slope angle

S_m = Sea steepness

R_2 = 2% run-up level

R_s = Significant run-up level

L = Wave length

L_p = Wave length for peak wave period

L_o = Deepwater wave length

C_r = Coefficient of reflection

ACKNOWLEDGEMENTS

The authors are grateful for the assistance of their colleagues in the preparation of this paper. In particular they are grateful for the assistance of the Single Layer Armour Research Club supported by :-

Coode Blizard
Kirk McClure & Morton
G.Maunsell & Parters
Posford Duvivier
Shephard, Hill & Co., Ltd.

Soil Structures International
States of Jersey
Hydraulics Research
Plymouth Polytechnic
University of Bristol

REFERENCES

1. WILKINSON, A.R. & ALLSOP, N.W.H. "Hollow block breakwater armour units." Proc Coastal Structures '83, ASCE, Arlington, 1983.

2. BARBER, P.C. & LLOYD, T.C. "The 'Diode' wave dissipation block." Proc ICE Part 1, November 1984.

3. BROWN, C.T. "Seabees in service." Proc Coastal Structures 83, ASCE, Arlington, 1983.

4. ALLSOP, N.W.H. "Concrete armour units for rubble mound breakwaters and sea walls: recent progress." Report SR 100, Hydraulics Research, Wallingford, March 1988.

5. ALLSOP, N.W.H. "The Shed breakwater armour unit, model tests in random waves." Report EX 1124, Hydraulics Research, Wallingford, April 1983.

6. ALLSOP, N.W.H. & HETTIARACHCHI, S.S.L. "Wave reflections in harbours: design, construction, and performance of wave absorbing structures." Report OD 89, Hydraulics Research, Wallingford, April 1987.

7. HETTIARACHCHI, S.S.L. "The influence of geometry on the performance of breakwater armour units." PhD Thesis, Imperial College, University of London, May 1987.

8. HETTIARACHCHI, S.S.L. & HOLMES, P. "Performance of single layer hollow block armour units." Poster paper to Breakwaters '88, ICE, Eastbourne, 1988.

9. STICKLAND, I.W. "COB units - Report on Hydraulics model reseach." Ref No H/334, Wimpey Laboratory, 1969.

10. READ, J. "Discussion on papers 3 & 4." Proc conf Breakwaters '85, ICE, London, October 1985, pp 130-132.

10. Examples of design and construction of various breakwaters in Japan

H. SATO, M. YAMAMOTO and Dr T. ENDO, Hydraulic Laboratory, Nippon Tetrapod Co. Ltd, Japan

SYNOPSIS. In Japan breakwaters are usually constructed using caissons, and in most cases these are protected by armour blocks in order to decrease the wave forces acting on the caissons. In recent years breakwaters have been placed at deeper and deeper locations further from the shore in large wave areas. In our paper we introduce examples of both "caisson" and "composite" breakwaters and present the results of experimental studies on stability and relationships between waves and damage to breakwaters after construction.

BREAKWATERS IN JAPAN

1. Breakwaters are constructed to protect harbours from wave action. Typical breakwaters used in Japan are classified as the following types. (1) Rubble mound breakwaters, (2) Caisson breakwaters and (3) Combinations of armour blocks and caissons (hereinafter referred as composite breakwaters) (see Fig. 1).

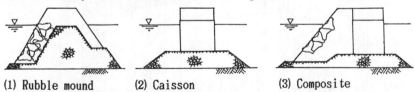

(1) Rubble mound   (2) Caisson   (3) Composite

Fig.1 Typical types of Japanese breakwater

Normally all three types are considered for design with regard to their merits and demerits for design conditions, construction methods and the economical view point with the most suitable type being selected. Table 1 shows their merits and demerits as tabulated in the Japanese design manual. In Japan composite breakwaters are generally chosen with the construction of rubble mound breakwaters being confined to shallow water where construction work by barges is difficult. The scarcity of rubble mound breakwaters in Japan is attributed to the insufficiency of large-sized rocks of good quality and the requirment for quick execution to avoid damage during construction.

2. Over 1300 breakwaters not including fishing ports were constructed in major Japanese ports for the 21 year period, from 1965 to 1985. Fig. 2 represents the percentage of individual types of breakwaters to the total number in ranks of design wave heights and water depths(ref. 1). As the design wave height becomes higher and the design water depth becomes deeper, the number of caisson type or composite type increases. In particular the number of composite

Table 1 Merit and demerit of each breakwater type

Type	Merit and demerit
Rubble mound breakwater	Merit: · lowest crest elevation · low reflected wave · slow progress of damage and easy repair Demerit: · large quantity of construction material · long construction period · high risk of damage during construction · large bottom width
Caisson breakwater	Merit: · quick construction · lowest risk of damage during construction · wide usage inside a harbour Demerit: · high crest elevation · high wave pressure · scouring of toe section of foundation · requirement of caisson building yard · difficulty in repair work
Composite breakwater	Merit: · low crest elevation · quick construction · low risk of damage during construction · wide usage inside a harbour Demerit: · requirement of caisson building yard · slightly higher construction cost

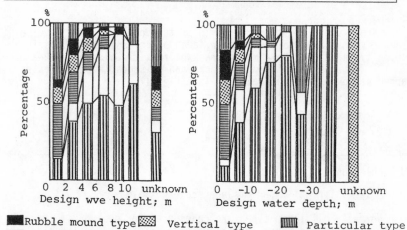

Design wve height; m

Design water depth; m

■ Rubble mound type ▨ Vertical type ▥ Particular type
▥ Caisson type ▢ Composite type ▤ the others

Fig. 2 Percentages of various types of breakwaters used

breakwaters increases remarkably in the large wave range. Typical cross sections of the caisson and composite breakwaters are shown in Figs 3-4. Furthermore, examples of large-size armour blocks used for the composite breakwaters are tabulated in Table 2.

3. In recent years, breakwaters have been constructed at deeper locations in order to meet various demands. When the ordinary type of caisson is applied in deeper areas, the cross section of the breakwater needs to be much greater so economic efficiency becomes lower and a significant problem arises in execution. As a result, new type caissons to reduce wave pressures have been developed for application in areas of large waves or deep areas. One type is the slanting surface caisson shown in Fig. 5. Two other types of interest are known as curved slit caissons and multi-cellular caissons.

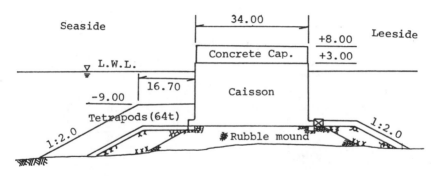

unit; m

Fig. 3   Typical cross section of caisson breakwater
(Hedono Port)

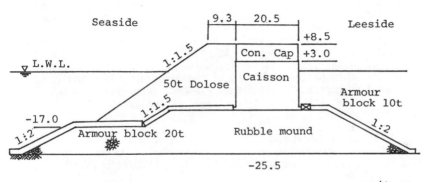

unit; m

Fig. 4   Typical cross section of composite breakwater
(Naha port)

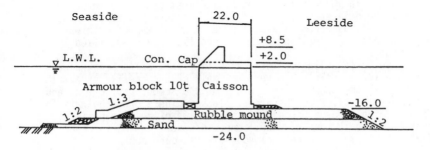

unit; m

Fig. 5   Typical cross section of slanting surface caisson
type breakwater (Niigata port)

191

Table 2 Examples of large-size armour blocks used for composite
breakwaters

Port	Armour blocks		Design water depth (m)	Design wave	
	Type	Mass(t)		$H_{1/3}$ (m)	$T_{1/3}$ (s)
Hososhima	Tetrapod	50	- 23.5	8.3	14.0
Rumoi	Tetrapod	50	- 18.0	7.6	10.0
Hitachi	Tetrapod	64	- 11.8	7.6	14.0
Ajiro	Tetrapod	64	- 18.0	8.4	14.0
Murotsu	Tetrapod	80	- 17.0	11.8	16.0
Naha	Dolos	50	- 28.0	10.7	15.1

DESIGN METHOD FOR COMPOSITE BREAKWATERS

4. In this section we introduce some calculation formulae of wave
forces acting on vertical walls and items for stability calculations.

Calculation of wave force

5. In stability of breakwaters calculations the following external forces should be taken into account. (1) wave forces, (2) hydrostatic pressures, (3) buoyancy, (4) dead weight and (5) dynamic
water pressures due to earthquakes. Wave forces acting on vertical
walls are classified into standing wave forces, breaking wave forces
and wave forces after breaking. As wave forces vary with the wave
conditions as well as tidal levels, water depth, sea bed topography,
profile and form of the face line of the structure, they should be
calculated appropriately with these elements taken into account.
Typical formulae for calculation of wave forces normally used in
Japan are shown in Table 3.

Table. 3 Formulae for calculation of wave force

Region	Caisson breakwater		Composite breakwater
Standing wave region $h \geq 2H_{1/3}$	Sainflou formula and its modification	Goda formula	Modified Goda formula
Surf zone	Hiroi formula		Morihira formula

6. Wave force acting on vertical wall of caisson breakwater.
In calculating the wave force acting on a vertical wall of a caisson
breakwater, the Goda formula has proved most practical in Japan. Goda
assumed the existence of a trapezoidal pressure distribution along a
vertical wall as shown in Fig. 6 regardless of whether the waves
broke or not. Factors and marks in the figure are specified in
reference 2.

7. Wave force acting on vertical wall of composite breakwater.
The wave force acting on a vertical wall of a composite breakwater
varies with the crown height, width of the armour protection, and
the characteristics of the armour blocks. Therefore, the wave force
should be calculated based upon the results of model experiments
conforming to the design conditions. However, the Morihira or modified Goda formula can be employed as a standard design practice for
the calculation of wave forces for cases such as where (1) the crown
level of the armour protection is approximately the same as the
crown height of the vertical wall and (2) the stability of the
armour blocks is insured against wave action.

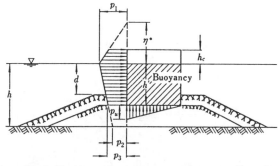

Fig. 6   Distribution of wave pressure according to Goda

The Morihira formula is applicable only in the region where the design depth of breakwater is shallower than that at which the significant wave height would become a maximum. It is assumed that the average wave pressure intensity according to equation (1) is exerted uniformly up to a height of $1.0\ H_{1/3}$ over the still water level or the crown height of the vertical wall from the bottom of the vertical wall, depending on whichever is lower (see Fig.7).

$$p = 1.0 w_o\ H_{1/3} \tag{1}$$

where w_o : specific weight of sea water
$H_{1/3}$: significant incident wave height

It is also assumed that the buoyancy acts on the entire section of the wall so that the uplift is included therein.

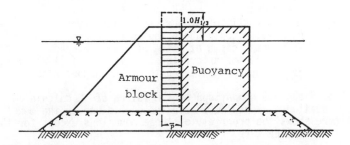

Fig. 7   Distribution of wave pressure according to Morihira

Tanimoto et al. improved on the Goda formula to apply it to the calculation of a wave force acting on a vertical wall of a composite breakwater (ref.3). This formula is useful in the range from a wave standing to breaking continuously. The maximum wave force acting on the wall and the uplift are calculated by using equations (2)-(8) respectively (see Fig. 8).

$$\eta^* = 0.75\ (1+\cos\beta)\ \lambda\ H_{max} \tag{2}$$

$$p_1 = 0.5\ (1+\cos\beta)\ \lambda\ \alpha_1\ w_o\ H_{max} \tag{3}$$

$$p_2 = \frac{p_1}{\cosh(2\pi h/L)} \tag{4}$$

in which $\quad p_3 = \alpha_3\ p_1 \tag{5}$

$$\alpha_1 = 0.6 + 0.5 \left\{ \frac{4\pi h/L}{\sinh(4\pi h/L)} \right\}^2 \tag{6}$$

$$\alpha_3 = 1 - \frac{h'}{h} \left\{ 1 - \frac{1}{\cosh(2\pi h/L)} \right\} \tag{7}$$

$$p_u = 0.5 (1 + \cos\beta) \lambda \alpha_1 \alpha_3 w_o H_{max} \tag{8}$$

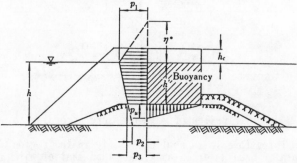

Fig. 8 Distribution of wave pressure according to Tanimoto et al.

λ in the equations represents the reduction ratio of the wave force. The value of λ should be decided based on results of model experiments conforming to the design conditions. However the value is usually taken to be approximately 0.8.

Recently Inagaki et al. showed that the reduction ratio of the wave force (λ) could be represented by a parameter which expresses block mound characteristics (K) based on dimensional analysis (refs 4-5).

$$\lambda = 1.2 \exp [-0.63K] \tag{9}$$

$$K = \left(\frac{1-\varepsilon}{\varepsilon} \right)^{1/3} \frac{H \cdot hd}{h^2} \frac{1}{B} \frac{L}{L_o} \tag{10}$$

where ε : porosity of armour protection layer
         h_d : water depth above the rubble mound
         h : water depth at a toe of armour blocks
         1 : width of armour layer along the still water level
         B : representative length of the armour block

Fig. 9 shows a comparison of the values of λ calculated utilizing equations (9) and (10) with those obtained in the model tests. The figure shows the propriety of these equations except for the case of a breaking wave.

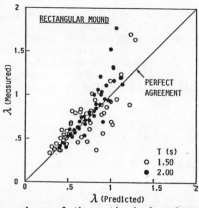

Fig. 9 Comparison of theoretical λ value with experimental

Items for stability calculation

8.   Stability analysis is generally classified as follows: (1) Stability calculation of upright section of breakwater, (2) Stability
of rubble mound, (3) Stability of the entire structure of breakwater
and (4) Stability of head and corner of breakwater.

9.   Stability calculation of upright section.   Sliding and overturning of structures and the bearing capacity of the foundation are
examined. With regard to the stability calculation for overturning
and the bearing capacity of the foundation, the dynamic water pressure due to earthquakes is also taken into account. Safety factors
are taken as the following values. (1) 1.2 against sliding, (2) 1.2
or greater against overturning due to wave forces and (3) 1.1 or
greater against overturning due to seismic forces.

10.   Stability calculation of mound. Entire and partial slipping
of the rubble mound and the weights of materials are examined.

11.   Stability calculation of the entire structure of breakwater.
When breakwaters are constructed on soft ground, circular slips and
settlements of the structure are examined. If the stability is insufficient, soil stabilization is employed. A seismic force is
taken into consideration in examining the stability. The seismic
force acting on the structure is calculated using the seismic coefficient method. The seismic force is calculated by the following
two equations, and whichever proves more disadvantageous to the structure can act on the centre of the gravity of the structure.

(1)   Seismic Force= Dead Weight×Design Seismic Coefficient
(2)   Seismic Force= (Dead Weight+Surcharge) ×Design
                     Seismic Coefficient

Problems on design of composite breakwater

12.   We introduced the calculation method of wave forces and items
on the stability calculation of composite breakwaters. However, in
applying them to the design, the following matters arise as significant problems.

(1)   In the Goda formula mentioned previously there are many factors which influence wave pressures. However phenomena of wave
pressures are so complicated that it can be dangerous to use the
same formula to calculate every wave pressure. Moreover the wave
force calculated sometimes takes on a significantly different value
according to the formula.

(2)   Extreme high wave pressure referred to as impulsive breaking
wave pressure often occurs on caisson type in cases where the sea
bed slope is steep or mound elevation is high. However as this
phenomenon is affected by many factors in a complex and sensitive
manner, it is difficult to clarify a criterion for its occurrence.

(3)   Armour block protection is often used to decrease reflected
waves and wave forces acting on a vertical wall as well as to reduce
danger due to generation of impulsive breaking wave pressures. The
wave force acting on a vertical wall of a composite breakwater is
more complicated than that acting on a vertical wall of a caisson
breakwater. So λ, mentioned previously, changes according to test
conditions.

(4)   Weight of armour blocks is generally calculated using the
Hudson formula. However this has some limitations such as, it does
not include effects of incident wave period etc.

(5)   Required weight of coating materials for the mound is affected by many factors such as wave height, water depth, the configuration of rubble mound foundation and the coating materials themselves.

A general method for calculating the weight of coating materials has not yet been established.

With the problems listed above in mind, in the case of important structures, hydraulic model tests are generally conducted to examine the stability of breakwaters.

EXAMPLES OF HYDRAULIC MODEL TESTS

13. The authors will introduce three examples of hydraulic model tests on stability of breakwaters in three ports, which were conducted in the company. Design conditions for breakwaters in the three ports are shown in Table 3. Brief explanations of the three examples follow below.

Table 3 Design conditions for three breakwaters

Port	Design water depth (m)	Design wave		Gradient of the sea bottom slope
		$H_{1/3}$ (m)	$T_{1/3}$ (s)	
Sonae	- 13.0	11.3	13.0	1/8 ~1/30
Kashima	- 15.5	7.36	15.0	1/500
Hedono	- 28.0	9.76	13.2	1/10~1/50

14. Hydraulic model test on stability of composite breakwater in Sonae Port. Sonae port is situated in Yonaguni Island which is located in the south-west of Japan. Wave conditions in this area are so severe that the construction period is restricted from April to October and even in this period some typhoons occur, causing huge storm waves. The sea bed is very complicated and the bottom slope is steep.

A two dimensional hydraulic model test was performed in order to estimate the stability of the caisson, armour blocks and rubble mound in conditions after .completion and while under construction.

Test cross sections were decided in accordance with the execution stages as follows. (1) Rubble mound without leveling, (2) Installation of caisson, (3) Concrete capping, (4) Partial installation of armour blocks and (5) Completion etc.

Summary of test results are as follows.

(1) Impulsive breaking wave pressures were generated under construction in the case of a 10 year return period wave ($H_{1/3}$=7.9m, $T_{1/3}$=10.8sec) when the crown height of the rubble mound was -7m. As a result -9m was decided as the crown height of the mound.

(2) The caisson after completion was stable against the design wave but slid slightly during construction in the case of the 10 year return period wave. So it proved desirable to place armour blocks as soon as possible after the installation of the caisson.

(3) Armour blocks (50t Dolose) were stable against the design wave.

(4) The weight of block for mound protection needed to be 50 tons to be stable against the 10 year return period wave.

(5) The rubble mound proved to be, for the most, stable against winter waves ($H_{1/3}$=6.1m, $T_{1/3}$=10.8sec) enabling construction to be carried out one year before placement of caisson.

From the test results the typical cross section of the breakwater was decided upon as shown in Fig. 12. Furthermore a three dimensional hydraulic model test was conducted to examine the armour blocks at the head section and mound protection blocks leeside of the breakwater etc. The actual construction of the breakwater will start in 1988.

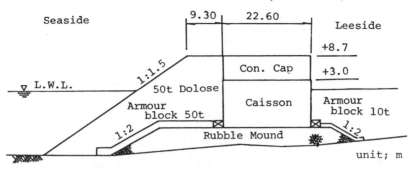

Fig. 12 Typical cross section of west breakwater (Sonae Port)

15. Hydraulic model test on wave forces acting on a caisson wall during construction in Kashima Port. Kashima Port faces the Pacific Ocean and the sea bed consists of sand. Armour blocks covering caissons have settled due to large wave action. Therefore a construction method where armour blocks were placed at a certain elevation and allowed to a settle in layers, was adopted. A hydraulic model test was performed in order to estimate the stability of the caisson while placement of armour blocks was going on.

The test result is shown in Fig. 13 as a safety factor varying with elevations of armour block berm. The figure suggests that armour blocks should be placed quickly, up to a height of +6.0m.

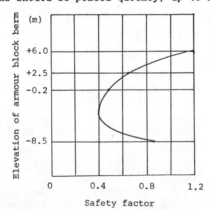

Fig. 13 Safety factors varying against elevations of armour block berm

16. Hydraulic model test on mound protection blocks in Hedono Port. The sea bed topography around Hedono Port is very complicated and the slope of the sea bottom is very steep. If breakwater were set-up, a large wave would hit it violently in the form of a plunging breaker. Therefore it was necessary to obtain data for the design of the breakwater using a hydraulic experimental method. In this design, mound protection blocks proved to be important and after consideration the protection by two-layers of 64 ton Tetrapods was employed. The stability of this protection was confirmed in hydraulic model tests.

The optimum cross section design based on the results of the test was previously shown in Fig. 3. This cross section, a caisson breakwater, was decided upon in order to avoid damage to the breakwater during construction as much as possible.

EXAMPLES OF DAMAGE TO COMPOSITE BREAKWATERS

17.   In Japan several instances of failure of breakwaters have been reported. Fig. 14 represents a correlation of the relative wave heights when the breakwater damage occurred (significant wave height during damage to critical wave height with respect to armour blocks) along with the actual number of occurrences of damage to armour layers of composite breakwaters (ref.6). From analyses of damage to the armour blocks concerned, in cases where they moved or broke due to wave action, the design weight of the blocks was lighter than that actually needed or the design wave heights had been taken lower than those which would cause damage to the blocks.

18.   In recent years, we have conducted stability analyses of armour blocks for composite breakwaters with probablistic approaches, so-called "Reliability analysis of breakwaters".

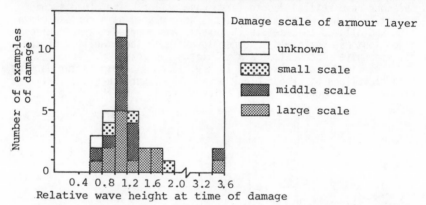

Fig. 14 Correlation between relative wave height and number of examples of damage to armour layers of composite breakwaters

REFERENCES

1.   KATAOKA S. and SAITA K.   Complications of breakwater structure. Technical note of the Port and Harbour Research Institute, Ministry of Transport, Japan, No. 556, June, 1986, 3-11. (in Japanese)
2.   GODA Y.   Random seas and design of maritime structures, 114-118 University of Tokyo Press, 1984.
3.   TANIMOTO K. et al.   An investigation on design wave force formulae of composite-type breakwaters, Proceeding of the 23rd Japanese Conference on Coastal Engineering, Japan Society of Civil Engineering, 1976, 11-16. (in Japanese)
4.   INAGAKI K. et al.   Reduction ratio of wave force using armour protection and parameters expressing block mound characteristics, Proceeding of the 33rd Japanese Conference on Coastal Engineering, Japan Society of Civil Engineering, 1986, 362-366. (in Japanese)
5.   INAGAKI K. et al.   Theoretical study on reduction ratio of wave force using armour protection, Proceedings of the 34th Japanese Conference on Coastal Engineering, Japan Society of Civil Engineering, 1987, 536-540. (in Japanese)
6.   KASHIMA R. et al.   Analysis of damaged breakwaters and seawalls, Proceedings of the 33rd Japanese Conference on Coastal Engineering, Japan Society of Civil Engineering, 1986, 626-630. (in Japanese)

Discussion on Papers 7 – 10

PAPER 7

MR A. G. BRINSON, Rendel, Palmer and Tritton, London
The marina at Brighton proved to be uneconomic on account of
the high cost of the breakwaters.

The marina at Eastbourne, which is based on the different
idea of forming a channel through the beach, has been studied
for about 15 years. However, developers have still not been
persuaded to proceed with the project, presumably because of
the capital cost of the breakwaters and, in particular, the
high cost of maintenance dredging of the entrance and beach
replenishment of the east side where erosion will occur. Do
the Authors believe that such a project will be developed?

MR C. DINARDO, Dinardo and Partners, Paisley
Could the Authors indicate what the seabed substructure
conditions are like beneath the proposed breakwater?

MR H. LIGTERINGEN, Fredric R. Harris (Holland), The Hague
Could the Authors explain what studies have been made to
investigate the littoral transport rate in a quantitative way?
Both for maintaining a minimum depth in the enhanced channel
and for the measures needed to remedy erosion of the shore
east of the harbour, it is of importance to have the best
possible estimate of deposition and erosion.

MR J. D. METTAM, Bertlin and Partners, Reigate
The Authors have spoken of littoral drift between 5000 and
50 000 m^3/year. Presumably, shingle will build up on the face
of the breakwater and must then be transported across the
entrance. This will not be a stable movement but surely a
sudden spilling of quite large quantities into the channel
during storms from west of south.

What depth do the Authors expect to be able to guarantee the yachtsmen who wish to use the entrance?

MR J. SIMM, Rendel, Palmer and Tritton, London
I think it is most important to distinguish between sand transport and shingle transport. Therefore, it would be helpful if the Authors could explain how they distinguished between the two and also what the effects of the breakwaters were on each one.

MR H. L. WAKELING, Consulting Engineer, Esher
My question concerns littoral drift and maintenance of the approach channel. Mean rates of drift might be representative if derived over a period of 20 years or more. However, wide variations from the mean occur from year to year and it must be expected that a lot of dredging may be required in some years.

Has any consideration been given to the form of breakwater to be adopted? A vertical face breakwater may cause reflection which can affect the rate of drift.

MR LACEY and MR WOOD (Paper 7)
In reply to Mr Brinson, it is indeed true that Brighton Marina proved uneconomic on account of the high cost of construction in total, and because it ran into a very poor economic climate. The original Eastbourne scheme of 1973 was based on the capacities of Brighton Marina, that was to be the biggest and the best. However, it too failed to materialise owing to the lack of funds.

It is not true to say that developers are not going to proceed with the project, as the scheme has indeed started and an auxiliary Bill is now going through Parliamentary procedures.

The capital cost of the breakwaters has been drastically reduced by forming a channel through the beach and by deleting the outer basin of refuge. The cost of maintenance dredging and beach refurbishment is down to very small limits and there is no doubt in the Authors' minds that the tests carried out by Hydraulics Research Ltd, Wallingford, have indicated that we have the correct design.

The Authors do believe that the complete project will be developed, and construction of the road and retail sites will begin on 1 June 1988.

In reply to Mr Dinardo, as Appendix A indicates, considerable site investigation work has been carried out primarily during 1973. Boreholes were taken both on land and in the areas indicated for breakwater construction. The sea bed under the breakwater is composed of 10 m of recent sands and gravels with the clay lenses overlying Pleistocene deposits in turn overlying lower Greensand.

In reply to Mr Ligteringen, hydraulic tests have been carried out by Hydraulics Research Ltd, Wallingford, as indicated in Appendix A. The storm basin tests showed where storm waves of long duration would be expected to erode the beach material, and the beach transport study was undertaken to ascertain the amounts of shingle transport.

The beach transport study was based on beach profile readings taken over a period of some nine years by the Southern Water Authority. The study indicated the possible movement of materials equivalent to some 20 000 m^3 of material per year moving along the beach.

The Authors agree that it is most important to obtain good estimates of deposition/erosion. However, they do not believe that this can be done accurately at the present time and with the present estimating methods.

In reply to Mr Mettam, it is expected that shingle will indeed build up along the face of the breakwater up to the stage where certain amounts will be transported across the entrance channel. However, we cannot, at the present time, forecast exactly when this will happen, but it will certainly be some time after the next decade.

The hydraulic tests did not show that individual storms would move larger quantities of shingle into the entrance channel but that considerable storm waves of considerable duration were needed.

With regard to Mr Mettam's second question, the design is able to provide sufficient depth in the entrance channel for the yachtsmen.

In reply to Mr Simm, the hydraulic model tests were not required to distinguish between sand transport and shingle transport, and the roles are based on the movement of the coarse part of the material. We agree that it is important to distinguish the predominant material, and at Eastbourne beach it is shingle.

In reply to Mr Wakeling, it is quite correct that large movements of sediment may well occur either annually or over a period of years. The behaviour of the sea is difficult to forecast. It is also true that the quantity of dredging required per annum will vary according to the storm patterns and this is the case in most marinas.

The form of breakwater to be used was considered during the feasibility studies of 1973, and it was considered that a rough rock armoured breakwater would be both aesthetically pleasing and functional. Thirteen years on we are looking closely at single layer armour options.

PAPER 8

PROFESSOR L. FRANCO, Politecnico of Milano
Could the Author give more information about the artificial seaweed which is being introduced to arrest the erosion of Victoria Beach: the height, depth of planting and so on? What

are the likely effects of this seaweed, as indicated by the model?

MR H. LIGTERINGEN, Fredric R. Harris (Holland), The Hague
Could the Author comment on the design of the offshore breakwater; does he envisage a submerged breakwater?

In relation to this, I should like to point out that submerged breakwaters have been successfully applied along long stretches of beaches on the Adriatic Coast of Italy. The advantage of providing an unspoilt view and the important cost reductions were the main reasons for selecting this solution.

MR J. D. METTAM, Bertlin and Partners, Reigate
The Author mentioned that the west breakwater had recently been extended to reduce the movement of sand into the entrance channnel, and thus to save dredging.

Taking the problem of the harbour entrance and the Victoria Beach together, would it not have been a better solution to dredge from the toe of the Lighthouse Beach and dump the dredgings off the Victoria Beach?

In time, the Lighthouse Beach will no doubt build up again, and perhaps dredging should be adopted to slow down this process and also to feed Victoria Beach.

DR O. J. SAYAO, F. J. Reinders and Associates Canada Limited, Brampton, Ontario
With reference to the solution to the Victoria Beach erosion problem, incorporating two offshore breakwaters and artificial weeds, did the Author test the synthetic weed as well as the breakwaters in the hydraulic model? As Victoria Beach is a source of those sediments which are necessary to restore the dynamic equilibrium downdrift, the breakwaters will decrease wave action on the beach and consequently the beach erosion. Therefore, the weeds may have proven not to be efficient owing to the lack of littoral drift to trap.

MR ONOLAJA (Paper 8)
In reply to Professor Franco, the totally flexible weed is made up from a non-woven polypropylene fibre with a buoyancy chamber, impediment fronds and anchorage containment. It will be arranged in such a way that the width of each strip will be about 20 m, and it will extend from 25 m to 45 m from the high tide shoreline. The lengths of fibres are between 1.5 m on the landward site and 2.5 m at the seaward limit of the field.

It was observed during the model studies that immediately after the most easterly breakwater, which is about 3 km east of the East Mole, severe erosion for about 0.75 km eastwards had taken place. The use of the synthetic seaweed will therefore promote on-shore sediment transport, reduce bottom surface current velocity and induce accretion.

In reply to Mr Ligteringen, submerged breakwaters are not envisaged on account of the fact that the area is close to the main harbour entrance and these would therefore pose a danger to shipping.

At the western extremity of the beach (close to the East Mole), waves are subject to refraction on the shallow areas and diffraction on the East Mole. These phenomena will create permanent instability on the future shoreline. In addition, there is concentration of energy in this area. For these reasons, it is essential to provide for a protective breakwater linked with and approximately perpendicular to the East Mole. Such a breakwater will also provide a protected area suitable for bathing.

For the first 250 m, the breakwater is flattened (with a wide berm at the top) so that the view from the beach should not be spoiled. For the remaining length, the breakwater is classical with a high profile and raised protective wall.

In reply to Mr Mettam, the fine nature of the sand accreting at the Lighthouse Beach makes it unsuitable for sand replenishment work at the Victoria Beach.

During the last sand replenishment exercise for Victoria Beach in 1985, three million cubic metres of sand were pumped, using trailer hopper suction dredgers, with the borrow pit located at between 10 to 13 nautical miles from the East Mole.

In reply to Dr Sayao, the two offshore breakwaters as well as the synthetic seaweed were tested in the hydraulic model. The source of sediment is not Victoria Beach itself but west of West Mole (through the harbour entrance offshore) and along the West African Coast. The breakwater will surely decrease the wave action on Victoria Beach and, consequently, reduce the beach erosion. The synthetic seaweed is to be installed immediately after the most easterly breakwater.

PAPER 8a

DR F. B. J. BARENDS, Delft Geotechnics
The Seabee block does not show lateral porosity, but a large orthogonal porosity, which makes it different from many artificial armour units. Is this aspect favourable with respect to the actual behaviour? Any lateral drag inside the armour is translated to an increase in interblock forces, increasing friction and thus improving stability. However, the absence of lateral porosity causes a larger impact on the underlying filter layer and probably causes a higher saturation of the core. Is there any information on this impact? Has the Author thought about the possibility of using these armour units on a relatively impervious base, and eventually with a high cohesion (clay, geotextile) material?

MR M. E. BRAMLEY, Construction Industry Reseach and
Information Association (CIRIA), London
It would appear necessary to stress the importance of
effective design of the underlayer to the Seabee, in view of
the fact that slope-parallel flow is not possible within the
Seabee-armoured layer. This presumably means that
considerable concentration of flow into, and out of, the
sublayer exists at the base of the unit at various stages in
the wave cycle.

Could Mr Brown please comment on the design of the sublayer,
not only in relation to the use of Seabees as armour to
relatively porous rubble-mound structures where the potential
for major flow already exists in the body of the structure,
but also to the use and cost of the Seabee system as armour to
relatively impermeable and potentially more erodible earth
embankments such as sea defences.

MR C. TUXFORD, Ardon International Limited, Tunbridge Wells
It seems to me that apart from rock armour, the Seabee System
provides for many aspects of the design process for single
layered armoured systems which other systems do not,
particularly those systems which are of a single size.

Would Mr Brown comment on this, particularly in the context
of advising the design engineer and justifying to this
engineer the need for lateral porosity within the primary
armour layer.

MR BROWN (Paper 8a)
It appears from the perspective of the questions of Dr Barends
and Mr Bramley that the essential variability of geometry
implicit in the linear or blanket theory, and as explicitly
realised in the development of the Seabee, has not got across.
There are some factors in the design of an armour unit that
are essential and are controlled by the requirements of
stability under the incident wave climate, and other factors
that are engineering or arbitrary options. In this Author's
mind, matters affecting stability are essential, but those
affecting run-up, reflection and cost are optional and are to
be weighed in the balance for each project.

In the case of armour systems reliant on mass rather than
ground anchors, the slope normal thickness is an essential
requirement for stability as is the slope normal porosity for
armour layers on a permeable underlay.

Slope parallel porosity, surface roughness, legs, etc. are
all intuitive means of affecting and refining service
functions, and may also have advantages or disadvantages as
regards stability. In most circumstances, the requirements
for a project are to achieve stability, durability and an
adequate level of hydraulic performance at a minimum cost or
optimum cost/benefit ratio, and jobs are generally awarded to
the systems that achieve either the optimum ratio or the
lowest cost, given always the subjective weighting applied to

the various systems according to their length of service history.

All armour systems are a compromise between hydraulic performance, structural integrity and economics. With most armour systems there are no options and what you see is what you get. In the case of the Seabee, this armour system was deliberately designed to take advantage of the inherent flexibility of geometry implied by the linear equations described in the Paper. As initially envisaged, the Seabee was endowed with lateral porosity, more to diminish impact and shock pressures than to allow rapid flow parallel to the slope. However, as an option, should it be economic? Also available are legs, which achieve something of the same effect, but do allow slope parallel flow at the interface, surface roughness and shaping, interlinks, cabling and shear keys. In the market place, no one has been found to pay for these extras, most being satisfied with the behaviour of the basic unit. In many cases, the increased toughness of the simple unit has been more sought after. The rapidity of manufacture of the simple unit also gives very good economics. As regards hydraulic behaviour, I would refer to the results of Dr Hettiarachchi quoted in the text of comparative tests of Seabees, Sheds and other voided units, demonstrating very similar behaviour over the tested ranges. In our own tests, alternative designs based on Seabees have generally met the standard design criteria established for various projects, with both increased margins of structural stability and decreased costs.

Dr Barends states that the slope parallel flow within the unit may increase inter-unit friction; but it also increases the downslope component of force and will lead to an increased potential for sliding and a requirement for toe stabilisation structures.

The subject of impact pressures on both the armour unit and the filter/underlayer was an early consideration in the development of the Seabee system, particularly with regard to the use of geotextiles. At the critical condition, when the wave jet is nearly normal to the slope, the impact occurs both on the surface of the armour unit (increasing the intergranular pressure) and on the surface of the underlayer within the void. There is a degree of interference between these two flows.

The shock pressures are superimposed on the slope drainage pressure field from the proceeding wave, such that a dynamic rotor is set up at the boundary. (In the 1978 paper (references 10 and 11), this effect was likened to a free form hydraulic jack.) This is partly the reason that failure generally occurs at units either side of the impacting jet. The balance of these forces is determined by both the minimum slope normal porosity and the entry co-efficient. The Author would welcome any evidence of the effect of lateral porosity on the jet and the dissipation of impact pressures.

Equally important is the structural load imposed on the

elements of the armour units by the same impact forces.
Therefore, the choice of porosity of the Seabee surface varies
according to location, wave size, strength and underlayer
grading; where appropriate, the porosity of the impact zone is
kept low to meter the rate of water ingress.

Testing in respect of Seabees has examined the effect of a
variety of monochromatic and polychromatic wave conditions,
with various porosities, aspect ratios, roughnesses (super and
sub-armour) and the effect of the underlying grading. The
further implications of these effects on post-loss behaviour
of the armour layer have always been of prime importance in
the development of the Seabee, as the loss of a single or a
few units must not lead to catastrophic collapse at exposures
below the critical or yield level.

Thus an early criterion for the design of a Seabee slope has
been the evaluation of the threshold of armour motion at which
the rate of damage increases with loss, and at which grading
the underlayer material is lost through the resulting hole. A
factor of safety or load factor is applied to these dimensions
(net surcharge, void diameter) or ratios (H/R, D/d_{50}).

There is also the question of the use of Seabees and similar
armour on relatively impermeable materials with geotextiles.
In the Author's early papers on Seabees and Blanket Theory,
the topic of wave amplitude attenuation through the various
layers of a revetment was discussed and the concept of the
hysteresis volume raised.

The resistance to wave action is the sum of energy
reflection, storage and degradation, each having a bearing on
structural design, secondary effects and economy. The use of
porous, planar armour systems directly on earth or clay banks
with only a geotextile underneath is not one with which this
Author concurs as either the soil surface or the geotextile is
subjected to a dynamic phreatic oscillation, which will either
break down the structure of the soil or pump and abrade the
cloth on the underside of the armour. This may lead to
clogging of the geotextile, pumping and understreaming or
disruption and failure of the filter.

At this time, Seabees are always specified to be placed on
an underlayer of graded material to allow both slope parallel
flow and slope normal attenuation of the wave dynamics.
Depending on grading, the practical variation of thickness is
between 0.6 and 2.0 times the armour layer thickness. The
practice of choking small revetments with rocks can be seen as
increasing the attenuation rate and decreasing the inflow of
water, but is confined essentially to low wave climates at
this time, on account of the heavy structural consequences of
the mobility of such material under larger waves. It is an
area requiring more work, especially at higher wave climates.

With regard to Mr Tuxford's questions, Fig. 3 of the Paper
shows a comparison of results of various single layer voided
units, and the simple Seabee compares well with the more
complicated geometries as regards run-up and reflection.

In our early tests in the late 1970s, experiments were made

on surface roughness, lateral porosity, legs, variable plan
aspect ratios and underlayer grading. These tests were
carried out in relation to a number of contract related
projects; in economic terms, the basic results were as
follows.

- Simple units are much quicker and cheaper to make (e.g. 18
  units/form/day at Abbot point, 12/form/day at Cronulla).

- Simple units are more rugged and have good post-fracture
  behaviour.

- Lateral porosity dissipates some shock and allows internal
  flows, but these effects are variable and depend on the
  registration of the rows and columns of units and the
  construction tolerances. The lateral porosity also allows
  fine particles to be carried through the units and may
  cause more rapid abrasion.

Any benefit from the use of lateral porosity is lost the
flatter the wave (high Xsi values). Run-up control would
appear to be affected as much by underlayer void volume and
the use of berms as by the use of highly voided and complex
units.

PAPER 9

MR C. T. BROWN, Seabee Developments, Bedford
While the Paper is entitled 'Single layer armour units', it
might more properly be called 'Hollow cubical armour units',
as it does not deal with any other type, except in passing,
and it does not rehearse the variations.
The Authors' attention is drawn to the placed stone at
Umpqua jetties: uniform and pattern placed Tribars, Svee
blocks, Gobi blocks and numerous revetment systems. The Paper
states that Seabees have no lateral porosity. While there are
no installations of such Seabee units, as simple geometry
units their performance is almost as good as that of the
aforementioned units, and they are much cheaper. Mr Allsop
may recall our early correspondence where it was pointed out
that lateral porosity was an option – not an essential –
depending on the economy.
Do the Authors consider that lateral porosity must be the
same as the entry porosity? Has there been any experience of
damaged or broken units?
How economic are Sheds compared with rock in direct cost per
tonne?

MR J. D. GARDNER, Sir William Halcrow and Partners, Swindon
I would like to comment briefly on the scheme at Kaoliköy,

Istanbul, where our design incoporated Shed units. The
project comprises five artificial beaches enclosed by ten
artificial headlands. Sheds were used for the headlands, for
which the design criteria were the following.

1. The visual impact of the headlands to be considered.
2. The headlands to be designed for a 100 year wave of H_s =
   3.5 m, with the reflection coefficient to be less than
   0.25 to prevent loss of beach material.
3. The wave transmission through the headlands to be less
   than 0.1.
4. The crest elevation to be about 1.5 m above high water
   level so that the headlands are at the same level as the
   promenade.
5. The maximum overtopping on to the beach to be less than
   0.2 m^3/sec per metre run.

Two particular problems which we experienced in the design
resulted from the use of Sheds. These were

(a) providing a level toe in breaking wave conditions so that
    the bottom row of Sheds was secure
(b) providing adequate venting in the headland crest to
    handle wave run-up during storm conditions.

MR M. C. FARROW, Excess Insurance Group, London
Could the Authors expand on the problems experienced during
construction, as it seems to me that, during construction, the
Shed wall is highly susceptible to storm damage - even though,
as was pointed out, the end could be closed. It seems to me
that the stability of any one block is reliant on the
completed wall. For example, there will not be much stability
in the case of a cube on a length of armouring.
   The Authors seemed to pass over the 'cracking' of the blocks
as not being a problem. Surely the way waves work would erode
or attack any weakness and once one block was lost, the
stability of the others would be undermined.
   In view of the method of construction and the relative
instability of the Shed wall during construction, how suitable
would this method be in a place where there is continual
swell/surf, such as West Africa?

DR S. S. L. HETTIARACHCHI, University of Moratuwa, Sri Lanka
Could the Authors state the maximum depth at which Cobs and
Sheds have been used so far in breakwater construction, and do
they expect any specific problems if the units were to be
placed at greater depths?
   The Authors mentioned that cracks were observed on Cobs and
Sheds. Could they state, very approximately, the number of
units (perhaps as a percentage) subjected to cracking, and its
intensity and distribution; and comment also on the positions

208

of the cracked units.

In the Authors' opinion, are these cracks due to dynamic wave impact and the resulting hydraulic impact forces, collisions, etc.; or are they due to the material properties of the concrete itself and its relationship to the geometrical configuration of the unit?

MR C. TUXFORD, Ardon International Limited, Tunbridge Wells
During the discussion, Mr Wilkinson made some reference to the current state of thinking with regard to lateral porosity and whether or not it was a positive or a negative aspect of a single layer armour system. Would Mr Allsop express an opinion on this, particularly as Mr Wilkinson did say that, on balance, it was thought that lateral porosity was a positive aspect of a protection system. Perhaps Mr Allsop would also comment on Mr Brown's hypothesis with regard to the ways in which a breaking wave hitting a slope is absorbed in a cyclical fashion within the underlayer and core of the breakwater.

MR DUNSTER, MR WILKINSON and MR ALLSOP (Paper 9)
Mr Brown suggests a change to the title of the Paper. The term 'single layer armour' was intentionally drawn as wide as possible by the Authors to encompass many types of concrete armour unit laid in a regular pattern in a single layer. As Mr Brown correctly notes, the research club has concentrated its resources on the most efficient type of single layer unit – the hollow cube. These units offer high porosity in each principal direction and would appear to approach most closely the idea of a network of holes held firmly together.

In discussing the influences of porosity and permeability on hydraulic performance, as raised by Mr Brown and Mr Tuxford, it is important to note that nearly all design procedures are based on long-crested wave attacks at normal incidence, and the model tests that support them are based on the wave/structure interaction as a two-dimensional problem.

At an armoured rubble structure, wave action will lead to flows over, through and within the armour, through and within the underlayer, and in the core. The relative magnitudes of these flows will depend on the hydraulic properties of the different layers, and on the principal wave parameters. Impermeable underlayers will lead to wave flows being concentrated within and over the armour layer. Wave run-up levels and reflections will be high and the armour will be subjected to severe forces. Where more permeable underlayer and core construction are used, the overall hydraulic performance will improve and forces on the armour will reduce. In theory, it might be possible for a designer to tune the transmissivity of the armour, underlayer and core to achieve optimum hydraulic efficiency. In practice, the specification of underlayers and cores is constrained by many other factors, including availability and cost. The designer's flexibility is generally restricted to the armour, and sometimes the primary underlayer, and it is by adjusting the properties of these layers that peformance can be optimised.

The most sophisticated hollow cube unit is the Diode, previously discussed by Barber and Lloyd (references 1 and 2). This unit offers a large open area at its upper surface while restricting flows within the unit at lower levels by perforated cross-walls. Barber (reference 2) has argued that an open upper face will allow as much of the wave into the structure as possible, thus maximising the energy dissipation of the different layers. Porosity and permeability normal to the slope will influence flows into the lower layers, but flows within the armour and underlayers will be governed mainly by permeability parallel to the breakwater surface. Observations made by the Authors in the laboratory and the field confirm the importance of allowing flows into and within the armour layer, emphasising the importance of controlled permeability in each main direction of flow. The hollow cube units considered in the Paper offer high porosity in each main flow direction.

At many structures, wave action may often be short-crested or may strike the structure at an oblique angle. Therefore, the main run-up flow will tend to be oblique, and it is important that the armour is sufficiently permeable parallel to the breakwater surface to maintain as much of the wave flow as possible within the armour layer. This requires relatively high porosity in each direction.

To date, Cobs and Sheds have had equal porosity on all faces, as this is convenient and appears to perform satisfactorily. However, in certain circumstances it may be advantageous to vary the aperture sizes not only to produce lateral porosity, which differs from the normal porosity, but perhaps also to produce normal porosity which has one value at the upper surface with a different value on the surface adjacent to the underlayer.

Mr Brown, Mr Farrow and Dr Hettiarachchi all expressed interest in the experience of cracked and damaged units. Cracks have been observed both in Sheds and Cobs. The greatest number of cracked units has occurred in the Cobs on the La Collette breakwater in Jersey. This breakwater is the first to have been constructed using hollow cubes, and it is also the breakwater subjected to the most exposed and severe wave action. About 12% of units on this breakwater exhibit some signs of cracking, many of which are fine hair cracks, after about 13 years in service. While some of the cracked units have been filled with concrete to ensure their integrity, it is the Authors' opinion that such filling is not, in fact, necessary.

The cracks may have a number of causes, such as early thermal and shrinkage strains, handling and placing stresses, settlement of the breakwater structure, irregular contact with neighbouring units and the underlayer, and dynamic wave forces. With the development of novel formwork, concrete mix design and other techniques, the incidence of cracking is being reduced.

Owing to the fact that hollow cubes are packed tightly together, cracking of a few individual units is not thought significantly to affect overall structural integrity; the elements of a unit which has broken right through are generally held in place by in-plane compressive forces. There has been no evidence to date of wave action causing chafing and progressive widening of cracks.

Means of reinforcing units cheaply with non-corrodible material is possible and is being investigated.

It should be noted that model tests and site experience indicate that a number of units may be removed without affecting the stability of the breakwater. The precise number which may be safely removed is dependent on not only the sea conditions but also the bonding pattern used and the size of the underlayer.

Mr Brown also raises the question of the economics of Sheds compared with rock. Clearly, it is not possible to give a simple answer to this question as there are so many influences. The most obvious influence is the availability of suitable size rock armour. Some indication of the possible savings is given by considering the size of Cob and Shed that would be needed for a given sea condition and comparing it with the size of rock needed for the same condition. Two-tonne Cobs or Sheds may be used on a $1:1^{1}/3$ slope to withstand 7 m waves, whereas 35-tonne rock would be needed at a slope of 1:2 when designing using the Hudson formula. The obvious size difference is not the only saving, as, depending on the height of the structure, considerable savings can be obtained by the reduction in core quantities, which is made possible by using steeper slopes. Steeper slopes and the size of unit being handled obviously have a great influence on site handling, particularly on craneage.

The particular problems experienced by Mr Gardner have previously been encountered and tested at Wallingford. Mr Read has described details of the design of a Shed armoured slope for Fort Jalali, Oman (reference 3). Large vented crest slabs were developed to restrain the crest of the armour, set particularly low for non-hydraulic operational reasons. It should be noted that three alternative crest slabs were tested by Shuttler (reference 4). Despite significant venting, the first two designs were rejected on account of movement of the slab under very heavy overtopping flows. It is most unlikely that this instability would have been discovered, and overcome, without appropriate model tests.

Again, at Fort Jalali, where the toe was exposed to severe wave action, precast toe beam sections, 3 m long, were used, which were restrained from sliding by driven piles (reference 3). At Pyrgos Marina, where the toe was founded at around -5 m, trapezoidal precast toe blocks, 1.3 m long, were used. A wide berm of 0.5-2 t rock was laid against the toe units to provide support and protection against possible scour. The design was tested under random waves for conditions up to H_s = 3.9 m, giving a maximum wave height, H_{max} = 5.5 m, limited by depth (references 5 and 6).

The Authors are surprised that Mr Gardner did not model test his design to refine the toe and crest details, and to confirm the performance of the design under extreme conditions.

In reply to Mr Farrow's concerns about storm susceptibility during construction, the Authors would point out that this problem is covered in paragraph 23 of the Paper, and they would further record that, to date, there have been no major construction problems on sites where these blocks have been used.

Another question raised by Mr Farrow is the suitability of Shed blocks in areas of continual swell/surf. Stability of individual blocks during construction stages can be ensured by selection of appropriate size blocks for the conditions.

Turning to Dr Hettiarachchi's questions, the maximum depth at which Cobs and Sheds have so far been used is on the Al Inshirah breakwater, Muscat, Oman, where the lowest armour unit was approximately 7.5 m below low water level.

The Authors do not expect problems in placing units at greater depths. Once the water is deep enough to require divers for placing, then increasing depth is of little concern except in so far as it affects the required reach of cranes. At depths below the area of significant wave action, one might expect to see a reduction in size of armour required.

References

1. BARBER P. C. and LLOYD T. C. The 'Diode' wave dissipation block. Proc. Instn Civ. Engrs, Part 1, 1984, 76, Nov., 847–870.

2. BARBER P. C. Discussion on 'The 'Diode' wave dissipation block.' Proc. Instn Civ. Engrs, Part 1, 1986, 80, Oct., 1401–1419.

3. READ J. Discussion on Papers 3 and 4. Developments in breakwaters. Thomas Telford, London, 1986, 119–140.

4. SHUTTLER R. M. Revetment at Fort Jalali, Oman: a laboratory sea wall study. Hydraulics Research, Wallingford, 1985, Feb., Report EX 1276.

5. ALLSOP N. W. H. Discussion on 'The 'Diode' wave dissipation block.' Proc. Instn Civ. Engrs, Part 1, 1986, 80, Oct., 1401–1419.

6. ALLSOP N. W. H. and BRADBURY A. P. Pyrgos marina: hydraulic model tests on the main breakwater. Hydraulics Research, Wallingford, 1984, Jan., Report EX 1180.

PAPER 10

MR G. CRAWLEY, Fairclough Howard Marine, Chatham
In Fig.12 of Paper 10, showing a typical cross-section of a
breakwater, a hatched rectangle has been included at the two
toes of the caisson. Could the Authors explain what these
objects are and what purpose they serve?

MR H. LIGTERINGEN, Frederic R. Harris (Holland), The Hague
The Authors have given a detailed and interesting account of
breakwater construction in Japan.
    In paragraph 12 of the Paper, it is stated that Goda's
formula for wave pressure on vertical faced caissons does not
apply in the case of impulsive breaking wave pressures. Could
the Authors explain which other methods are employed in Japan
to determine these pressures and to analyse the dynamic
stability of the caisson subjected to short duration high
amplitude forces?

MR SATO, MR YAMAMOTO and DR ENDO (Paper 10)
In reply to Mr Crawley, the objects in question are called
foot-protection blocks. When a breakwater is constructed in
Japan, a few rows of foot-protection concrete blocks are
usually provided just in front and at the rear end of the
upright section shown in Fig. 12 of Paper 10. These
rectangular blocks for foot-protection usually weigh from 10
to 40 tons, depending on the design wave height.
    The upright section reflects most of the incident wave
energy; therefore, turbulence occurs in front and scours the
rubble mound foundation. Small-sized rubble stones are used
at the crown just under the caisson, because the crown of the
rubble mound under the caisson has be be levelled in order to
install it. Therefore, because the rubble mound foundation
under and just near the caisson is easily scoured, foot-
protection blocks or armour blocks are provided as protection.
If the rubble mound foundation at the toes of the caisson were
scoured, the caisson itself might receive damage. For this
reason, foot-protection blocks are placed at the front and
rear of the caisson.
    The reason for the rectangular shape is that we can install
them more accurately than other shapes, and thus can secure
the foot of the caisson.
    In reply to Mr Ligteringen, as an impulsive wave pressure is
inversely proportional to its duration, the impulse which is
exerted on a vertical wall is rather small. Furthermore, a
rubble mound and the ground around a prototype breakwater are
elastically deformed under conditions of an impulsive breaking
wave pressure, and this softens the wave impact on the upright
section. A computation of such elastic deformation of the
foundation indicates that the effective wave pressure which
causes the sliding of an upright section is at most twice to

three times the normal wave pressure corresponding to the wave height, when the wave pressure is averaged over the exposed area of wall. The Goda formula does apply in calculations of this effective wave pressure. The reliability of a calculation method for wave pressure on a vertical breakwater is judged by the accuracy of the prediction of breakwater stability. From the analysis of scale model tests and actual examples of sliding of breakwaters under storm waves, as well as the examples of breakwaters which have withstood the attacks of high waves without experiencing damage, the Goda formula has been found to be practical.

The effective impulsive pressure due to breaking waves is much greater than the pressure usually employed in breakwater design, and there is a danger that impulsive breaking wave pressure will be generated under certain conditions. However, several factors cause the conditions of impulsive wave pressure to be quite sensitive to changes; for this reason, whenever there remains a suspicion of the generation of impulsive breaking wave pressure, the wave force should be examined through hydraulic model experiments, using a large-scale model to confirm particular conditions.

It is desirable to avoid the use of a cross-section or structure that tends to generate such high wave pressures. When an impulsive wave pressure is unavoidable because of a sea bed with a steep slope, it is desirable that the structure should be designed to alleviate the wave force by providing an appropriate means of wave dissipation.

The above-mentioned was discussed in Goda's book (reference 2 of the Paper), which should be referred to for further information. The Authors would also recommend the use of wave force dissipating structures against any situation where impulsive pressure is suspected.

11. The constructability of a breakwater

J. S. MOORE, Kier Limited, Gatwick Airport

SYNOPSIS

Much research is under way into the hydraulic, geotechnical
and structural stability of breakwaters which may lead to more
sophisticated designs emerging. This paper is written to
emphasise the point that breakwaters must nevertheless be
designed to be built in similar conditions to those they are
to withstand. Some common practice construction techniques
are described to illustrate the continuing need for constructability.

INTRODUCTION

1.   The client requires a functional breakwater at an
economic total cost. A factor which will contribute to this
is the breakwater's constructability, that is, the degree to
which the design is planned for construction.
2.   Since a breakwater is located to give protection from
the sea, it is probable that the structure will experience
significant wave attack many times before its completion.
Survival of the partly built structure in those conditions
may depend upon its constructability.
3.   Designing a breakwater with its construction in mind
calls for an appreciation of the supply problems, the
environmental conditions and the probable techniques. Some
of these are discussed below, the emphasis being on rubble
mound works.

PLANNING FOR A RUBBLE MOUND BREAKWATER

Use of plant

4.   The main plant requirements are for armour placing,
core shaping, core spreading and general transport and
distribution.
5.   Floating craft is readily positioned but needs a wide
spread of working and storm moorings. Operating close to a
rock slope on a falling tide is not ideal and in addition
there is extensive weather downtime and poor placing accuracy.

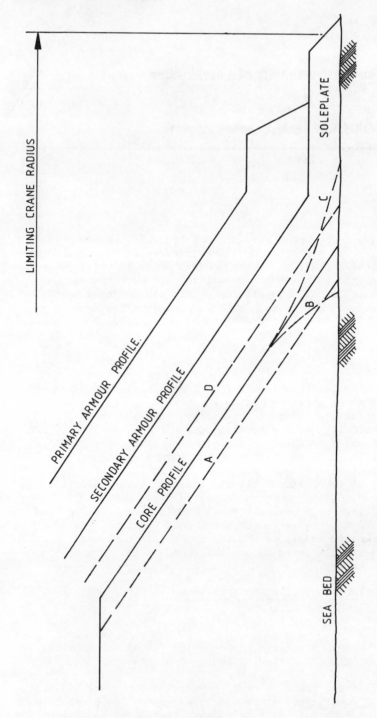

Fig. 1. Some probable deviations from specified core profile

Floating cranes are therefore best avoided except when they provide the only viable solution.

6.   Land based cranage, which should ideally cover the entire working area, is usually provided in the form of crawler cranes. These machines are designed for uneven ground conditions and can readily be travelled out to safety in bad weather. If they can have the versatility of crane, grabbing and dragline duties this is an advantage although the time taken in changing duties is sometimes counterproductive.

7.   When considering the loads which have to be placed out to the lower slopes advantage can be taken of the immersed weight. However, at low states of tide there may be difficulty in achieving full submersion of the load within safe radius. This reduces the effective working period. A remote release flotation tank may also be of assistance in increasing effective radius.

8.   For core spreading there is a choice between blade dozers and front end loaders. The latter is the more versatile but the ultimate choice of either need not be a design consideration. A small hydraulic backacter is a useful addition for local trimming.

9.   Core shaping is a key operation, the effectiveness of which may determine the whole speed of advance. End tipped core material will slump to about 1 in 1.25. Over time this will flatten out, but unevenly so.

10.   For the flatter slopes large quantities of material have to be transferred to the toe area and this can be done by large grabs, rock trays or nets. An alternative method is by calculated overtipping. The tendency is for the toe to fill up under sea action. Pulling back the upper reaches can then be done by dragline provided the seabed is not dotted with reefs. In much shallower depths a large hydraulic backhoe may fulfil this purpose.

Specification tolerances and acceptance

11.   An "as drawn" profile is neither easily achieved nor easily proved at the rougher end of the workable weather range i.e. the so called moderate seas with a wave height of 1.25 - 2.5 metres. A tight specification with low tolerances for slope, thickness and profile may put an additional constraint on progress in this weather. As an example Fig 1 shows some likely deviations from specified core profile which may call for corrective action. Profile A would probably be unacceptable whereas B may only require the filling of the depression with coarse or armour type rock. This solution may be acceptable to the contractor unless armour rock is in short supply. C might be acceptable with the provision of the correct thickness of rock underlayer on top but if this results in putting the subsequent work out of radius then it is not a good solution. The same conclusion would apply to D.

12.   In calm weather soundings are easily carried out from a small captive boat. The weather window is reduced if slack

water conditions are required. In worse conditions a profile may be obtained by using a long crane - suspended dipstick. This is a slower operation and removes a crane from production, possibly leading to a difficult choice having to be made.

13. When the weather is unsuitable for both types of checking it may be necessary to accept the evidence of the placing control equipment. This is not always considered acceptable but failure to do so may result in increased risk.

14. The frequency and timing of diver's inspection will also be limited in this weather.

Rock source investigation

15. Thorough investigation of rock sources, as to quality and quantity, is essential during the design period. The time available to the contractor for preparing his bid is very limited and this will not permit a comprehensive investigation of local and other sources.

16. In recent years deliveries of Scandinavian rock in 20,000 tonne barge loads have become available. These are of particular application to the South and East coasts of the UK. Planning of the transfer from barge to breakwater is necessary.

17. Above all, adequate quantities of armour sized rock must be available to meet the temporary and permanant works requirements of the construction. A strategic surplus is also necessary for compensation purposes (See Fig 1, Profile B), for emergency repairs and for a progress boost when conditions are favourable. A shortage of armour rock will certainly delay progress and may endanger the structure.

Materials haul

18. By definition, substantial quantities of rock are required for a rubble mound breakwater. If the job is close to its rock source it is possible that a private haul road may be constructed suitable for large off-road dumptrucks.

19. In other situations, the transportation may be by rail or public road. By their very nature, quarries which are suitable for rapid expansion are likely to be in rural areas with a roads system to suit. Any necessary upgrading of the roads should be put into effect well in advance of the need to exploit the quarry.

20. Public resistance to substantial increases in traffic concentration may be met. This should be anticipated and early measures taken to resolve the problems.

21. If large quantities of rock are to be transported by rail, as for example at Port Talbot, it may be necessary to provide a special siding.

Operating levels and protection

22. There are basically three stages of construction:

1) Firstly core material is end tipped and armoured from a level just above H.W. This is the primary working level.

2) Secondly the core and armour are raised to capping level.

3) Finally the construction of the capping and wave wall.

Stage 1

23. At any one time during normal working there may be three or four operations in progress, namely end tipping, core shaping, underlayer and soleplate placing and armour placing. The ideal of each of those operations proceeding at the same steady pace is rarely achieved.

24. Available core width at the working level is critical. The dumptrucks must be able to pass the tail swing of the largest crane and have room to turn near the scar end. If both weather and leeside slopes are being progressed in tandem this must also be taken into account. Temporary additional width can be obtained by overtipping as described earlier but this is lost once the core is being shaped.

25. Protection of the core is very important and is achievable in stages. A large and well graded mix, say 0 - 3 tonne, has some intrinsic stability. Nevertheless it is subject to selective washdown in quite gentle waves and over a period of two weeks' winter weather may assume a slope of about 1 in 6. Over a shutdown or breakdown period the scar end can be permitted to do this if the material at the toe will still remain accessible and within the payline. As a rudimentary form of self protection this is quite useful.

26. The side slopes should be protected immediately they have been finally shaped and accepted. Prior to that they are best left to their own devices.

27. The most vulnerable part, the tidal zone face should be covered first with a single layer of armour rock. This provides a degree of immediate protection and restricts washdown onto the lower approved slope. The remainder of the first layer can then be worked up from the toe, followed by the soleplate and the second layer. In the case of a stabit protected breakwater, such as Douglas Harbour (Fig 2), the concrete units will follow close behind the rock but laid back at a 45° angle. This will mean that if all operations are proceeding smoothly and strictly in train there can be a distance in excess of 100 metres between the scar end and the stabits in working level. This is protected only by the underlayer which should be adequate for the purpose.

Stage 2

28. This stage, to the underside of the capping, usually follows closely behind. The raising of the structure brings

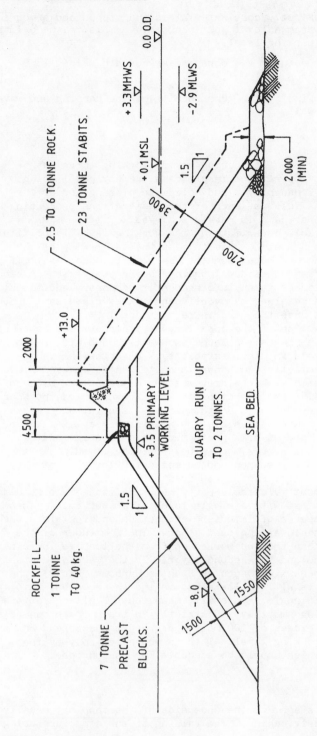

Fig.2. Typical offshore cross section of Douglas Breakwater

an obvious reduction in vulnerability to overtopping and it
will be a further advantage to build a temporary wave wall on
this level, say to 5m above HW.

29. It should be recognised in the design that stage 2
brings about a considerable reduction in available working
width which must still be adequate for plant movement and
operation.

Stage 3

30. In designing the capping, it should be remembered that
access is needed for its construction (the contractor will
probably work his way back from the end). A simple shape is
desirable requiring only one or two lifts. If a wave wall is
incorporated which is required to support heavy armour units
there must be adequate width, at least on the lower pour, for
the placing crane. Remember also that a crane placing say 25
tonne concrete units is not rigged in the same way as that
which will be placing 6 tonne underlayer rocks by grapple.

31. It is an advantage if the capping can be constructed
within the protection of the temporary wave wall, re-using
that material for permanent work. A plane level formation is
desirable to allow the cranage to get close up. The width is
now very restricted and there may not be manoeuvring room for
different items of plant.

Side slopes and reach

32. In paragraph 24 reference is made to critical width at
the primary working level. The other element which must
receive serious consideration is reach. Assuming 20 metres
depth, a slope of 1 in 2 requires a crane radius of 55 metres
to place material at the outer edge of a stabit soleplate.
However, a reduction of the slope to 1 in 1.5 will lead to an
improvement in the radius of 10 metres, a significant change.
For this reason the side slopes should be designed to be as
steep as possible.

33. Measures such as building projecting rock platforms
at frequent intervals along the working level can assist in
achieving reach but they are costly and disruptive. For
placing stabits the flotation tank mentioned earlier worked
quite well, but in heavy swell or strong currents there were
problems of accuracy due to bounce or drift.

The Environment

34. The degree of constructability planned into a design
should be influenced by the weather and sea conditions anticipated at the site during the construction period. For
reasons of cost and continuity the contractor will usually
wish to carry out work in all the seasons although his
detailed programme will be drawn in such a way as to minimise
the risk. If at all possible the design should reflect this.

35. Studies of the wind roses and the wave climate will

already have indicated the likely conditions, and an early
decision can be made on what structural integrity the partly
built structure should have. It seems reasonable to suppose
for example that in a location where the design Hs is 8 metres
there will be frequent occasions when the site experiences
3 - 4 metre waves. Model tests can be carried out on several
anticipated construction stages at the same time as the design
model tests are in progress. The effect of headlands, beach
gradients etc. can be investigated as well as the effect of
lack of full protection at the forward end.

36. The feasibility of closing down the forward advance
during the stormy season can also be investigated but from
the contractors' point of view a closedown is not a good
option unless he can usefully deploy his resources elsewhere.
Also, protection of the scar end will be a costly and time
consuming exercise not to be undertaken lightly.

37. Besides wave action there are other environmental
factors which should be considered.

38. A rocky shore or a shelving beach will each have their
own influences on the construction methods necessary to get
clear of the surf zone and into deep water. The tidal range
in relation to them will determine how much of the early
construction can be carried out in visible conditions, thereby
affecting the learning curve. The timing and strength of the
currents at the site, whether tidal streams, outfall discharges
or estuarial flows will certainly have an effect on load
placing accuracy, underwater visibility, divers' bottom time
and manual soundings.

Geotextiles

39. At the 1985 conference there was clearly a "constructability gap" when it came to discussing these membranes. In
the Geotechnics workshop "there was fairly strong support for
the increased use of geotextiles in breakwaters" whereas the
Construction workshop expressed quite a different view.

40. It is suggested that the following questions should be
posed:

(i)   Is there a clear design advantage in specifying a
       geotextile membrane?

(ii)   Can the contractor be expected to instal the membrane
       in accordance with the requirements in all workable
       conditions?

(iii) Can the Engineer's Representative establish that item
       (ii) has been met?

41. It is desirable that the answer to all these is
affirmative, in which case a fourth question may be asked.
Will the use of a membrane restrict the contractor's flexibility, increase his vulnerability and lead to significantly
higher cost. If so, is there an alternative?

Innovative concrete armour

42. The Seabee and Shed are two types of voided concrete unit which are easy to cast, transport and place. The natural wedge action of these blocks enables more economic side slopes to be constructed. The hexagonal and cubic shapes are not difficult to place and do not require any great precision. The Seabee unit also has the advantage that a simpler capping, detail is possible by filling the voids. The Seabee's versatility in size allows a particularly economic design to be achieved.

CONCLUDING REMARKS

43. The paper has described some aspects of rubble mound construction which the author suggests should be considered during the design period. Individual contractors have their own favourite methods but it is felt that much of what has been said can be considered common practice.

44. The concentration has been on rubble mounds but the need for constructability (and reduced vulnerability) applies equally well to other forms of breakwater such as caissons, cofferdams, openwork, heavy precast blockwork and composite structures.

ACKNOWLEDGEMENTS

45. The author is grateful to several collegues for helpful comments and advice, and in particular to Mr R.Malyon.

12. The development of a design for a breakwater at Keflavik, Iceland

W. F. BAIRD, W. F. Baird & Associates Coastal Engineers Ltd, Ottawa, and K. W. WOODROW, Bernard Johnson Incorporated, Bethesda, Maryland

SYNOPSIS. A breakwater design was developed to build a breakwater using local quarried rock, 100% quarry utilization and relatively simple construction methods. The design significant wave height is 5.8 m and the maximum depth of water is 24 m. The final design consists of a wide layer of 1.7 to 7 tonne armour stones in place of a traditional two layers of armour stones, which would have required 30 to 40 tonne stones. The core of the breakwater contains the remainder of the quarry yield, that is, stones weighing less than 1.7 tonnes.

The development of the design was supported by an extensive series of physical model tests. The tests demonstrated that by varying the geometry of a breakwater cross-section a design can be prepared that makes full use of the yield of a quarry and may use armour stones weighing five times less than the stones required for a conventional design. At this location this design approach achieved cost savings, compared to that for a conventional design, in the order of 40 per cent.

AKNOWLEDGEMENTS
1. The authors wish to acknowledge Phil Smith of the U.S. Navy for assembling the successful construction team, and for the opportunity to participate in the post design period of this project. The conscientious work of Steve Emrick and Ken Trotman of the Navy Atlantic Division project management and construction staff respectively, Lieutenant Commander Tom Hilferty, Arnfinnur Bertelsson and Julius Arnorsson in the Iceland ROICC office, Barry MacMillian of Tarmac, and the Icelandic Prime Contractor and Nupur Construction staffs are acknowleged for making this a successful project.

INTRODUCTION
2. In 1981 the U.S. Navy, as part of an on-going design contract with Bernard Johnson Inc. (BJI), required a protected harbor at the U.S. Naval Station, Keflavik, Iceland. The harbor facilities

required a breakwater to provide protection from
wave action.   In 1983, Bernard Johnson Inc.
retained W.F. Baird & Associates (WFBA) to assist
in the development of the design for the
breakwater.
   3. Helguvik Bay is located on the east side of the
Reykjanes peninsula and is exposed to waves generated
within the Flaxafloi, Figure 1.

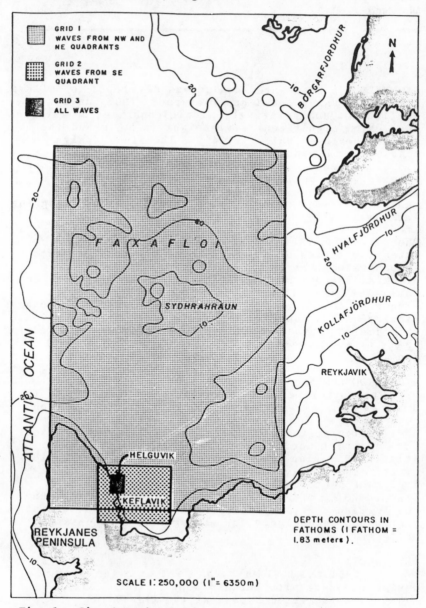

Fig. 1. Site location map

4. The maximum depth of water at the location of the breakwater is 24 m, Figure 2. The tidal range is approximat ly 4 m.

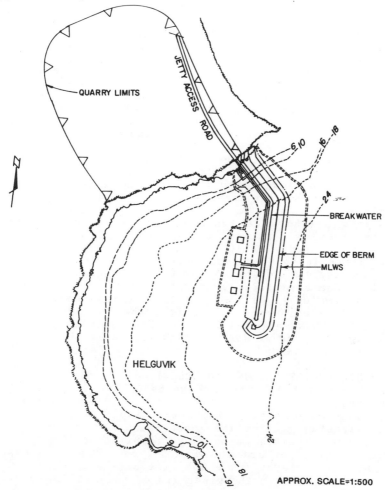

Fig. 2. Harbor layout

5. The design wave conditions were determined from a wind-wave hindcast study based on winds recorded at Reykjavik and Keflavik airports, adjusted to represent overwater wind speeds. The breakwater was designed to successfully survive storms with a return period of 50 years. The maximum wave conditions in this storm were determined to be as follows:

North east (waves attack at 45° to the breakwater centre line)

Hs = 5.8 m, Tp = 9.6 s

227

East (waves attack perpendicular to the
breakwater centre line)
        Hs = 4.3 m, Tp = 7.2 s

INITIAL PROJECT DEVEOPMENT
6. Initially, consideration was given to a
conventional breakwater design based on the U.S. Corps
of Engineers Shore Protection Manual. This design
would have required two layers of 32 tonne armour stone
on a slope of 1:2. Armore stone of this size could not
be produced from the local basalt quarries, so
consideration was given to the use of concrete cubes.
This design was of concern since initial estimates of
the cost for this design exceeded the available
funding.
7. At this time, another option was investigated to
utilize available natural armour stone only. WFBA had
previous experience with the design of a "berm"
breakwater where the traditional armor layer is
replaced by a large mass of smaller sized stones. It
was possible for a design to be developed to make use
of the locally quarried rock having a maximum stone
size in the order of 10 tonnes.
8. Three independent cost analysis were done to
compare the two potential options. The estimates
indicated a savings of approximately 40% for the "berm"
design placing it well within the program funding
limits. Based on these results, the "berm"
breakwater was selected for complete design
development, and extensive model and quarry blast
testing were begun.

BERM BREAKWATER DESIGN AND THEORY
9. The berm breakwater design using smaller stones
then required in a two layer armour system, is based on
the observation that if the armour layer is built to
significantly greater thickness than that of two
stones, much smaller stones are required to provide
stable protection against wave action. The thickness
of the armour layer for a specific breakwater is
determined based on the gradation of the available
armour stones and the incident wave climate.
10. The relatively high porosity of the mass of armour
stones allows the waves to propagate amongst the stones
and dissipate their energy over a large area within the
wide armour layer. In a conventional two stone armour
layer, the flow produced by the incident wave is
restricted by the relatively impermeable filter and
core and, consequently, there are large velocities
produced by the wave uprushing or downrushing within
the narrow armour layer. In the berm the flow has a
larger area into which it can move and as a result
localized velocities are greatly reduced therby
decreasing the external hydrodynamic forces applied to
the stones. A considerable increase in stability is

achieved as a consequence of this dissipation of wave energy within the permeable mass of armour stones.

11. A benefit of the mass of armour stone is that it increases stability as a result of wave action. Wave action causes consolidation and a resulting increase in shear strength of the mass of stones. Motion of some stones at the surface results in "nesting" of the surface stones. This nesting process also results in an increase in the frictional restraint on individual stones. Depending on the size of stones available and the design wave conditions, movement of stones on the outer surface may occur to varying degrees. Movement takes place during the early stages of exposure to wave action. The stones eventually find a geometrically similar space in the berm surface into which they nest. The result of this process is a natural armouring of the outer layer of the stones. A typical armoured profile is illustrated in Figure 3.

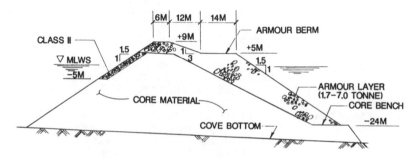

Fig. 3. 'Berm' cross-section

MODEL TEST PROGRAM

Program Overview

12. A program of model tests were performed to assist the development of the design and to demonstrate its' performance.

13. The tests were completed at a geometric scale of 1:35 in an 8 m wide wave flume at the National Research Council of Canada. Waves occurring from the north east and east were generated. For the tests with waves from the north east the bottom topography in front of the breakwater was modelled. The full profile of the design storm, having a total duration of 40 hours, was simulated by nine segments in each test and irregular waves used for each segment of the storm.

14. The full breakwater, including the head of the breakwater was built in the wave flume. The armour layer was an accurate representation of the assumed prototype gradation. Each stone in the armour was individually weighed and sorted to provide the required gradation. A total of ten complete tests with four additional demonstrations were completed. The tests

were designed to evaluate the overall stability and
performance of the structure when built using different
stone gradations and subjected to various wave
conditions and water levels. The information obtained
as the tests proceeded were used to change and revise
the breakwater design.

The ten tests are summarized below:

(a) Test 1   Initial design with the armour
              layer built using two gradations of
              armour stone subjected to the design
              storm from north east at water level
              of +4 m.

(b) Test 2   Initial geometry retained and built
              with armour stone of 1.7 to 7 tonnes
              throughout.

(c) Test 3   Model breakwater tested in Test 2
              subjected to a) a second design
              storm, b) 24 consecutive hours of Hs
              = 5.8 m, Tp = 9.6 s (maximum wave
              conditions) and c) 14 hours of Hs>7
              m and Tp = 9.6 s.

(d) Test 4   Crest elevation reduced, water level
              increased to +5 m.

(e) Test 5   Test of breakwater at low water
              level, i.e. +0 m.

(f) Test 6   Test to evaluate the effect of
              localized placement of armour stones
              not meeting the gradation. One area
              contained 1.7 to 3 tonne stones, the
              other contained 6 to 7.5 tonne
              stones.

(g) Test 7   Width of the armour layer reduced.

(h) Test 8   Breakwater subjected to design storm
              from the east.

(i) Test 9   Repeat of Test 8.

(j) Test 10   Breakwater with reduced armour stone
              sizes to evaluate the influence of
              the armour stone gradation on the
              overall stability of the breakwater.

Berm Armouring Performance

15. It is useful to describe the performance of the
breakwater during the design storm. No change to the
profile of the breakwater occured during the first two
segments of the storm (Hs = 1.8 m and Hs = 3.05). Some
motion of stones was observed but was limited to
"rocking-in-place". During the third segment (Hs = 4.3
m) some rounding of the outer edge of the horizontal
berm occurred. This resulted from some stones
initially placed in a relatively unstable position
being rolled a distance down the slope during the
downrush of the largest waves.

230

16. In the fourth segment (Hs = 5.2 m) more intense motion was observed. The stones that were removed were rolled to an elevation of between 0 m to -10 m. However, as these stones were rolled out of position the net effect was to leave behind a more compact, nested and therefore stable outer layer of armour stones.

17. During the peak of the storm (Hs = 5.8 m), the observations were similar to the previous segment with the motion generallly restricted to "rocking-in-place".

18. During the sixth segment (Hs = 5.2 m) and later segments the motion was considerably reduced compared to that observed with the same wave heights on the "up side" of the storm. When the design storm was repeated the structure was observed to be considerably more stable - in the sense that less motion was observed than noted in the first storm. Each subsequent storm produced similiar effects, with the profile becoming more static and more stable.

19. It is important to note that unlike conventional two layer armouring, movement of stones caused by storm action does not produce failure or weakening of the breakwater protection. It can have the opposite effect, with an increase in the stability of the armour layer.

Observed Safety Factors:

20. Upon completion of several of the test segments, the model was subjected to storms more severe than the 50 year design storm. With flume limitations the most severe wave climate achievable was Hs = 17.5 m. During this storm there was limited overtopping and conserably more movement in the armour layer, however exposure of the core or damage to the near side of the breakwater did not occur. The resulting armour profile would still provide protection for future storms. These tests as well as the tests using smaller than specified stones, indicate that this berm breakwater possesses a higher safety factor and greater flexibility than a conventional two layer design.

Test Conclusions

21. Following the completion of the model studies the following conclusions were drawn:

(a)   The breakwater will successfully survive repeated occurrences of storms with a return period of 50 years.

(b)   Tests were completed at water levels of 0, +4 m, and +5 m. The breakwater was stable for all water levels. WAve runup was predictably greater during the tests at high water. The maximum extent of wave runup during the high water tests was approximately +8 m to +9 m.

(c) Tests were completed for two wave
directions. The response of the breakwater
was most critical for wave attack
perpendicular to the centre line of the
breakwater.

(d) An armour stone gradation of 1.7 to 7 tonnes
was recommended. Some reserve stability
exists with this design and this was
demonstrated using both lighter stones and
reduced width of armour layer. This
gradation was selected to allow full use of
the quarry yield.

(e) The head of the breakwater was carefully
observed and its performance recorded in all
tests. It was recommended that the full
width of the horizontal berm of armour
stones be placed around the head through 135
degrees.

(f) Localized placement of uniform sized stones,
whether smaller or larger than the specified
size did not affect the stability of the
structure. It is anticipated that limited
segregation of storms, as could occur during
construction would not affect the
performance of the structure.

QUARRY TEST PROGRAM

22. A blast test program to establish quarry gradation
curves from the basalt flows adjacent to the breakwater
site was undertaken concurrent to the model test
program. Since the design goal is for 100% quarry
utilization the quarry tests ideally would be completed
prior to the model tests, however design schedule did
not allow this to occur. As a result, another
Icelandic basalt quarry was used as the basis for the
initial model test, with the results of the blast test
either confirming the assumed gradation, or requiring
additional model tests.

23. The quarry test program consisted of developing
three test quarries at locations close to the
construction site. Yield volumes and size distribution
were determined from these tests quarries, with a graph
of the composite results shown in figure 4. Quarry
yield estimates were then made for the entire potential
quarry site using the test results, extrapolations of
boring information, and geologic mapping of the cliff
face.

24. The final test results confirmed that the modelled
armour gradations were achievable, however the design
would not fully use the quarry.

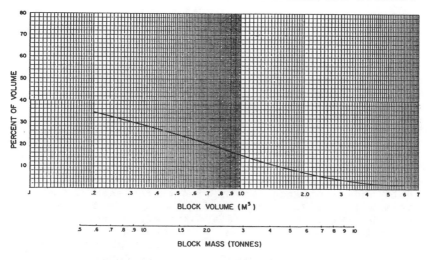

BLOCK VOLUME (M³)

BLOCK MASS (TONNES)

Fig. 4. Helguvik test quarry combined exceedance plots

25. A minor adjustment was then made to the breakwater profile adding a bench at elevation - 24 m, see Fig. 4. The elevation was determined to have no effect on the armour stability, and allowed for a projected 100% quarry utilization.

FINAL CONTRACT DOCUMENTS/DESIGN
26. Upon completion of the test programs, the final breakwater plans and specifications were prepared. An essential part of this work was to ensure that the breakwater was built to the same specifications as the model. The following are several key items which were emphasized in the documents:

(a) Durability, geometric aspect ratios, and minimum size were specified for the armour stones to maintain proper porosity.
(b) Gradation curves and tolerances were delineated for the armour stone.
(c) Berm width was allowed to vary based on the natural in-place slope, however volume of armour at any section was fixed.
(d) Testing and quality control requirements were delineated.

27. A final important design service provided was a pre-construction presentation to the construction and inspection staff. This allowed for explanation of the berm concept and emphasis on those key contract document requirements that could adversely effect the design. The presentation and subsequent discussions were extremely valuable in eliminating early construction deficiencies.

CONSTRUCTION

28. The breakwater was built during 1986 and 1987 to an elevation of 3.4 m. A temporary head was built in November 1986 to protect the breakwater during the winter of 1986-87. It is worthy of note that no downtime due to wave action was experienced during construction nor was there any damage to the partially built structure. The contractor found that the partially built berm of armor stone absorbed the wave action and minimized spray and overtopping that for a conventional structure would have caused downtime and damage to the unprotected core. A particular good example of this construction benefit was in November 1986, a storm from the North east with winds of 55 knots and significant waves of 4 m, representing a storm with a return period of greater than 5 years. The above water portion of breakwater extended approximately 150 m of its total length of 350 m, to an elevation of 3.4 m. With the exception of wind blown spray, there was no overtopping and no damage to the construction. The partially completed berm armour mass was able to dissipate the wave energy.

29. The armour stones were individually placed by a crane, with occassional assistance from an excavator equipped with a rock grab. Careful and precise placement of each stone was not required with the armour mass, and the final placement was not the limiting factor in the construction process. This method also allowed a final opportunity at the breakwater to reject any substandard armour stone.

BENEFITS OF THE BERM DESIGN AT HELGUVIK

30. Design factors are unique for any breakwater site, and therefore requires site specific evaluation. This evaluation enabled us to select the berm breakwater, with the following benefits. Some of these benefits would apply to any berm breakwater design, some apply to this project due to site specific factors:

(a) An approximate 40% construction cost savings was achieved.

(b) The design targeted a 100% utilization of all quarried stone.

(c) Armour stone size was drastically decreased allowing easier handling with smaller equipment.

(d) The breakwater was less susceptible to storm damage during construction, and less construction downtime was encountered.

(e) A significantly higher stablity safety factor was achieved for the armour layer compared to traditional designs.

(f) The breakwater was awarded within funding limits.

(g) The design allowed for simple construction procedures without close tolerances for armour stone placement.

13. The control of rubble-mound construction, with particular reference to Helguvik Breakwater in Iceland

J. READ, G. Maunsell & Partners

SYNOPSIS. Two aspects of breakwater construction are examined, the quarrying of the rock and the means to ensure that it is correctly placed in the breakwater. The recent construction of Helguvik breakwater in Iceland is used to illustrate the way in which the quality and size of rock quarried may be monitored and also the method of survey adopted. Other survey methods are described and discussed.

INTRODUCTION

1.   This paper addresses the problem of ensuring that the intentions of the designer of a rubble mound breakwater are achieved during construction.

2.   Control must be applied to the production of the rock to achieve a product complying with the specification with the minimum of waste. When quarried the stones must be placed in the breakwater in zones allocated by the designer and it must be shown to the satisfaction of the Engineer that this has been done. Thus there is the need to provide positional control over rock placing and separately to provide means to survey the placed rock.

3.   The techniques used to control quarrying operations at Helguvik will be described together with the survey system which satisfied both survey requirements. In addition a brief review of other survey methods has been included to provide perspective.

THE QUARRY

4.   Helguvik is situated on a shield volcano composed of rock known as olivine tholeite; a type of basalt. This comprises both highly fragmented scoriaceous basalt lava and more massive compound unit flows. The compound unit flows tended to predominate in most of the designated quarry area but in the south west quadrant highly fragmented lavas were present to below 10m above sea level. A particularly thick flow unit frequently formed the highest level of the lava sequence and extended down to sea level in the south eastern and north western quadrants. Fig. 1.

Fig. 1.   Thick lava flow unit

5.   The top 2 to 3m of rock in the quarry area was
moderately weathered with very closely spaced fractures. It
was frequently overlain by about 2m of silt and fine sand
with thicker soil overburden in topographical depressions of
the ground surface. Below the moderately weathered zone the
thick flow unit had closely spaced fractures extending from
3m to 8m below ground level. Fracture spacing in this flow
increased with depth until below depths varying from 7m to
12m below ground level the fractures were widely spaced.
Flow units other than the very thick one had much closer
spaced fractures.

6.   An individual lava flow usually comprises a
scoriaceous and highly vesicular base, up to 1m thick, with
as much as 2m of similar material at the top of the flow,
the remainder being a core of relatively massive and only
slightly vesicular rock. Typically the relative density of
the massive material ranged between 2.75 and 2.85 with only
7% of the rock having a relative density less than 2.69. By
comparison the relative density of 60% of the rock in the
rest of the layer was less than 2.75. A preliminary
estimate of the mix of stones which might be produced from
the two main grades of basalt together with the overall
yield is given in Table 1.

7.   In general it was expected that most of the material
for the core would be obtained from the upper levels of the
quarry with a lower bench providing armour stone. An aerial
photograph of the quarry forms Fig 2.

Table 1. Estimated quarry yield

| | Fracture spacing | | Total yield |
	Wide	Medium	
Class 1.	20%	–	14%
Class 3.	60%	80%	66%
Waste	20%	20%	20%

Fig. 2. The quarry and breakwater

Table 2. Classes of rock

Class	Stone weight range	Mean stone weight
1	1.5 - 7 tonnes	3.2 - 4.2 tonnes
1a	greater than 7 tonnes	–
2	0.5 - 1.7 tonnes	1 tonne
3	less than 1.7 tonnes	–
4	Stones between 6 in and 1 ft in size	–

DESIGN OF BREAKWATERS

SPECIFICATION REQUIREMENTS

8.   The contract documents identified five classes of stone, shown in Table 2, and they specified that the Class 3 material was to contain no more than 5% by weight of material less than 6 inches in diameter, and that the Class 1 rock was to lie within the grading shown in Table 3.

Table 3. Grading of armour

Size (tonnes)	Percent finer by weight
7	95-100
5	75-85
3	35-50
1.7	0-10

9.   Trial blasts confirmed that adequate quantities of armour stone would be produced but that the core material (Class 3) would contain more small material than was specified. Typical results showed that, without screening, the proportion of material less than 4 inches in size might average 30% of the total volume of Class 3. As the contract had been let on the basis that no screening would be required the matter was referred to the client and it was finally agreed to relax the specification so that an average of 30% by weight of Class 3 material might be less than 4" in diameter provided that in no case should an individual truck load contain more than 40% by weight of small material.

SELECTION AND QUALITY CONTROL
Class 1: Selection

10.   Armour stone was extracted by excavator from heaps of blasted material and transported by wheel loader to a stockpile for that particular weight category. Weights of stones were assessed visually by the loader driver by reference to previously weighed sample stones. During training, and less frequently when experienced, operators would send stones to be weighed as a check upon their own accuracy of selection. Occasionally stone would be loaded at the quarry face for direct transport to the breakwater but it was more usual to draw stone for the breakwater from previously assembled stock piles. Fig. 3.

11.   Stones were sorted into three categories and, in order to monitor the number of stones, counters for each category were installed in the wheel loaders. As each stone was loaded the relevant counter would be updated and the status of the counters would be checked and recorded at intervals of about two hours. Correct grading in the breakwater was maintained by ensuring that each batch contained a suitable number of stones in each category. A batch size of 300 stones was chosen for convenience, and the

Fig. 3. Loading armour stone

maximum, minimum and optimum number of stones from each
category was specified (See Table 4.). The weight of each
truck load was recorded and the total compared with the
recorded number of stones for the batch to check the mean
stone weight. Each batch of 300 stones weighed about 1000
tonnes.

Table 4. Mix of armour stones

Size	1.7-3 tonnes	3-5 tonnes	5-7 tonnes
Optimum No.	174	93	33
Range of numbers	153 - 195	81 - 105	24 - 39

12. Class 1, Quality Control

Quality control included visual assessment of the soundness
of the rock including breakages, planes of weakness and
cracks and porosity together with the shape of the stones.
In addition check weighing was required and tests were made
for density and absorption, graduation and average stone
weight. The frequency of testing is given in Table 5.

13. For check weighing of individual stones, selection
was at random from the source in use at the time of
weighing. The weighed stones were returned to the
stockpiles and marked clearly as additional reference
stones. When a truckload of stones was checked the status
of the loader counters before and after the truck had been

loaded would be checked against the total weight of the
stones in the truck. Each stone was subsequently weighed
individually. Results from the check weighings would be
used to modify selection as necessary. In general, greater
emphasis was laid on check weighing the smaller sizes
because operators found it more difficult to assess their
weight correctly.

14. Daily checks were also made upon the records kept by
the operators of the numbers of each size placed in the
breakwater. It was verified that the numbers of each
category were within the acceptable limits and from the
total weight of the armour placed a check was made upon the
average weight of the stones.

Table 5. Frequency of tests on quarried stone

Test	Intervals between tests		
	Class 1	Class 2	Class 3
Weighing individual stones	1000 tons	as Class 1	-
Weighing truck load of stones individually	15000 tons	-	-
Mean weight	4190 tons	1000 tons	-
Grading	4190 tons	1000 tons	-
Density and absorption	4190 tons	as class 1	-
Test screening	-	-	1500 tons (plus any truck load with high % of small sizes)

(Note: units are US tons)

Class 2

15. Class 2 stones ranged in weight between 0.5 and 1.9
US tons (0.5 - 1.7 tonnes) and were used as protection to
the rear face of the breakwater and for scour protection
around the berthing caissons. Tests upon this category are
also listed in Table 5. In this case the mean weight of the
stones was checked by counting the number of stones in a
particular truck load and comparing that with the
weighbridge weight.

Class 3

16. Test screening amounted, in practice, to about six
tests per day. A test load would first be weighed in the
usual manner and then passed over a small grizzly, the fines

being collected by a conveyor and reweighed at the weighbridge. As the selection of core was by eye there was appreciable variation from load to load and a moving average for a sample of ten tests was therefore calculated as a check on any trend in fines content. If a truck load appeared to have an unusually large proportion of fines it would be sent for test screening to check that it did not contain more than 40% and this could have affected the moving average significantly. However in practice only about 10% of test values were in excess of 30% and more usual test values of fines lay between 10 and 30% with the moving average varying between 19 and 28%. It would appear that the core contains about 25% of material smaller than 4" (100m) and about 5% of sand sized grains.

17. Where the core was end tipped no mixing of the core took place but a very large part of the core was placed by barge. Some mixing of that material occurred as it was first end tipped into a stockpile at the jetty from where it was loaded into the dump barge by frontend loader. Selection from the tipped material was used to correct the segregating effects of end tipping and to distribute fine material as evenly as possible in the barge.

CONTROL OF CONSTRUCTION

18. Three commonly used methods of construction were used at Helguvik: placing by crane, end tipping and dumping by barge. Although a small amount of armour was placed by barge most was placed by crane so that, effectively, the crane was used for armour and the other two methods for the core.

19. The much larger thicknesses of armour than is usual simplified armour placing. It proved possible to control this satisfactorily by repeated surveys at fairly close centres. It was not necessary to devise a placing pattern instead; if a section was found to be deficient, the crane driver was instructed to place the next stones (giving an approximate number) between markers painted on the berm. One was not, of course, attempting to build the final stable surface and movement of stones under storm attack was expected.

20. Core construction was chiefly by barge as about 85% of the total volume of the core could be placed in that way. Again control was by repeated survey allied to close control of the barge's position. Trimming of the core face was not required. For each barge load a position and heading was specified and a check was obtained on the actual dump location because the readout of the display unit froze as the dump began allowing a record to be made. Errors in position varied widely; in one day ten loads were dumped at distances varying from 0.5m to 19m from the positions specified. The average error was 7m. Distances from the centreline of the breakwater generally showed less error than the position along the breakwater, comparable figures

for smallest, largest and average errors for the two directions in the same sample of dumps being 0.5m, 7m, 2.2m and 0.25m, 17.7m, 6.2m respectively. On that particular day a total of 7450 tonnes was placed in 18 hours.

21. Core placed by end tipping was delivered directly from the quarry and was tipped as evenly as possible over the whole width of the scar end. Trimming of the faces was necessary to achieve the specified slopes.

SURVEY

22. Survey essentially comprises the determination of the location of a point in plan and of the level of that point above or below a datum. A description of the various methods available may be divided conveniently between these two tasks.

Determination of Location

23. Traditionally, methods for the determination of location have been divided into the simple alignment of markers by eye for control of construction and slower instrumental measurement for inspection and permanent record purposes. The accuracy of construction was therefore limited and the more accurate surveys took too long to be useful in controlling the works except where gross errors had been found. Methods now available may be applied more flexibly and can provide accurate control of construction as well as enabling the more rapid completion of pre and post-construction surveys.

24. Radio position fixing is long established and provides a very rapid method of location the results of which can be manipulated by computer without intermediate processing. Both range-azimuth and range-range methods of location are in use but only the latter can provide an accuracy approaching that which is required for engineering works. Range-range methods may be subdivided into those with active shore stations and those with the active station afloat. Although the former system can control several vessels the accuracy possible is no better than $\pm$ 5m. The latter system can achieve an accuracy of $\pm$ 1m in the estimated position by addressing at least four slave stations on shore and comparing and weighting the results by computer. This method has the disadvantage that a master station and computer are needed for every vessel requiring such control.

25. When allied with a very sophisticated method of depth measurement which will be described later an accuracy of $\pm$ 1m would probably be sufficient but if used with a simple echosounder the resulting survey of a breakwater would probably exhibit errors of the same order as the tolerance permitted for construction.

26. Better accuracy over the limited field of a breakwater can be provided by a combined theodolite and electronic distance measurer (EDM) working in a

Fig. 4a. Display unit in target mode

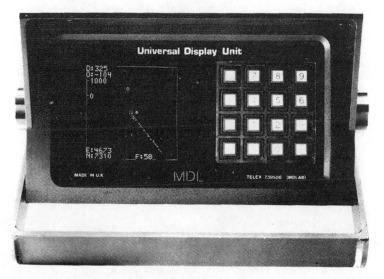

Fig. 4b. Display unit in track mode

range-azimuth mode. Both range and bearing are output in
digital form admirably suited to radio transmission and
computer processing. Methods of operation range from simple
manual tracking to fully automatic systems which will seek
and lock onto their target. Both range and bearing are
transmitted by telemetry to an onboard processor which
includes a display unit to inform the surveyor or coxswain
of his position relative either to a desired track or to a
specified position. Fig. 4 shows the two modes of display.

243

27.   The accuracy of the equipment used at Helguvik gave a
potential accuracy of $\pm$ 25mm (distance) $\pm$ 10mm (bearing) at
1000m. Manual tracking proved to be very simple as the
operator had only to turn a continuous tangent screw, the
readings being taken automatically. Fog would undoubtedly
prevent the use of this system but tracking at night is
possible if a navigation light is set immediately above or
below the reflector prisms. Fig. 5 shows the survey launch
and mounting of prisms and transducer.

28.   Where the prisms are not mounted directly over the
echosounder transducer in a survey launch or the centre of
the hopper in a barge a correction to the calculated
position must be made to allow for the separation. A gyro
compass is usually fitted in such cases to give the heading
of the vessel so that a correction can be computed
automatically. As the gyro can be reset frequently when the
barge or survey vessel is lying alongside with a known
heading, the gyro compass chosen can be a relatively
inexpensive instrument.

Depth Measurement
29.   Dimensional control of breakwater construction
essentially reduces to the measurement of profiles of the
mound. The spacing will generally be governed by the speed
with which a profile can be measured, the degree of
disruption to the works which measurement will cause and the
maximum spacing which will satisfy the engineer that his
specification is being met. At Helguvik while the bulk of
the core was being placed, a spacing of 50 feet (15m) was
adopted but for later stages this was halved.

Fig. 5. Survey launch

30. Methods using a heavy plumbing rod fitted with a basket at its bottom are slow and usually require the use of a crane. Very often the crane will be that used to place armour so that profile measurement directly influences production. Remote sensing methods therefore have considerable potential both to minimise interruption and to increase the density of measurement. Those currently available are all electro-acoustic in character and are classified by their intended use and the frequency of the transmitted signal. Echosounders usually operate at frequencies of between 30 and 210 kHz whereas sonars often use frequencies of up to 1 MHz. In consequence the range of sonar is shorter than that of an echosounder and will be unlikely to exceed 100m in clear water.

Echosounders

31. Those used for navigational purposes are fitted with transducers which emit radiation over an included angle of about 45°, intended to enable the instrument to function when the vessel is rolling heavily. The footprint of such a beam is too large to be used to survey a seabed which slopes steeply. The systematic error arising when measuring over a slope can be reduced by reducing the spread of the transmitted pulse. Narrow beam units have therefore been developed to minimise this problem, but are restricted to the higher frequency pulses. A transducer with a 3° beam angle is currently available for a pulse frequency of about 210 kHz whereas the beam angle for a 38kHz transducer may be about 8° (ref 1).

32. The potential for improved accuracy of a narrow beam will only be achieved if the alignment of the transducer can be kept truly vertical. A restriction of survey work to calm weather when both roll and heave of the survey vessel were at a minimum proved satisfactory at Helguvik but elsewhere gryostabilisation has been adopted to isolate the transducer from pitch angles of $\pm$ 10° and roll angles of $\pm$ 15°. Eleotronic compensation for roll, pitch and heave is also offered by at least one manufacturer. Both mechanical and electronic stabilisation are expensive and are comparable in price to alternative measurement techniques which are described in subsequent sections.

33. Typical accuracy of a hydrographic survey sounder will be about 5cm at 30m depth but resolution will depend upon the type of record and environmental conditions. Practical accuracy is unlikely to be better than 30cm (1ft) at any depth but proved sufficient at Helguvik to control the finished surfaces of both core and armour. Comparison of duplicated soundings showed differences normal to the slope of up to 4ft although the average difference was 1.3 ft. Fig. 6.

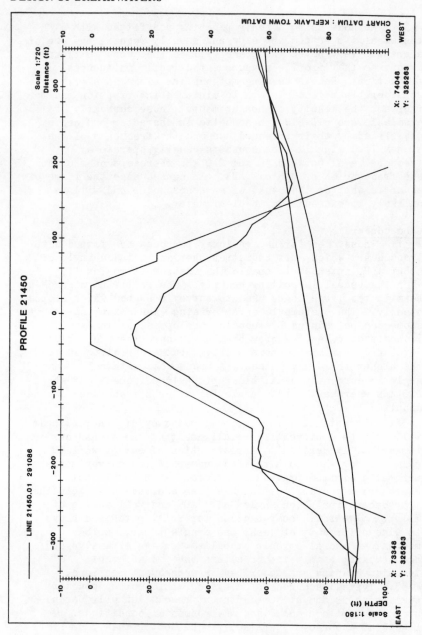

Fig. 6a. Typical measurement profile during core
construction

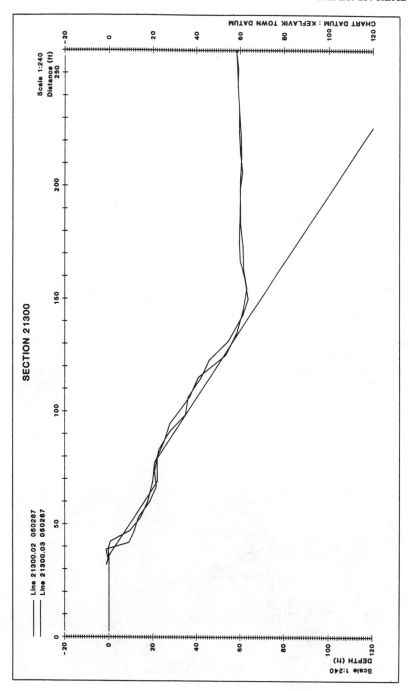

Fig. 6b. Comparison of duplicated measurements of one
profile

Sonar

34. A sonar with a narrow beam (1°) has been used with a mechanical stepping device to record profiles of pipe trenches by making a series of discrete measurements. For those purposes it was mounted upon either a boat, trenching plough or ROV and produced computer plots of the profiles. It has not, so far, been used to measure a breakwater and would require further development for this to be possible. In particular the heading and attitude of the sensors would need to be known with some accuracy.

35. Two methods mount transducers in a fish towed behind the survey craft. One, developed in Norway can scan a 90° sector beneath the fish, while the other can scan more than 180°. The limitation of the scanned sector, combined with its extremely high cost is unlikely to commend the first to civil engineers but the second could provide the basis for a very rapid means of accurate survey and monitoring of construction. The principles of operation and the results of trials of the swathe-sounder have been described by Cloet (ref. 2,3). In use at the breakwater site the fish would be towed close astern of the survey vessel at the minimum depth necessary to be in unaerated water. The frequency of 300 kHz is relatively low so that ranges in excess of 100m are obtained and the largest breakwater structure could be surveyed in one pass. The fish would be towed parallel to the breakwater axis and five profiles would be measured every second, each profile comprising about 200 measurements of depth and distance. At a practicable speed of six knots profiles would be spaced at 0.6m (2ft) centres, Fig. 7.

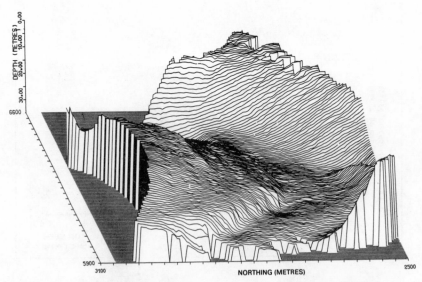

Fig. 7. Drake Channel and Vanguard Bank, Plymouth Harbour. Surveyed by BATHYSCAN from m.v. Catfish (Plymouth Polytechnic)

36. The accuracy claimed using radio position fixing of the survey vessel has been of the order 25cm (10 inches) in depth when used for hydrographic survey. This degree of inaccuracy results from both position fixing errors and those associated with the measurement of range, fish attitude and heading. Errors of position could be reduced considerably by the adoption of the optical/EDM system described earlier. The system also lends itself to compensation during computation and plotting. The plotted relative positions of robust permanent seabed markers could be compared with positions established by other survey techniques. Plot scale could then be adjusted to minimise the differences between the two methods. Such a system could provide a very rapid means of checking the whole surface of a breakwater rather than at isolated cross-sections.

Processing and Presentation of Results
37. On-line processing is offered by most manufacturers and will generally be installed on survey vessels operated by specialist companies. At a construction site, off-line processing will probably be satisfactory as any need to re-survey can quickly be satisfied. For off-line processing, positional and depth data would be collected by a small data-logger on board the survey vessel, tidal data would be recorded separately and all would be processed by a computer on shore. At Helguvik after correcting the survey data for the state of tide, the software permitted the inspection of each profile for erroneous readings with an option of substituting an average of the readings on each side of that in doubt or the rejection of the profile. The latter was necessary when shoals of fish suppressed the echo from the seabed. Each profile was checked before printing on a drum plotter. The same equipment was also used to prepare a plan of the survey to show the location of each profile.

CONCLUSIONS
38. The production for a breakwater of rock of the correct size and grading is a process which greatly depends upon the integrity and efficiency of the work force. It would be quite impracticable to weigh each rock, even for armour, and control is essentially statistical. The paper describes the main features of the system used to ensure compliance with the specification.
39. Survey methods adopted at Helguvik have been automated to a considerable degree although remaining simple in principle. No special precautions proved necessary to compensate for boat movement as calm weather was sufficiently frequent during the construction season to limit breaks in surveying to about two to three days. The resulting records of breakwater profile and location were used to control construction and to provide the final measurement of the structure.

40. Where breakwater design calls for primary and secondary armour to be placed in thinner layers than were required at Helguvik a closer spacing of survey sections would be desirable especially where visual observation is difficult or impossible. In such cases the recently developed swathe-sounder would seem to offer a means to maintain the essential degree of control without extending the time taken for survey.

ACKNOWLEDGEMENTS
The author has pleasure in acknowledging the help given in the preparation of this paper by the Iceland Prime Contractor and their consultants Tarmac-Maunsell who advised upon the planning and management of the project. He also wishes to thank the Naval Facilities Engineering Command of the United States Navy for their permission to publish this paper.

REFERENCES
1. Simrad EA 200 Product Information
2. CLOET R.L., HURST S.L., EDWARDS C.R., PHILLIPS P.S. and DUNCAN A.J. A sideways-looking towed depth measuring system. J. Roy. Inst. Nav. Sept 82 pp 411-420.
3. CLOET R.L. and EDWARDS C.R. The bathymetric swathe sounding system. Oceanology International Conference March 1986.

14. Main breakwater repair at the Port of Ashdod, Israel

L. STADLER and M. RADOMIR, Ports Authority, Israel

SYNOPSIS. The Ashdod Port Main Breakwater was designed in the late fifties and its construction completed in 1965. The 1976-77 comprehensive survey of the structure and subsequent partial surveys executed, showed that major repair works should be undertaken in order to ensure the structure's capacity to withstand storms, whose characteristics are considerably higher than those assumed in the original design. The repair design included three main aspects namely, armour, crownwall and geotechnical stability.

EXISTING HARBOUR AND ITS MAIN BREAKWATER

1. The existing port is an open sea port, located in the southern part of Israel's Mediterranean coast, protected by two breakwaters, Fig. 1. It was designed in the late fifties and constructed during the period 1962-65. The main breakwater is of the rubble mound type, composed of different rock layers and armoured with concrete tetrapods of 10 and 16 cubic meters in the trunk and roundhead, respectively. The upper tetrapods are supported by a mass concrete vertical wall monolithically connected to a concrete road, cast on the mound's crest, Fig. 2. The overall length is 2200 m and the original maximum water depth was 15 meters below Land Survey Datum (LSD).

Original design

2. <u>Environmental data</u>. Waves, tides and winds were measured over a period of one and a half year. Wave heights were determined at fixed hours (08hoo, 11hoo and 14hoo), by visual measurement of the highest breaking wave during five to ten minutes of continuos observation of the elevation of the breaker heights relative to sea level, wave directions were measured three times a day by visual observation, using binoculars and a compass. Tidal data was gathered at the adjacent cooling water basin by means of a mareograph.

Basic design

3. A desk design was prepared, primarily based on the recommendations of the tetrapod's unit patent holders and subsequently tested in 2D and 3D models.

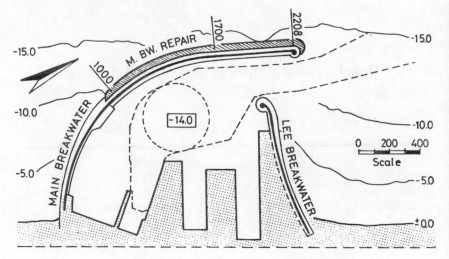

Fig. 1. Existing port

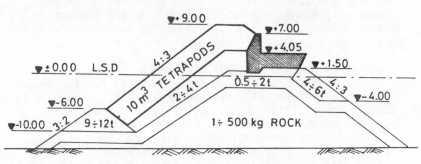

Fig. 2. Typical cross section

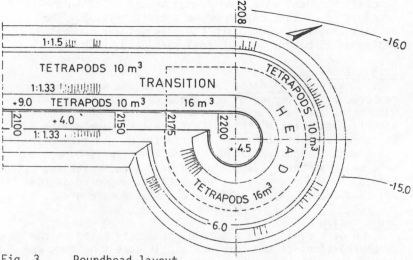

Fig. 3. Roundhead layout

The "design criteria" set down by the designers was:

(a) No damage and no overtopping with wave heights of up to 6.50 m.
(b) Repairable damage for waves of 8 m height and 13 sec period.

4. Two dimensional model tests. The model tests were done by Safona, France in a wave flume at a scale of 1 in 50. Three alternative cross sections were tested in order to reach the selected design. The testing conditions consisted of regular waves with constant period for different wave heights, combined with two water levels, -0.30 and +1.20m L.S.D. The recommended cross section was subject to the following tests:

Table 1. Two dimensional model tests

Test	Wave height m	Wave period sec	Duration hrs	Water level m
A-a	5.5-8.2	13	30.5	+1.20
A-b	3.0-8.0	13	20.5	-0.30
B-a	4.0-8.2	11	19	-0.30
B-b	3.0-8.1	11	27.5	+1.20

Evaluation of the structure's behaviour was of a qualitative character, based on observations during testing.
The conclusions drawn by Safona were that "The stability of this cross section appears to be complete up to a wave height of 8 m, above which only some light incidental damage was observed" (ref. 1).

5. Three dimensional model tests. The wave tank tests included testing of a typical cross section, situated in the curved part of the breakwater and the end section together with the roundhead at a scale of 1 in 50. For the first one, a section located in water depth of 10 to 12 metres was selected with the cross section arrived at in the flume tests but for an increased thickness of the concrete slab by 0.50 m.

Table 2. 3-D model tests programme, typical cross section

Wave height m	Wave period sec	Duration hrs	Water level m
4.85-7.00	9	21.5	-0.30
5.50-7.40	9	23	+1.20
4.10-7.70	11	42.5	-0.30
6.35-8.35	11	43.0	+1.20
3.70-7.50	13	16.0	-0.30
4.40-8.30	13	57.8	+1.20

Table 2.   3-D model tests programme, typical cross section

Wave height m	Wave period sec	Duration hrs	Water level m
4.85-7.00	9	21.5	-0.30
5.50-7.40	9	23	+1.20
4.10-7.70	11	42.5	-0.30
6.35-8.35	11	43.0	+1.20
3.70-7.50	13	16.0	-0.30
4.40-8.30	13	57.8	+1.20

Upon completion of the above series, apparently carried out without any rebuilding of the section, the conclusion drawn was that... "The section was quite stable up to wave heights of 8m and periods of up to 13 sec. Although the breakwater was not damaged, these were the extreme conditions under which it would remain stable" (ref. 2).

The roundhead and end section layout tested was different than the constructed one, Fig. 3. The roundhead section tested had an armour layer at a slope of 4 in 3 composed of 16 cu m and 10 cu m tetrapods, Fig. 4. This final design was tested in only one test under waves of 6.60 to 8.0 m height and a constant period of 13 sec with an overall duration of 17.5 hrs prototype. Safona's summary states: "Satisfactory stability for waves up to 7.50 m.   A limited number of tetrapods lost without this becoming aggravated for 8.0 m waves. Repairable damage" (ref. 2).

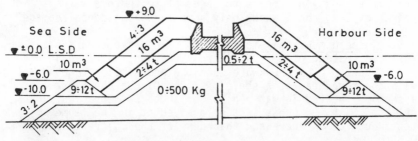

Fig. 4.   Roundhead cross-section

REVIEW OF THE EXISTING BREAKWATER

6.   The performance of the existing breakwater until 1978 was considered to be satisfactory. This evaluation was based on the state-of-the art at that time, confirmed by the findings of more or less regular monitoring of tetrapods displacements. These were determined by comparing elevations of colour marks painted on tetrapods resting against the crownwall. The fact that some tetrapods showed progressive movements and additional tetrapods units were placed during the 12 years since completion of construction, was accepted as normal "wear and tear".

7. <u>1977 Survey</u>. The Coastal and Marine Engineering Research Institute (CAMERI), Haifa carried out in 1977 a detailed survey of the breakwater (ref. 3) whose main findings are presented in the following paragraphs.

8. Ten full profiles measurements of the seawards slope were executed from station 600 to 2200 that were classified, by shape, into three types as shown in Fig. 5. The upper layer of tetrapods was visually inspected and the number of broken tetrapods was evaluated as shown in Table 3. The stone berm was shown to have lost, on the average, 30 per cent of its mass with a maximum of 50 per cent between stations 1400-1600.

Table 3. Tetrapods breakage and movement

| STATION | MOVEMENT AT CREST | | BROKEN TETRAPODS | | | DISPLACED TETRAPODS ($^\circ/_\circ$) |
	NUMBER	AVERAGE (cm)	ABOVE L.S.D(N°)	ABOVE LSD($^\circ/_\circ$)	BELOW LSD($^\circ/_\circ$)	
900-1100			13	4	27	0
1100-1300	4	35	15	4	36	3
1300-1500	6	42	20	6	44	9
1500-1700	4	52	16	5	17	6
1700-1900	11	50	15	4	80	13
1900-2100	17	59	20	6	56	9
AFTER 2100	6	51	21	6	74	2

9. Harbour slope profiles were surveyed at 50 m intervals which show slump to S-shape deformations, as well as cavities under the crownwall (Fig. 6.).

10. Several levelling surveys of the crownwall road were carried out since completion of construction, showing a small overall settlement as shown in Fig. 7.

11. <u>1980 damages and maintenance</u>. Damage to the end section and roundhead was caused by the March 1980 storm ($Hmo,o = 5.4m$, $Tp = 12sec$). Maintenance repairs were executed in October 1980 at station 2175, where 26 units were placed (7 underwater); at the roundhead 35 units (26 of 16 cu m and 9 of 10 cu m) were placed aiming to fill the gaps and achieve a reasonable degree of interlocking. Inspection of the repaired sections carried out in 1981 (ref. 4) showed extensive breakage of replaced tetrapods Fig. 8, that was caused by a December 1980 storm ($Hm,o = 6.30m$, $Tp = 12sec$).

Geotechnical stability assessment

12. Slope stability analyses were carried out, using both Bishop and Normal (Fellenius) methods of slices. Loading cases considered included:

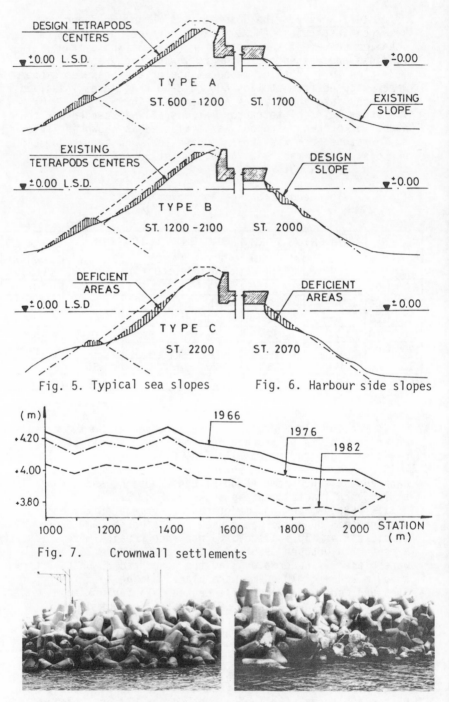

Fig. 5. Typical sea slopes

Fig. 6. Harbour side slopes

Fig. 7.   Crownwall settlements

Fig. 8.   Existing roundhead and transition

(a) normal water level (+0.00 LSD)
(b) different down rush levels.
(c) overtopping related high water level
(d) seismic loading

Table 4. Seawards slope minimum factors of safety

Water level LSD m	Bishop	Normal
+0.00	1.39	1.34
-2.00	1.18	1.19
-4.00	1.01	1.07

Table 5. Harbour slope minimum factors of safety

Method	Water level +0.00	Overtopping	Earthquakes
Bishop	1.28	0.93	-
Normal	1.22	1.00	1.085

REPAIRS DESIGN

13. Basic approach. The basic approach adopted by the Ports Authority was to look for a design solution that would not involve a reconstruction of the structure but repair works, at lowest possible capital costs, accepting future maintenance and fulfilling the following:

(a) reducing the wave height directly impinging on the existing tetrapods and thus increasing armour stability
(b) increasing the seawards slope geotechnical stability to an acceptable factor of safety
(c) securing the concrete crownwall stability for higher sea states
(d) improving the harbour side geotechnical stability

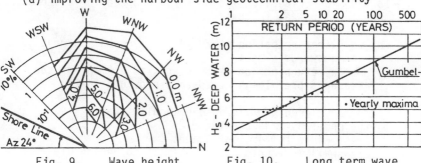

Fig. 9.   Wave height   Fig. 10.   Long term wave
directional distribution   height distribution

14. Environmental data. Wave, wind, current and tide data were collected since 1957 by different methods. A detailed description of the data collection and analyses is given in (ref. 5), Figs. 9 and 10.

15. Geotechnical data. Although extensive soil exploration studies were carried out prior to the construction of the existing harbour and its subsequent phases of development, limited information was available on soil layout underneath the breakwater, studies conducted for a future major extension (ref. 5) served as soil mechanics input data.

16. Seismic data. Seismic activity in Israel varies considerably from relatively high intensities along the Jordan Valley rift to low activity, along the Mediterranean coast. Probabilities of occurence of seismic ground peak accelerations were determined based on historic and measured data (ref. 6).

17. Design parameters. The values adopted for the different design parameters are presented in Tables 6, 7.

Table 6. Deep water significant wave heights, m

Return period - years		20	25	50	100
Wave Direction	W	7.4	7.6	8.2	8.8
	WNW	6.4	6.6	7.1	7.7
	NW	4.6	4.7	5.2	5.6

Table 7. Extreme water levels and seismic acceleration

High water	+1.00 m LSD
Low water	-0.50 m LSD
Ground peak acceleration	0.075 g

18. Design criteria. The criteria adopted for acceptable damage level was as follows:

(a) for a return period of 20 years, minor adjustment of the seawards slope.
(b) for a return period of 50 years, adjustment of the seawards profile is acceptable provided that:

- the developed profile is dynamically stable
- the underlayer is uncovered
- crest width remains sufficient
- crownwall movements are very small to negligible

MODEL TESTING

19.   Several alternatives were considered aiming primarily
to comply with the hydraulic stability criteria as mentioned
in paragraph 13. Upon consideration of possible concrete
armour units solutions, it was decided to use for the
different alternatives rock as armour. This decision was
based on past experience in Israel as to rock availability
up to 12-13 ton and to future maintenance considerations as
to crane capacity, easily available in the country.
Three alternatives were selected for model testing namely an
underwater wide berm, a submerged breakwater and a
sacrificial rock armour layer, Fig. 11.

20.   The model testing sequence adopted consisted of the
following stages:

(a) quick tests of the alternatives in an existing 3-D
    model
(b) study of most promising alternative in a 2-D model
(c) detailed testing of selected solution in a 3-D model,
    including roundhead
(d) wave forces on crownwall in a 2-D model

It should be pointed out that in all three stages a
"reference" test of the existing breakwater, as per the
original design, was carried out. The purpose of these tests
was to obtain a correlation between model and prototype
tetrapods damage percentages, thus confirming that the
damage definition adopted is realistic and enables us to
overcome the fact that concrete units structural strength is
not to scale in the model.

21.   Quick 3-D tests. The model was built in a irregular
wave basin 27m long, 24.5m wide and 0.90m deep to a scale of
1 in 65. Waves were generated in 30m water depth according
to the EM Jonswap spectrum, W'ly, WNW'ly and NW'ly waves
were generated in the model (ref. 7).

Table 8.   Quick 3-D tests programme

Test	Description	Testing conditions
T-1	Underwater berm, 50m	W,+1.0 LSD, Hmo,o=3-10m, 72hrs.
T-2	Submerged breakwater	W,+1.0 LSD, Hmo,o=3-11m, 75hrs.
T-3	Submerged breakwater	WNN,+1.0LSD, Hmo,o=3-11m, 75hrs.

Damage of tetrapods was determined by a cine-technique for
rocking and by photographs, using an overlay technique, for
displacement. Tetrapods rocking over 1/3 of the time or
displaced by more than half their height, were counted as
damaged. For the rock armour, profiles were measured and thus
a developed cross section was established.

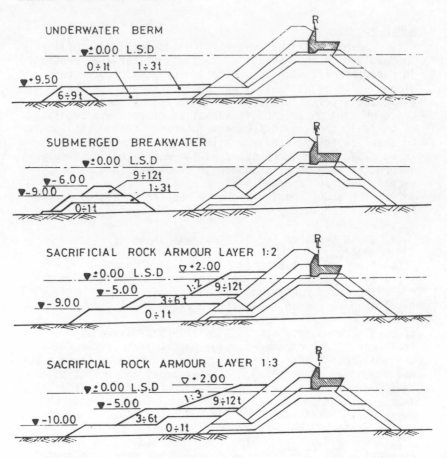

Fig. 11.   Alternatives tested,trunk

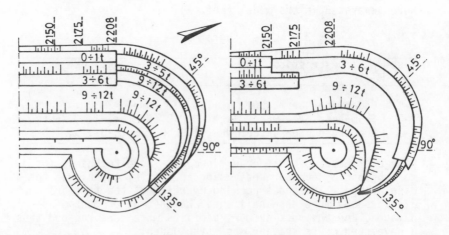

Fig. 12.   Alternatives tested,roundhead

22. 2-D tests. The model was built in an irregular wave flume 0.60m wide, 27m long and 1.25m deep to a scale of 1 in 65. Waves were generated in 40m mater depth according to the EM Jonswap spectrum (ref. 8).

Table 9. 2-D tests programme

Test	Description	Testing conditions
ASH-1	Reference cross section	WNW, +1.0 LSD, Hmo,o=3-8m, 66 hrs.
ASH-2	Submerged breakwater	WNW, +1.0 LSD, Hmo,o=3-10m, 72 hrs.
ASH-3	Sacrificial rock armour 1:2	WNW, +1.0 LSD, Hmo,o=3-10m, 82 hrs.
ASH-4	Sacrificial rock armour 1:3	WNW, +1.0 LSD, Hmo,o=3-10m, 72 hrs.
ASH-5	Sacrificial rock armour 1:3	WNW,-0.5LSD,Hmo,o=3-10m,72 hrs.

23. Detailed 3-D tests. The tests were carried out in the wave basin as described in paragraph 20 (refs. 7-9). The two alternatives for the roundhead are shown in Fig. 12.

Table 10. 3-D detailed tests programme, trunk and roundhead

Test	Description	Testing conditions
T-4	Sacrificial rock armour 1:3	WNW, + 1.0 LSD, Hmo,o=3-11m, 75hrs.
T-5	" " " "	WNW, +1.0 LSD, Hmo,o=3,4,8,8,8,9,10,11m, 62hrs.
T-6	" " " "	WNW, +1.0 LSD, Hmo,o=3-11,7m, 85hrs.
T-7	" " " "	W, +1.0 LSD, Hmo,o=3-11m, 75 hrs.
T-8	" " " "	W, +1.0 LSD, Hmo,o=4,8,8,8,9,10,11m, 49hrs.
T-9	" " " "	W, -0.5 LSD, Hmo,o=3-11,7m, 85hrs.
A-1	Reference roundhead	WNW, +1.0 LSD, Hmo,o=3-8m, 66hrs.
A-2	Alternative 1	WNW, +1.0 LSD, Hmo,o=3-10m, 72hrs.
A-3	" "	W, +1.0 LSD, Hmo,o=3-10m, 72hrs.
A-4	Alternative 2	WNW, +1.0 LSD, Hmo,o=3-10, 7m, 82hrs.
A-5	" "	W, +1.0 LSD, Hmo,o=3-10,7m, 82hrs.
A-6	" "	W, +1.0 LSD, Hmo,o=3-8m; WNW, -0.5 LSD, Hmo,o=7,8m, NW, +1.0 LSD, Hmo,o=4-6m; W, +1.0 LSD, Hmo,o=9-10m, 125hrs.

24. Wave forces 2-D model. The model was constructed in a wave flume, 0.75m wide, 1.10m deep and 32m long to a scale of 1 in 65 at the National Research Institute for Oceanology, South Africa (ref. 10).

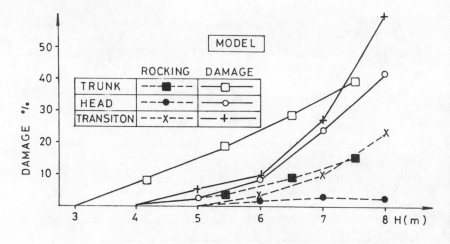

Fig. 13.   Reference cross sections damage

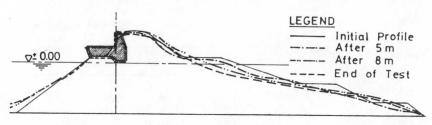

Fig. 14.   Development of final profile, trunk

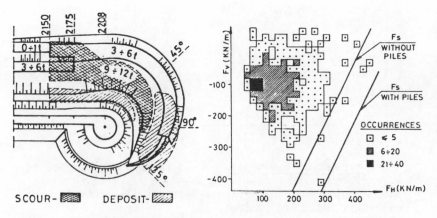

Fig. 15. Roundhead damage

Fig. 16.   Crownwall
forces and safety factors

Table 11. Wave forces on crownwal, 2-D tests programme

Test	Description	Testing conditions
A	Sacrificial armour layer	+1.0 LSD, Hmo,o=4,6.5,8,9.5m, 8.3hrs
B	Sacrificial armour layer with tetrapods removed down to +4.0 LSD	+1.0 LSD, Hmo,o=4,6.5,8,9.5m, 36.3hr
C	Sacrificial armour layer, developed profile	+1.0 LSD, Hmo,o=6.5,8,9.5m, 35hrs.

The waves forces were measured on a 5m crownwall test section
that was attached to a steel frame consisting of a vertical
and a horizontal lever arm, which were fitted with strain
gauges. Thus, bending moments could be measured at four
positions and horizontal and vertical forces determined
as well as the position of the resultant force.

Model tests results

25. Reference section. 2-D and 3-D damage percentages
(rocking and displacement) obtained for both trunk and
roundhead sections compared reasonably well with prototype
values up to Hmo,o=7m. Fig, 13. It should be bear in mind
that the highest storm recorded, during service of the
structure, was Hmo,o=6.5m.

26. Trunk section. The underwater berm and submerged
breakwater alternatives were shown to provide inadequate
protection. For the first one, 20 per cent of damaged
tetrapods was evaluated for Hmo,o=6m; during Hmo,o=7m the
underlayer and crownwall were exposed leading to total
collapse for Hmo,o=8m. The underwater breakwater showed
lower damage percentage for sea states up to Hmo,o=6m but
later on the values obtained were similar to those for the
existing "reference" section.
The sacrificial rock armour layer was shown to meet the
design criteria. In all tests the cross section remained
unchanged in the lower sea states, developing into a stable
composite slope between Hmo,o=5 to 7m consisting of a 1 in 6
middle slope between -2 to -10m, 1 in 3 above - 2m and 1 in
4 below -10m, Fig. 14. This composite slope remained
unchanged for wave heights up to 10m but for a slight
steepening of the upper slope (1 in 2.5).

27. Roundhead. The first alternative Fig. 12 provided
acceptable protection for westerly waves but damage caused
by WNW'ly waves was unacceptable. The second alternative,
incorporated a westward orientation of the transition zone
between trunk and head as well as a widening of the armour
crest at the roundhead. This design remained unchanged up to
Hmo,o=5-6m, where as for Hmo,o=7-10m, rocks were gradually
scoured from the outer quadrant and deposited in the inner
one, Fig. 15.

28. Crownwall stability. During the first 3-D tests the crownwall collapsed at Hmo,o=8m. Following tests were carried out with brass rods, simulating piles through the crownwall road into the mound.

29. Wave forces on crownwall. The maximum forces obtained for the designed sacrificial armour and its developed profile were similar for Hmo,o=8m. The vertical forces are of the order of 340 KN/m, the horizontal forces are between 360 and 430 KN/m and the maximum overturning moment is between 2150 KNm/m and 2820 KNm/m. With the tetrapods removed down to +4m LSD similar vertical forces and overturning moment were obtained whereas the maximum horizontal force is 560 KH/m, Fig. 16.

FINAL DESIGN

30. The sacrificial rock armour layer (1 in 3 slope) tested in the models, was shown to comply with the hydraulic stability criteria and except for minor adjustments, was accepted as final design, Fig. 17.

31. Crownwall stability. The maximum forces and moments obtained were used to compute the factors of safety for the existing and a pile reinforced crownwall. Drilled, 60cm diameter, 15m long concrete piles, reinforced by a steel profile and at 2.50m center to center distance were shown to yield an acceptable factor of safety, as shown in Fig. 16.

32. Geotechnical stability. The stability of the sea slope is largely increased by the relatively mild slope of 1 in 3. The harbour slope condition is improved by the stabilizing effect of the piles and the addition of flow divertors, that will prevent any further damage due to overtopping.

33. Armour rock specifications. The physical requirements are a minimum bulk specific gravity of 2.68 ton/cu m, 2.0 per cent maximum water obsorption and a 25 per cent crushing rate.

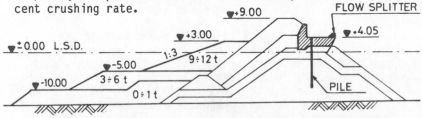

Fig. 17.   Final typical cross section

ACKNOWLEDGEMENTS

The excellent cooperation of CAMERI, Israel and NRIO, South Africa who assisted in the work and the comments and suggestions of Mr. Dan Halber, Chief Engineer, are acknowledged with thanks. Special thanks are owing to Mr. J.A. Zwamborn, whose comments and advise were invaluable.

REFERENCES
1. SAFONA. Port of Ashdod. Flume investigation into the cross-sectional stability of the main and lee breakwaters, 1959.
2. SAFONA. Port of Ashdod. Profile stability studies, wave tank tests, 1960.
3. Buslov V. and Bishop J. Survey of marine structures at the Port of Ashdod: main breakwater. Coastal and Marine Engineering Research Institute, Haifa, 1978, Report PN 25/78.
4. Buslov V. Inspection of damaged sections of Ashdod Port main breakwater. Coastal and Marine Engineering Research Institute, Haifa, 1981, Ref. No. 3045.
5. Halber D. et al. Design of Ashdod Port extension, coal unloading terminal for 150000 dwt bulk carriers. 26th. PIANC Congress, Brussels, 1985.
6. Shapira A. Earthquake hazards and peak ground accelerations maps of Israel. Proc. Symposium on earthquakes hazards in Israel, Jerusalem, 1983.
7. Rosen D.S. and Gottlieb O. Ashdod Port expansion, breakwaters stability three dimensional model study. Coastal and Marine Engineering Research Institute., Haifa, 1986, PN 180/86.
8. Gottlieb O. and Rosen D.S. Repair of the main breakwater trunk at Ashdod Port, two dimensional stability model study. Cameri, Haifa, 1986, P.N. 170/86.
9. Finkelstein A. and Gottlieb O. Repair of the main breakwater head at Ashdod Port, three dimensional stability model study, Cameri, Haifa, 1986, P.N. 181/86.
10. Scholtz D.J.P and van Tonder A. Ashdod harbour, wave forces on crownwall of repaired breakwater. CSIR, Stellenbosch, 1986, C/SEA 8645.

Discussion on Papers 11 – 14

PAPER 11

MR M. E. BRAMLEY, Construction Industry Research and
Information Association (CIRIA), London
I would like to comment on, rather than question, Paper 11 –
and refer specifically to paragraphs 15 to 17 concerning 'Rock
source investigation'. First, I would like to emphasize that
the question of rock armour for UK breakwaters should be seen
as one end of a spectrum of demand which stretches along the
coastline with rock-armoured revetments on coastal floodbanks,
up estuaries to riverbank protection, and on to protection to
the upstream slopes of earth dams (where there is currently
considerable activity in renovation and maintenance under the
Reservoirs Act, 1975).
CIRIA has had contacts with the UK construction industry in
all these fields through the use of, and current proposals to
update, CIRIA Report 61 – 'Design of rip-rap slope protection
against wave attack'. While the existing report relates
principally to the hydraulic design and sizing of rip-rap,
comment from practitioners confirms a general view that more
attention should be given to design for ease and practicality
of construction. In planning the revised CIRIA report, we are
giving particular attention to the specification and sources
of rock, of which two aspects need to be emphasised.
Firstly, where local quarry sites exist, the engineer needs
to be able to relate his design and specification more readily
to total quarry production – that is, to likely shape,
durability and size range of rock production.
Secondly, and more importantly for breakwater armour stone,
some standardisation needs to be brought into rock
specification if the UK quarry industry is to develop
economically its production of large armour stone. At
present, no two designer specifications are the same, so the
UK quarry industry is unable to match anticipated demand on a
planned basis. Commercial quarries are simply expected to
respond to demand by immediate production – this necessarily
entails high unit costs, which could be reduced significantly
if rock were produced to stockpile on a planned product basis
over a period of time.

There is no doubt that a significant market for rock armour exists in the UK - particularly with the future requirement for upgrading the East Coast sea defences. CIRIA is therefore bringing the UK quarry industry into the planning of the new CIRIA report, with a view to ensuring that the best possible framework is developed for future planning, design and specification of quarried rock protection.

MR A. G. BRINSON, Rendel Palmer and Tritton, London
From my own experience, I would agree that it is essential that design must take construction into account. Some additional comments to the Paper are as follows.

The core width at working level is critical, as is stated in paragraph 24, In fact, contractors always want it wider. One way of increasing the width is to place small stone (aggregate size) on top of the secondary armour to provide a wider running surface.

It is essential to minimise the distances between the leading edge of core, the secondary armour and the primary armour, in order to minimise storm damage. In my experience, it is the primary armour which becomes left behind. To minimise damage to the section with only secondary armour, it has been found that if secondary armour is temporarily placed on top of the core, damage is reduced to a minimum.

Weather forecasts of storms are available for anywhere in the world, but are these always reliable?

MR J. E. CLIFFORD, Consulting Engineer, London
This had been a most useful Paper on the problems of constructing a rubble mound. I was the project director for Douglas breakwater, and would like to make a few comments on the design.

We considered the option of a caisson breakwater, but an estimate showed it was more expensive, with problems both of stepping the foundation level over a rapidly changing rock sea bed, and of joining in with the existing breakwater, and also some doubt about whether we could build the caissons in the small harbour at Douglas.

There was good quality granite in the Isle of Man, and therefore we were firmly committed to a rubble mound breakwater at a fairly early stage. Quarry boreholes were made but trial blasting was not possible before design. We had hoped to get about 20% armour yield, but in the event the yield was about 12%. Therefore, the very large underlayer provided in the design, principally for protection during construction, was slightly reduced from 4-8 tonne to 2-6 tonne of rock. Similarly, the lee side armour was replaced by 7 tonne fluted concrete blocks laid in one single layer.

The design used very few classes of rock, to keep checking to a minimum, but it was recognised that the checking of soundings from the crane interrupted the contractor's work.

The number of cranes employed was quite large, but was
evidently necessary to ensure that a variety of work could be
carried out to suit all wave and other conditions.

Model tests caused us to raise the capping level to reduce
overtopping, but nevertheless the design considered a working
level for the contractor below the cap, to ease construction.

Mr Moore has described some of the severe conditions
encountered during construction. I believe, however, that the
specified tolerances were generally met. I would ask Mr Moore
to let the PIANC Working Group on Rubble Mound Breakwaters to
have any information on or recollections of tolerances
actually achieved.

MR C. DINARDO, Dinardo and Partners, Paisley
The Author referred briefly to a tendency for slumping to
occur. Could he give more details of this?

MR P. LACEY, Ove Arup and Partners, London
I refer to paragraph 35 of Paper 11, in which the Author
appears to be advocating model tests on anticipated
construction stages. This is most difficult at predesign and
design stages, as even on breakwaters different contractors
use different methods. I doubt whether the client body would
take kindly to extended tests at his cost. Perhaps this is an
area where consulting engineers should exert more pressure.

Paragraphs 39–42 are inconclusive and I would appreciate
further explanation of why they were included. Is the Author
of the opinion that Seabee or Shed units are the best for
contractors?

MR H. L. WAKELING, Consulting Engineer, Esher
In Paper 9, a placing rate for hollow armour blocks weighing 2
tonne was quoted as 50 per day. At the Humbolt Jetty
breakwaters, where 42 tonne Dolos were being used, a placing
rate of 50 per day was also claimed by the contractor. Could
the Author say what placing rate had been achieved on his
contract?

Also, the larger the armour unit the smaller the number that
had to be placed per unit area, and so the overall cost
increase was small for placing much larger blocks, probably of
the order of 3%. Could the Author comment on this?

MR MOORE and MR MALYON (Paper 11)
In reply to Mr Brinson, this method of obtaining additional
working width over the secondary armour is the one we used at
Douglas. The material size of the stone is relevant: if it is
too small, it vanishes into the voids; a middle size may be
washed down to clog the face of the voids, needing to be hand
picked out later. Some form of separation medium is useful.

The primary armour does tend to be left behind, particularly when the leading edge of concrete armour is laid back at 45°. Consequently, it is important that the secondary armour is large enough to be adequate protection for most purposes and that the contractor has adequate supplies readily available to him.

The desirability of keeping the leading edges of core and secondary armour close together can be overstressed. In rough weather the scar end is likely to suffer substantial movement leading to undermining, loss of scarce armour material and additional rebuilding. If the core (preferably up to 2 or 3 tonnes in size) is allowed to keep further ahead of this secondary armour, and is continually topped up, a self-protective transition back to the armour develops and the risk of serious loss outside the payline is small. Final shaping followed by placing the first layer of secondary armour is thus carried out on a well-compacted and relatively stable section.

As Mr Clifford has said, the tolerances were generally met. By developing the methods of placing and proving to suit the conditions, the site team were able consistently to produce work to the required standards.

In reply to Mr Dinardo, observations at Douglas confirmed that the degree of slumping, undercutting and general rock movement is time and condition related. The core material, as tipped and spread, formed side slopes of about 1 in 1.25. For 0-2 tonne core material, quite gentle waves plucked out head-sized boulders and rolled them down to the toe. Larger waves increased the rate of washdown and the size of boulder plucked out. Over a period of time, if left unattended, the upper material was cut back, a beach formed in the wave zone and the S curve began to form, i.e. steep and unstable above wave level, flatter and 'beachy' in the tidal zone, steeper below LW but with washdown material curving outwards at the toe.

During a two-week Christmas closedown, allowance was made for this situation to develop at the scar end, and it assumed an average final slope of about 1 in 6. Supplementary core material was stacked at the top of the slope backed by a barricade of secondary armour material some distance behind the anticipated intersection.

Two serious storms were experienced, totalling about five days of very rough weather with moderate to rough conditions in between. The precautions taken were adequate and our observations would suggest that deterioration beyond 1 in 6 would have been very slow.

In reply to Mr Lacey, we would advocate that the Client and his Consulting Engineer test intermediate stages of construction when carrying out model testing for a project. The results can be made available to tenderers in the form of a brief report and video tape.

On the subject of geotextiles, we accept that there are good technical reasons for using them in designs, and they have been successfully used in many rubble mounds. However, they are effective only if correctly positioned, and if, during a storm, they are displaced in the partly-built structure, they may cause more extensive rebuilding than would otherwise be necessary. Also, their presence in the design may put a costly constraint on the type of plane used and on the timing and execution of various operations.

Kier has been involved in the construction of several major breakwaters in the last 20 years and on each there have been problems in obtaining sufficient armour stone. We can therefore see the advantage to both client and contractor in using the voided concrete units of the Seabee and Shed type.

In reply to Mr Wakeling, the output rates quoted for 2 tonne hollow blocks and for 42 tonne Dolos seem perfectly feasible, provided that there is a sufficient area of work available and the preparation work is able to keep ahead.

At Douglas, the record for a one-night shift was 34 stabits placed. This merely means that on a good run, stabits can be placed at that rate, or marginally better, with a faster crane. The activity average of the whole job was significantly less, being limited by the linear rate of advance of the breakwater, long radius placing, and other factors. Rate of placing stabits was never in itself a constraint on the breakwater progress.

Placing costs for different weights of unit can be related only to the particular project. The 3% quoted (not unreasonably) could jump to 30% if the heavier unit demanded a substantially bigger crane or extra temporary works. Although the heavier units would reduce the hook time, they may not always reduce the time the resources are on site.

PAPERS 12 and 13

INTRODUCTION TO PAPER 13 BY J. READ
Information on the control of quarrying operations and the construction of rubble mound breakwaters is not easily obtained: maybe this is because control is considered to be a trifling matter or perhaps it is thought to be of sufficient commercial value for its circulation to be restricted. In either case this reticence is disappointing. Paper 13 attempts to give some insight into the methods adopted on one project, with the hope that the information will not only prove useful itself but will also prompt others to share their experience.

It is mentioned in the Paper that the quarry produced more armour than was required but relatively little core-sized material. Initial blasting tended to produce up to 40% of armour but one-third of the remainder was smaller than 4 in. The actual proportions of three typical trial blasts are given in Table 1.

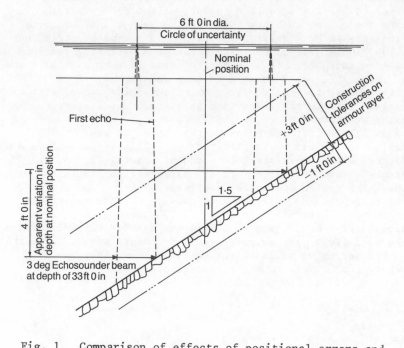

Fig. 1.   Comparison of effects of positional errors and
construction tolerances at Helguvik

Table 1. Blasting trial results

Explosives ratio	Armour >4 in	<4 in	Core	fines ratio
0.35 kg/cum	39%	40%	21%	34%
0.25-0.35 kg/cum	45%	42%	13%	24%
0.36 kg/cum	40%	41%	19%	32%

The actual proportions varied throughout the quarry, as was
to be expected when a number of different flow units were
being worked. The thick lava flow unit produced a
particularly high proportion of large stones, of which many
had to be broken into smaller sizes. In general, the yield
per shot hole was about 2200 m^3 from holes 3.5 in in diameter
and 10 m and 15 m deep.

As the Specification detailed only the small end of the
quarry run material, no grading tests were carried out on
stone sizes greater than 4 in. With that limitation, Table 2
shows a typical grading envelope of the core.

The uncertainty of the position of the survey vessel in plan
will influence the accuracy with which the surveyed surface is
known, especially when that surface is a steep slope. Fig.1
has been drawn for a slope of 1 to 1.5 and shows that an
uncertainty of position of 3 ft could result in an apparent
error in depth of $\pm$2 ft. At Helguvik, the tolerance for the
main armour thickness was +3 ft and -1 ft, so that a survey
with an order of error shown in the figure would not be very
useful. For that reason, a method which would limit that
uncertainty to something much smaller was adopted.

Some explanation is required about the accuracy of a
position meaured by the various methods described in the
Paper. In the case of microwave range-range systems, the
likely error is described quite clearly as a circle of
uncertainty of 2 m diameter. This measure takes into account
all the likely sources of error which may affect a system such
as Microfix or its equivalent. A comparable figure for radio
range and bearing systems such as Artemis, deduced from the
standard deviations quoted by the manufacturer, would include

Table 2. Grading of core

Size: in	Percentage passing
4	20-40
3	14-30
2	10-20
1	8-16
0.5	5-13

the effects of temperature, atmospheric conditions and counter clock stability. The error increases with range, and at 1000 m would approximate to a rectangle 4 m by 1.4 m, with the greater dimension in the range direction. There would be a 95% probability that the actual position would be within that rectangle. Polarfix, which is a self-tracking laser range bearing system, has a quoted accuracy of ± 1 m at 5000 m and ± 0.2 m at 1000 m.

Practical inaccuracies of the EDM/theodolite stem from the repetition rate of the measuring cycle of the instrument combined with the speed of the tracked vessel. The potential accuracy of the instrument quoted in the Paper is that which can be achieved when the target is stationary. In tracking mode, the receipt of a signal reflected from the target prism initiates a measurement, but the time lag between the measurement and its reception on the tracked vessel varies depending on the relationship between the time of initiation and the beginning of the measurement cycle. As the relationship is not measured, no allowance can be made for the ensuing time lag. In practice, this leads to an uncertainty of 0.5 m in position if the tracked vessel is travelling at 4 m/s (8 knots). At Helguvik, the speed of the survey boat was low – when measuring breakwater profiles it was only about 2 knots – so that the error resulting from the unknown time lag would have been between 0.1 and 0.2 m.

While it is true that the survey system worked well at Helguvik, its dependence on an unobstructed line of sight prevented tracking of the survey boat when to seaward of the breakwater once the crest of the armour layer had been placed. The arrangement of prisms at the top of the same spar that carried the echosounder transducer meant that the prisms were barely 2 m above water level when the transducer was lowered to its working depth. This arrangement avoided the need to correct for the different locations of tracked position and sounder, which would have needed a gyro-compass to supply the vector for the correction, but allowed no second look at the completed sections.

Before considering the costs of these systems, a point should be.made about the Bathymetrics side scan sonar. Its developers have some impressive examples of its capabilities both in home and foreign waters, but a situation was discovered recently where it gave no useful output at all. This was in that narrow and turbulent channel of the River Severn known as 'The Shoots'. It was hoped to survey the proposed site of the main piers of the proposed Second Severn Crossing, but the combination of turbulence and turbidity prevented any signal echo from being received. The reason for the failure is believed to be the turbidity caused by the intense turbulence stirring up the soft mud. High frequency sonars suffer reduced range in anything less than clear water, and Bathymetrics use a frequency intermediate between that of an echo sounder and sonars. It is of interest to note that on a subsequent occasion, but working on a smaller tide, it was

Table 3. Comparison of equipment prices

Position fixing equipment
(Two mobile stations except for Polarfix)

	£
Range-range microwave (Microfix $^+_-$1 m)	59 000
Range-bearing microwave (Artemis IV)	101 000
Range-bearing laser (Polarfix)	62 000
Auto-tracking EDM/theodolite (AGA 140T)	46 000
Manual tracking EDM/theodolite (AGA 140H)	24 000

Echosounding equipment

	£
Raytheon 719C with 3 deg transducer	5 600
Atlas Deso 2 frequency sounder with 33 kHz and 210 kHz transducers	25 000

Wide swathe sounding equipment

	£
Bathymetrics Bathyscan 300	0.25m
Simrad multibeam echosounder	0.25m
Echos XD multibeam echosounder	1.25m
Bennigraph multibeam sonar	1.5-2.0m

---

only possible to use an echo sounder at slack water. The character of the water changed quite suddenly as the current slackened and the water became clearer.

The equipment used at Helguvik, with today's prices, is as follows.

	£
1 AGA 140H Geodimeter	17 000
(a 2 second manually tracked EDM/theodolite)	
2 sets of prisms	7 000
3 Universal display units	7 500
4 No. 5 Channel Microtel transmitter/receiver	8 000
2 Seahawk gyro compasses	24 500
1 Hydrolog data logger	6 000
1 HP9122 Disc drive	1 050
1 Raytheon DSF 6000 with 3 deg transducer	5 600
(priced as Raytheon 719C)	
1 Tide gauge	4 600

Note: The Raytheon DSF 6000 is designed for a towed array of four transducers. The 719C is for one transducer only and is equivalent to the duty actually performed by the model installed.

The office equipment which was used to plot the survey information and to calculate quantities consisted of a desk-top computer, printer and drum plotter. Almost any of the many examples of such equipment would be satisfactory, and typical prices would be as follows.

	£
AT personal computer	1 500
Printer	500
A1 size drum plotter	7 500

The software provided with this computer was the most
expensive item at £15 000.

The main items of this kit compare favourably with others
mentioned in the Paper, as can be seen from Table 3.

MR J. BERRY, Bertlin and Partners, Reigate
Is a berm type breakwater designed so that it will be reshaped
under wave action?

At Helguvik, 'fines' is defined as less than 4 in in size.
Was there a restriction on dust?

MR J. E. CLIFFORD, Consulting Engineer, London
I remain somewhat unclear about the philosophy of movement or
non-movement of a berm type breakwater. Does a static or
dynamic stability occur in the long term?

It would seem that if dynamic stability applied then in a
case of angular wave attack there would be a lateral
component, analogous to a beach, which would cause the
breakwater gradually to migrate.

Should such a beach breakwater only be used where near
normal attack is more likely, and perhaps, therefore, not in
deep exposed conditions?

MR P. HUNTER, Sir Alexander Gibb and Partners, Reading
With reference to the design of the Helguvik Breakwater, Paper
12 refers to two rock gradations, while Paper 13 has five
gradations. Could the Authors expand on this? Concerning the
excess rocks weighing more than seven tonnes, did the
specification allow the contractor to put these in the core?

DR O. J. SAYAO, F. J. Reinders and Associates Canada Limited,
Brampton, Ontario
With reference to Paper 12, as model tests were developed
based on gradation curves from similar quarries in Iceland,
how did the Authors ensure that the berm was constructed with
the stone gradation as modelled?

With reference to Paper 13, what was the construction
methodology for the armour berm width? If a construction road
was placed on top of the berm, what measures were adopted to
prevent contamination of the berm with smaller (than
specified) material?

MR G. VIGGÓSSON, Icelandic Harbour Authority, Reykjavik
Since 1983, ten rubble mound breakwaters of the berm concept
have been constructed in Iceland under the supervision of the
Icelandic Harbour Authority. Four of these are new; three are
the result of the repair and construction of a berm on the
sea-facing side of old cassion breakwaters; and three are a
modification of existing conventional breakwaters. Their
sizes range up to 100 000 m^3 of rocks and core. The new
breakwaters have usually been constructed by local
contractors, whereas the repair and modification of
breakwaters have been undertaken by the Icelandic Harbour
Authority.

Berm breakwaters have a great advantage over conventional
breakwaters in Iceland, because the wearing of rock in some
places is high, and a shortage of funds means that the
construction period has to be extended over two years. A
partially completed bermed breakwater has functioned well
through winter storms.

The first berm breakwater in Iceland was built in 1983, in
Bakkafjördur. The construction technique consisted of
placement of rock by truck and dozer dumping methods. This
placement technique results in a high proportion of fines,
which eventually plug the voids and thus diminish the berm
permeability and wave energy dissipation. This method was not
acceptable at the outer end of the breakwater, however, and
the berm was constructed by backhoe. Today, the outer end of
the breakwater is partially reshaped, while the inner end is
almost completely reshaped, resulting in movement of smaller
rocks along the breakwater.

The Icelandic Harbour Authority has performed five stability
tests on berm breakwaters in three dimensions, using natural
irregular waves in a scale 1:45 to 1:60 (most often 1:60).
The breakwaters are designed in such a way that minor
reshaping is acceptable at a sea state a little lower than the
design wave situation. The size of the stones of the two
uppermost layers is increased to compensate for eventual
breakage of the rock. The size of these rocks depends on the
availability of larger rock in the quarry. One of the basic
ideas behind the berm breakwater concept is the utilisation of
all the fragments of material blasted in the quarry.

Today, the berms have to be constructed by crane or by
backhoe to avoid the accumulation of small stones in the berm.
The dimensionless wave height parameter $H_s / \Delta D_{50}$ (where D_{50} is
the nominal diameter of the stone and the relative mass
density) is chosen so that settlements and minor reshaping are
accepted at design level. Usually, $H_s / \Delta D_{50} \leqslant 3.0$ is used for
the berm, and $1.2 \leqslant H_s / \Delta D_{50} \leqslant 2.0$ for the two uppermost layers
of selected stones and also at the head of the breakwater.

MR BAIRD and MR WOODROW (Paper 12)
In reply to Mr Berry, some breakwaters, called berm
breakwaters, are reshaped by wave action. The Helguvik

breakwater was designed so that only a minor change in the profile would occur when exposed to the design wave conditions with essentially static profile after several severe storms. To date, after two winters, there has been no change to the profile of the breakwater.

There was no restriction on dust in the core specification; however, the material had to be produced from competent stone.

In reply to Mr Clifford, the term 'berm breakwater' has been used to describe a breakwater that has a relatively wide 'berm' of armour stones, as opposed to an armour layer with a width of one or two stones.

If a wide 'berm' of stones is used, the size (weight) of the stones may be significantly less than that required for a conventional structure. This may have significant cost advantages to the owner.

Clearly, there exists a wide range of 'berm breakwaters' that would be stable at a particular site. Within this family of berm breakwaters, structures can be designed where there is essentially no movement of stones after exposure to several severe storms. The Helguvik breakwater is an example.

Alternatively, structures have been designed where more significant movement is expected to occur. In the Authors' experience, this movement is expected to occur only during the initial reshaping of the berm, after which a stable profile exists and continuing movement of stones, during storms of less severity than the design event, does not occur.

In either case, a properly designed berm breakwater should not have continued movement, analogous to a beach, and can be used in deep exposed conditions as well as in structures subject to angular wave attack.

In reply to Mr Hunter, the breakwater was initially designed to have only two gradations of stone, Class I armour and Class III core. However, during the design development it was determined that the design would need to include small quantities of three other classes of stone. As can be seen from Table 1, 97.5% of the stone was in the two classes. The other classes of stones were included in the design for the following purposes.

Table 1. Helguvik stone quantities

Stone Class	Volume required: tons
IA	No limits
I	370 730
II	34 750
III	1 476 680
IV	11 175

Class IA

Stones weighing more than 7 tonnes were not allowed in the
berm. Class IA was developed to allow the contractor to use
stones which were produced weighing more than the Class I
limit. Although they were not required for design stability
the crest location for these stones gives added safety factor
since the crest stones must resist drag forces in the event of
overtopping waves.

Class II

The inside harbour face needed protection from minor wave
action, and scour from the vessel propellors. Since this
could be provided by using stones weighing less than the Class
I armour, it was decided to specify the 0.5-1.7 tonne stone so
that armour would not be expended where not required. As
anticipated, this stone was obtained as a by-product of the
quarry effort, with no significant additional sorting effort.

Class IV

Various structures and a roadway were constructed on various
portions of the breakwater. Since there was essentially no
limit to the size of the core material, the Class IV stone was
used as an intermediate filter layer between the core and the
bedding material for the structures. It should be noted that
the specification allowed any excess Classes I, Ia, and II
stones to be used as Class III.

In reply to Dr Sayao, the contractor was required to
estimate the weight of each stone as it was loaded for
transportation to the quarry. The correct gradation was
achieved at the same time by requiring the operator to load
the appropriate number of stones in the following size ranges:
1.9 - 3.3 tons; 3.3 - 5.5 tons; and 5.5 - 7.7 tons. The
operator used automatic counters in the loader to ensure that
the correct gradation was achieved.

The berm stones were placed by crane. A construction road
was not built on top of the berm. A pad was built for the
crane. This consisted of blasting mats covered with granular
material, which was easily removed on completion without
contaminating the berm stones with undersize material. The
specification required 100% removal of any temporary
construction roads or platforms on the berm.

MR READ (Paper 13)

In reply to Mr Berry, no limitation was placed on the amount
of dust in the core other than the overall requirement which
is given in paragraph 9 of Paper 13. Although there was no
contractual requirement to do so, the contractor did make a
number of analyses of the fines and concluded that those of
sand size and smaller formed about 5% of the core. The in
situ density of the core was estimated to be 1.85 tonnes/m^3
compared with the solid rock density of 2.8 tonnes/m^2.

The apparent contradiction between Papers 12 and 13 in the number of rock gradations, noted by Mr Hunter, is easily explained. Class 3 rock formed the core and Class 1 the armour layer. Class 2 was used for the lee-side armour and Class 1a was placed in the wave wall above the berm. Class 4 was used as a foundation immediately below a concrete retaining wall behind the wave wall and a concrete duct along the lee-side of the crest. Paper 12 refers only to Class 1 and Class 3.

In reply to Dr Sayao's questions, the implementation of the designer's intention to build an armour berm in which the sizes of stone were well mixed was not easy. The system adopted is described in paragraphs 10 and 11 of Paper 13. This system set out to ensure that each batch of 300 stones contained the proper proportion of the different stone sizes. What is perhaps not made clear is those paragraphs is that the stones were loaded at random so that it was unlikely that any individual truck load contained stones of only one size. Thus the stones were placed in the breakwater at random although the overall grading of the stones was closely controlled. The result appeared to be a well-mixed mass of stones.

A little of the armour stone was dumped by barge in the early stage of construction, but the great majority was placed by crane and no construction road was needed on the berm. Where the crane stood on the armour, blastmats were laid over the armour stone and smaller stone spread on the mats. Clearing the smaller material was then a straightforward operation.

Mr Viggósson's contribution is very interesting, giving as it does some quantitative information on the design of berm breakwaters. His values for the dimensionless wave height parameter are very similar to that actually achieved at Helguvik where $H_s / \Delta D_{n50} = 3.1$. No distinction in size was made between the uppermost stones in the berm or in the breakwater head and the stones in the main body of the berm.

PAPER 14

MR N. W. H. ALLSOP, Hydraulics Research, Wallingford
The PIANC Working Group 12 is interested in crown wall stability. Together with the tests reported in Paper 14, there have also been tests conducted to study possible extensions to the Ashdod breakwater. Could the Authors give details of the crown wall movements recorded in those models, the respective wave conditions, and how these test results can be used in calculations of wave force and crown wall movement?

MR W. F. BAIRD, W. F. Baird and Associates, Ottowa
What have the Authors learnt - i.e. what information can they pass on - about what went wrong with the original design of the Ashdod breakwater?

MR J. G. BERRY, Bertlin and Partners, Reigate
With reference to paragraph 31 of Paper 14, could the Authors
describe any practical difficulties which have arisen in
connection with the installation of the piles through the
crown wall?

PROFESSOR L. FRANCO, Politecnico of Milano
Following the description of the Ashdod breakwater, I believe
it would be of interest to mention some rubble-mound
breakwaters in southern Italy (enclosing small boat and
fishing harbours) which were damaged during a severe storm on
11 January 1987 (estimated $H_s \simeq 7$ m).
 It is clear that not only armour stability is important: the
crown wall is also a very delicate part of the structure.
Although it may be obvious, one lesson to be learnt is that
the wall face should not be exposed, and therefore the armour
crest should be higher to protect it. Moreover, the failure
can also start at the 'backdoor' on account of scour on the
rear side caused by heavy overtopping (as in the case of
Maratea Harbour where none of the first tetrapods broke).
 Another important aspect is the critical transitional
section between two different types of unit (or different
weight): this structural joint is always a weak part and must
be carefully studied.

MR H. LIGTERINGEN, Fredric R. Harris (Holland), The Hague
Mr Stadler mentioned the low geotechnical safety of the lee-
side slope of the breakwater in its present state. Could he
expand on the way this problem is going to be solved in the
rehabilitation works?

PROFESSOR M. A. LOSADA, University of Cantabria, Santander
What are the Authors' criteria for selecting a particular type
of breakwater?

MR STADLER and MR RADOMIR
In reply to Mr Allsop, the tests carried out for a future
extension to the Ashdod breakwater did not include direct wave
forces on crownwall measurements. They did include crownwall
movement measurements and analytical recalculations of its
stability. Since it would be lengthy to present all the
results, the Authors will point out only that the combination
of a low parapet and flow splitters on the crownwall's harbour
edge gave excellent results. This combination reduced wave
forces on the one hand, and protected the inner armour layer
on the other. More details will be presented to PIANC Working
Group 12.
 In reply to Mr Baird, the Ashdod Breakwater was designed in
1959 by Fredric Harris Inc., according to the state of the art

at that time. We now know that breakage of concrete units, geotechnical and crownwall stability should be carefully assesed during the design process. In common with many other breakwaters, Ashdod was designed on the basis of only 1 1/2 years of visual wave measurements, which leads to underestimated, extrapolated extreme values.

In reply to Mr Berry, the drilling was carried out through three different layers (concrete, 0.5 - 2 ton rock and quarry run). The drilling, using a Caldwell bit, was done through all layers; the major difficulty was that the borehole diameter varied from the designed 60 cm up to 1.5 m. This enlarged diameter was most probably due to high air pressures, used to air lift the drilled material. This difficulty was overcome by using a steel casing.

The Authors are of the opinion that this type of drilling and concrete casting should be carried out by means of a bentonite slurry, which will ensure the borehole diameter and will prevent high leakage of concrete into the quarry run mound during tremie pouring of concrete.

The Authors fully agree with Professor Franco's last comment. The vulnerability of a transition section between two different sizes of tetrapod (10 and 16 m^3), is well known in the Ashdod case. Heavy damage to the transition section was caused by a March 1980 storm (Hmo, o = 6.2 m). Replacing of tetrapods was extremely difficult and it is practically impossible effectively to re-establish interlock between armour units along the 'seams'.

In reply to Mr Ligteringen, the geotechnical stability factor of safety of the harbour slope will be improved by the introduction of piles and flow splitters. The first will transfer the horizontal forces to a deeper shear plane, while the latter will divert overtopping water flow, thus avoiding erosion of the inner armour.

In reply to Professor Losada, the Authors, as officials of the Ports Authority, are in the quite unique situation of being both designers and 'owner'. It is well known that risk and reliability versus capital and maintenance expenditure are the basic parameters in the breakwater design decision-making process. From this follows that the Authors would, by means of desk designs and model tests, as extensive as practically possible, present to the Ports Authority Board a reduced number of recommended designs, with their related confidence levels and required capital investment. Based on this information and other types of policy-making consideration, a decision about the final design is taken.

15. Slotted vertical screen breakwaters

J. D. GARDNER and I. H. TOWNEND, Sir William Halcrow & Partners Ltd

SYNOPSIS
The first half of the paper reviews the results from theoretical and experimental studies on wave transmission, reflection and loading on slotted vertical screen breakwaters. The second half provides a case history by reference to a recently completed marina breakwater at Plymouth, England

INTRODUCTION
1. The design of breakwaters for small boat harbours and marinas can present problems that are not encountered in "traditional" breakwaters. Marinas are often situated inside existing large harbours or deep bays where design wave heights are typically less than H_s = 2.0m but other factors can be equally important. These might include:-
(i)   space requirements prohibit the use of a mound breakwater.
(ii)   excessive wave reflection from the breakwater back into existing channels is not acceptable.
(iii)   wave reflection must be reduced to prevent seabed scour in front of the structure.
(iv)   soft sedimentary sea bed material offers poor foundation conditions and wave loading on the structure must be minimized.
(v)   wave transmission through the structure may be permitted but the level of transmission must be realistically predicted to ensure that the breakwater produces the required degree of shelter.

2. Vertical breakwaters with vertical or horizontal slots or a combination of slotted and solid breakwaters can be used to help solve the factors mentioned above.

DESIGN OF BREAKWATERS

3. The three main areas of interest to the designer are wave transmission, wave reflection and wave loading. There are theoretical methods for approaching each of these. problems and there are also case histories which provide useful experimental results on some aspects of the problem. However, for the designer faced with, say, the task of producing a conceptual marina design, there is a scarcity of information showing the relationship between the solution to each of these three areas of interest. For example a single slotted breakwater which has low reflection characteristics will have high transmission characteristics but reduced wave loading.

DEFINITIONS
4. Wave studies for a marine project will typically yield specified return period design conditions in terms of the significant wave height H_s for which corresponding mean zero crossing period T_z can be determined from the expression

$$\frac{2 \ H_s}{gT_z^2} = .05 \ (\text{ref 1}) \tag{1}$$

on the assumption that the design conditions correspond to a fully developed sea state.

5. Other terms are defined as follows:-

L = wavelength($=gT$)
H_{max} = average maximum wave height
H_I = incident wave height
H_R = reflected wave height
H_T = transmitted wave height
K_R = H_R/H_I = reflection coefficient
K_T = H_T/H_I = transmission coefficient
K_D = energy dissipation coefficient of slots
b = width of slot (a typical minimum value is 150mm because smaller slots are likely to be blocked by marine growth)
D = width of solid element
p = porosity or ratio of slot width to total width
= $b/(b+D)$
l = distance between front slotted screen and solid wall
d = water depth at the structure

6. The fundamental equation linking reflection and transmission is:- $E_R + E_T + E_D = E_I$ where $E_{I,R,T\&D}$ denotes wave energy incident, reflected, transmitted and dissipated respectively, now as $E \ H^2$ the expression can be rewritten as:-

$$K_R^2 + K_T^2 + K_D^2 = 1 \tag{2}$$

SLOT ALIGNMENT
Slots are vertical or horizontal. No examples have been found of slots arranged diagonally. The choice between

vertical or horizontal slots is primarily governed by the
structural design of the breakwater.   Vertical slots are
frequently formed by closely spaced piles tied together at
the top by a pile cap, for example as at Pier 39, San
Fransisco (ref 8).   Horizontal slots are formed by plank
units spanning between king piles, for example as used at
Plymouth Marina (ref 3) or tested for Shilshole Marina,
Seattle (ref 2).

7.   There is no evidence of a comparative study between
the performance of horizontal or vertical slots except for
the few tests carried out for the Plymouth design (ref 15).
From the Plymouth tests it was concluded that there was no
obvious hydraulic advantage of one arrangement over the
other.   The choice can largely be made on structural
grounds.

WAVE TRANSMISSION

8.   The amount of wave energy that is transmitted through
a slotted structure depends on the proportion of incident
wave energy that is reflected and the extent of energy
losses, in turbulence, at the slots.

9.   The value of the transmission coefficient is primarily
a function of the porosity of the structure and the incident
wave steepness, but water depth also becomes an important
factor in shallow water.   Energy losses at the slots depend
on shape factors of the slots such as rounded or square
edges,thick or thin structural units.

10.   Most existing theoretical work on prediction of wave
transmission is based on monochromatic wave theory.   However
work by Truitt and Herbich (ref 5) demonstrates that
expressions based on monochromatic waves can be used for
prediction of the transmission of a random sea.   The
characteristic wave height for the random sea, say H_s, is
substituted for the monochromatic wave height, H_T,and the
resulting H_T is the significant wave height for the
transmitted spectrum.   Other characteristic wave heights,
such as the maximum wave height could equally be used to
determine the maximum transmitted wave height.

11.   Some expressions available for estimating
transmission coefficients through a single slotted structure
are:-

$$K_T = p \quad \text{(ref 5)} \tag{3}$$

(ref 5 points out that measured results typically 25%
greater)

$$K_T = 4.(d/H_I). \quad E \; (-E + \sqrt{E^2 + H_I/2d}) \quad \text{(ref 5)} \tag{4}$$

where $E = C.p./\sqrt{1 - p^2}$
and $C = 0.9$ to 1.0

$$K_T = \sqrt{1/(k^2 c^2 + 1)} \qquad \text{(ref 4)} \tag{5}$$

where for flat plates $C = -\dfrac{(b+D)}{\pi} \cdot \log \sin(\pi p/2)$

k = wave number = $2\pi/L$

12. Table 1 shows a comparison between these three expressions and laboratory results (ref 15). For calculation purposes the incident wave has been taken as $H_I = 1.0m$, $L = 20m$, water depth = 5m and slot width = .15m.

Table 1 - Transmission coefficients for a single slotted breakwater

p%	Transmission Coefficient, K_T			
	(3) +25%	(4)	(5)	lab (ref 15)
5	.30	.25	.86	.18
10	.35	.43	.98	.40
15	.40	.57	.99	.54
20	.45	.67	1.00	.61
25	.50	.74	1.00	.66
30	.55	.80	1.00	

WAVE REFLECTIONS

13. Wave reflection from a slotted breakwater encompasses reflection from:- a single screen, a single screen backed by a solid screen and several rows of screens backed by a solid screen.

14. Kakuno (ref 4) proposed the following expression for estimating the reflective coefficient of a single slotted structure:-

$$K_R = \sqrt{k^2 c^2/k^2 c^2 + 1} \tag{6}$$

Where k and C are as defined in expression (5) which does not take account of K_D

15. Table 2 compares the results derived from (6) with the measured results. By using the corresponding measured values of K_T from Table 1 it is possible to estimate the value of K_D for the structure.

Table 2 — Reflection coefiecients for a single slotted breakwater.

p%	Reflection Coefficient, K_R		Energy Dissipation Coefficient
	from (6)	Laboratory (ref 15)	K_D for test structure (ref 15)
5	.31	.74	.42
10	.12	.52	.57
15	.06	.39	.56
20	.04	.32	.53
25	.03	.28	.49
30	.02		

16. When a solid wall is included behind the slotted wall the wave transmission past the structure is reduced to zero. The extent by which reflected wave energy is less than the incident wave energy depends on the amount of energy lost through wave interaction in the "box" between the two screens. The major variables which affect the hydraulic efficiency of the structure as an energy absorber are the porosity of the front wall, and the ratio of the distance between the front and rear walls and the incident wave length, l/L.

17. There are several theoretical approaches to the determination of K_R for a box type structure but not all are in a form suitable for practical use. K_R based on a selection of references is compared where possible with measured results (ref 15) in Table 3.

18. There are examples of structures either constructed, or planned with 2 or 3 slotted screens and an analytical solution to the problem in ref 13. Table 4 shows in summary form a comparison between the results based on ref 13 and selected case histories.

Table 3 - Reflection coefficients for "box" type breakwater with a single slotted front screen.

Front Screen Porosity p%	Ratio l/L	Reflection Coefficient, KR	
		Laboratory (ref 2,6,4)	Theoretical (ref 6,9,12)
8	.25	.56	
	.40	.61	
	.50	.69	
	.75	.64	
16	.25	.25	
	.40	.39	
	.50	.60	
	.75	.37	
20	.20		.21)
	.40		.78) ****
	.50		1.00)
	.75		−)
	.20		.62)
	.25	.35	.56) **
	.30		.52)
	.35		.49)
23	.25	.13	
	.40	.35	
	.50	.57	
	.75	.32	
25	.25		.50)
	.40		.85) **
	.50		1.00)
30	.25	.17	
	.30	.25*	
	.40	.40	
	.49	.71*	
	.50	.60	
	.75	.30	
34	.25		.62)
	.40		.92) ****
	.50		1.00)
	.75		−)

Note: * = Ref 2, ** = Ref 6; *** = Ref 9; **** = Ref 13.

Table 4 - Reflection coefficients for structures with 2 or 3 front slotted screens in front of solid screen.

Screen Porosities P% outer-inner	Ratio l/L	Reflection Coefficient KR
30-30	.99	.48)
	.61	.45) *
30-30-30	1.48	.25)
	.92	.19)
30-30	.53	.33 ***
20-20	.25	.14)
	.40	.37)
	.50	.27)
		) ****
34-20	.25	.20) Screens equally
	.40	.23) spaced
	.50	.27)

WAVE LOADING

19. The theoretical determination of wave loading on a slotted screen is very complex and there is no evidence of any existing theoretical solutions to this problem. Slotted screens have however been used where wave protection coupled with a reduction in wave loading is required, for example when a wave screen is built onto an existing weak structure. However in each case the resulting wave loading has been determined experimentally. A summary of measured or anticipated wave loading is given in Table 5.

20. The reduction in loading is given as a percentage reduction in the loading that would be experienced against a solid screen.

Table - 5 Wave loading on a slotted, or porous, screen.

Ref	Structure Description	Reduction in Loading
7	Composite breakwater, front face slotted	40%
8	Two rows slotted piling in front of solid wall	33%
11	Timber piling, p = 10%	60%

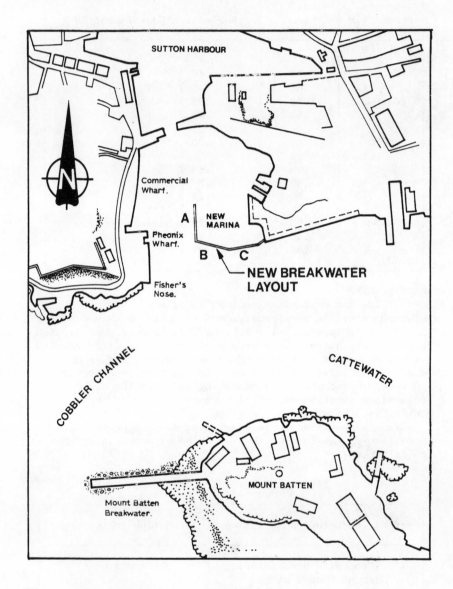

Fig. 1   Plymouth Marina – Location Plan

21. The opportunity of determining the wave loading experimentally is not always available to the designer, particularly during the conceptual design stage. In consequence wave loading has to be treated in a suitably conservative manner until a programme of load testing has been carried out to demonstrate that reductions in loading are permissable in the design.

A CASE HISTORY – PLYMOUTH MARINA

DESIGN CRITERIA

22. The marina is located in the north east corner of Plymouth Sound as shown in Figure 1. It is encompassed on the seaward side by two existing shipping channels which are frequently used by small boats. Foundation conditions comprise up to 30 metres of silt overlying bedrock. Space is at a premium – the seaward limit of the marina is fixed by the channel boundaries and the landward limit is fixed by the rock foreshore. These factors combined with a tidal range of 5 metres precluded the use of floating breakwaters or rubble mound and gravity structures. A vertical screen structure supported by piles was selected.

23. The major criteria governing the hydraulic design of the screens are:-

(i) Wave heights inside the marina should not exceed Hs = 0.3 metres for more than 200 hours per year.

(ii) With the exception of a 10 metre wide strip adjacent to the south west face of the breakwater wave heights in the existing shipping channels should not be increased, as a consequence of wave reflection from the breakwater, by more than 50% above the conditions existing before construction.

(iii) The breakwater should be designed to withstand the 100 year return period storm occuring at any tide level.

24. Numerical methods were used to predict the wave conditions that would exist after construction of the breakwater. The study concluded that the design criteria would be satisfied for the predominant wave periods expected at the site if the three sections of the breakwater met the following criteria:-

Section A = K_R 0.8; K_T = 0.4

Section B = K_R 0.5; K_T = 0

Section C – K_R = 1; K_T = 0

BREAKWATER STRUCTURE

25. Examination of the existing literature discussed in the first part of this paper indicated that Section A should be a single slotted screen; Section B should be a double

screen with the front screen slotted and the rear solid and
obviously Section C should be a single solid screen.
Preliminary design for the double screen section concluded
that the porosity of the front screen should be 25% and the
spacing between screens should be 10 metres.

26. The design wave height is Hs = 1.7 metres which
corresponds to H_{max} = 3.2 metres. Allowance was made for
the partial sheltering of Section A and here the wave height
was reduced to H_s = 1.5. It was decided that the most
effective way of transmitting wave loads to the bedrock was
by pairs of raking piles. Initially the idea was to attach
the screens direct to the raking piles but this scheme
produced unacceptable additional bending in the raking piles
and raised doubts about the practicality of driving raking
piles sufficiently accurately to form a framework to which
the screens could be fixed. The distance between the
screen supports was chosen as 3 metres as this produced
realistic member sizes for the screen both for construction
and future maintenance.

27. As the existing information on the transmission and
reflection characteristics of single screens was
insufficient for detailed design purposes, hydraulic model
tests were carried out as part of the detailed design stage.

MODEL TESTING FOR TRANSMISSION AND REFLECTION
28. Model testing was carried out at Hydraulics Research,
Wallingford in their 45 metre long wind wave flume. The
purpose of the tests was to measure the reflection and
transmission coefficients (K_R and K_T) for single and double
screens whilst changing the following variables identified
during the conceptual design stage:
(i)   Wave periods between 3 and 12 seconds
(ii)   Screen porosity varying between 8% and 35%
(iii)   Space between screens varying between 5 metres and 15
        metres
(iv)   Slots horizontal or vertical
(v)   The gap size between the bottom screen and the seabed
(vi)   The effect of changing the plank thickness from 300mm
        to 150mm
(vii)   Screens tilted or vertical
29. The model scale was 1:15 and random waves were used.
Wave measurements were made using three wave probes and
incident and reflected spectra calculated using an analysis
program. The method determines the reflection and
transmission coefficients for each frequency considered in
the incident wave spectrum.

30. As the model test progressed some changes were made
to the testing sequence taking account of the previous
results and the requirements of the structural design, which
was proceeding simultaneously, thereby avoiding tests
unlikely to yield a useful solution. In all, 54 tests were
carried out and the results are given in ref 15.

31.   The principal results are:-

For the single screen (Section A):

(i)   a porosity of 8% gave acceptable values of K_R and K_T

(ii)   the performance of horizontal slots was similar to
          that of vertical. Horizontal slots were therefore
          adopted because this orientation suited the proposed
          structural form

(iii)   a 1 metre gap beneath the screen gave unacceptable
          wave transmission results

(iv)   the hydraulic performance of 150mm planks was similar
          to that of 300mm planks

For the double screen (Section B):

(i)   a double screen with a front screen with 16% porosity
          spaced 8 metres from a solid back screen gave
          suitable reflection coefficients

(ii)   the performance of horizontal slots was similar to
          that of vertical. Horizontal slots were therefore
          adopted because this orientation suited the proposed
          structural form

(iii)   a 1 metre gap beneath the screens gave unacceptable
          wave transmission results

(iv)   the hydraulic performance of 150mm planks was similar
          to that of 300mm planks

(v)   tilting the screens did not show any improvement over
          vertical screens.

WAVE LOADING

32.   Wave loading measurements were not included in the
model test programme because the price of the breakwater
structure was not particularly sensitive to the small
reductions in wave loading that the model might have shown
to be permissable. The number of piles for example was
unlikely to be changed although some reductions in wall
thickness might have resulted. Another factor was the
additional time needed to set up the model for load testing
could not be accommodated in the tight design programme.

33.   Wave loading on the screens was calculated as though
the screens were solid and the methods contained in BS 6349
Part 1 were used. To determine the loading on the double
screen it was assumed that the arrival of the wave crest of
the transmitted wave on the internal screen occurred at the
same time as the arrival of the incident wave crest on the
external screen. The results showed that for the full wave
loading the peak landward load is 420 kN per metre run and
the peak seaward load is 240 kN per metre run.

34.   This approach is somewhat conservative but in
practice piles proved to be longer than expected and rock
conditions beneath the pile toe were quite variable. It was
therefore just as well that a conservative approach was
used.

STRUCTURAL DESIGN

35.   The typical structural details for the double screen

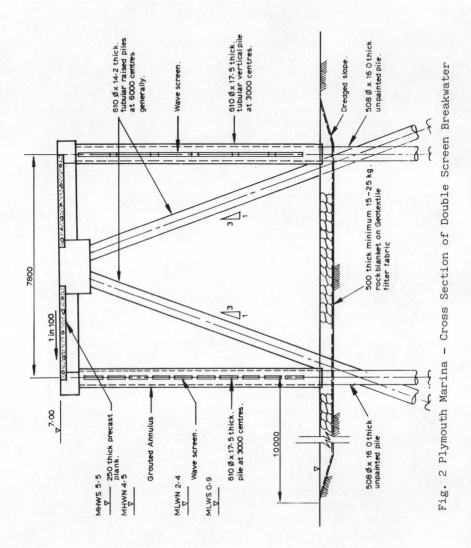

Fig. 2 Plymouth Marina – Cross Section of Double Screen Breakwater

are shown in Figure 2. The single screen is similar except that the rear screen and half of the deck is deleted.

36. The wave loads on the screen are transmitted to the bedrock by the pairs of raking piles. The maximum compression and tension loads are approximately 2500 kN and 1800 kN respectively. The high compression loads are taken by end bearing on the rock and for this reason the pile toes are plugged with concrete. The tension loads are taken by dead anchors drilled beyond the pile toes and grouted into bed rock.

37. The vertical pile is a composite pile with the upper section comprising an external sleeve grouted onto the inner pile after the inner pile has been driven to bed rock. The external pile is a prefabricated section with flanges for wave screen plank fixing. The use of the composite pile allowed the inner pile to be designed primarily for axial loads carried to bedrock, whilst the grouted sleeve above bed level provides the additional stiffness to carry the high moments induced by the wave screen. This arrangement also helped to eliminate alignment problems with the plank fixings.

38. The screen planking material is greenheart timber and the deck is concrete. Handrails and lights are provided on the deck to provide a public promenade and viewing area.

CONSTRUCTION

39. Construction was carried out by floating plant although the hydraulic backhoe dredger had spud legs for extra stability. The main piling barge was 18 metres by 33 metres and supported a 20 ton Butters Derrick.

40. The maximum pile lengths were 50 metres and these were pitched in one length. Pile driving was by hydraulic hammers, a BSP 357 at the start of the contract and an IHC S70 for the remainder. The facility to vary the energy of the hydraulic hammer was particularly useful for dealing with the soft overburden when the pile needed only low driving energy.

41. The reinforced concrete deck was designed to be cast in-situ but the contractor elected to use precast sections. This provided a high quality concrete finish but produced some problems where the piles connected to the deck and required non-standard units where the breakwater changed direction or cross section.

42. Anchors were inserted after construction of the deck. The deck provided a good working platform for the drilling rigs.

43. The fixing of the screen planks took place after anchor installation to avoid the risk of wave loading on the un-anchored structure.

44. Figure 3 shows a section of the slotted screen under construction.

45. The cost of the breakwater, excluding pontoons and

Fig. 3 Plymouth Marina – Fixing Planks on the Front Slotted Screen

other marina items, is approximately £1.8 million, of which approximately 65% is the cost of the tubular steel piles.

46. The Contractor for the project was Dean and Dyball (Western) Ltd from Exeter.

CONCLUSION

47. The Plymouth project required Parliamentary approval before commencement. As the Act included the criteria on wave reflection mentioned above the provision of a structure with a known reflection performance was of paramount importance for planning approval in addition to the provision of an effective marina for operational purposes. It is probable that as more marinas are required on the boundaries of crowded waterways similar structures will be needed in these locations.

48. Current design practice relies on the use of hydraulic models to determine wave transmission, reflection and loading characteristics and it is unlikely that final design will be able to proceed without such tests. However, it is possible that there is sufficient published work to enable feasibility studies and preliminary designs to be successfully completed without physical tests. Hopefully the information in this paper, and the discussion generated, will help bring this work into the form where preliminary design, at least, can proceed with a greater degree of confidence.

ACKNOWLEDGEMENT

49. The authors are indebted to the City Engineer, Plymouth for his permission to publish this paper.

REFERENCES
1. British Standard Code of Practice for Maritime Structures, Part 1. General Criteria
2. WECKMAN J, BIGHAM G N and DIXON R O, Reflection characteristics of a wave absorbing pier. Coastal Structures '83.
3. GARDNER J D, TOWNEND I H and FLEMING C A, The design of a slotted vertical screen breakwater. Coastal Engineering '86.
4. KARUNO S, Reflections and transition of waves through vertical slit type structures. Coastal Structures '83.
5. TRUITT C L and HORBICH J B, Transmission of random waves through pile breakwaters. Coastal Engineering '86.
6. RICHLEY E P and SOLITT C K, Wave attenuation by porous walled breakwater. Journal of the Waterways, Harbours and Coastal Engineering Division, August 1970.
7. ONISHI H and NAGAI S, Breakwaters and seawalls with slitted box-type wave absorber. Coastal Structures '79, Vol I
8. NOBLE H M, Low wave reflection construction at San Fransisco. Coastal Structures '79, Vol I.

9. KAKUNO S, ODA K, IBA T, YOSHIDAT, Wave reflection by a slitted breakwater as a boundary-wave problem. (Abstract only)

10. SEELING W N, Wave reflection from coastal structures. Coastal Structures '83.

11. HUTCHINSON P S and RAUDKIVI A J, Case history of a spaced pile breakwater at Half Moon Bay Marina, Auckland, New Zealand.

12. CHILDS K M, Perforated wave barriers in small boat piers. Ports '86.

13. KONDO H, Analysis of breakwaters having two porous walls. Coastal Structures '79, Vol II.

14. SJELIN G W, YAMASHITA T, ROLSTON J W, FOTINOS G C, Fisherman's Wharf breakwater design and construction. Ports '86.

15. ALLSOP N.W.H. and KALMUS D.C. Plymouth Marine Events Base. Performance of wave screens, Report No. EX 1327. Hydraulics Research, July 1985.

16. Some North American experiences with floating breakwaters

Professor R. E. NECE, University of Washington, Seattle,
E. E. NELSON, US Army Corps of Engineers, Seattle, and
C. T. BISHOP, National Water Research Institute, Burlington, Ontario

SYNOPSIS. This paper discusses field experience with floating breakwaters during the past 10–15 years on the East and West Coasts of North America. Regional preferences exist; on the East Coast the most common type is the floating tire breakwater, while on the West Coast concrete units (single pontoon or catamaran) are prevalent. General features of both categories are discussed, and one example of each is presented. East Coast experience is divided between freshwater and saltwater sites; saltwater sites are most common on the West Coast.

INTRODUCTION
   1.   Floating breakwaters must be sited with care. The local wave climate must be assessed to see if a floating breakwater is indeed feasible and then the breakwater must be sized to provide adequate wave attenuation. The degree of wave protection sought is somewhat subjective and varies with type and size of craft. Guidelines have been proposed for wave conditions in sheltered harbors (ref.1). For example, for head seas and a design wave period between 2 and 6 seconds (more than spanning the range of application of floating breakwaters) a "good" wave climate is one for which 0.3m waves are exceeded no more than once per year.
   2.   Until the early 1980's, operational problems with floating breakwaters were often the rule rather than the exception. A number of problems are listed here; many are common to both types of breakwaters:
   -   Inadequate buoyancy, including effects of biofouling.
   -   Connections between modules and/or individual components of the breakwater.
   -   Mooring and anchoring systems.
   -   Corrosion of connections and mooring lines.
   -   Accumulation of litter and debris.
   -   Boat-wake transmission, diffraction, and reflection; boat waves may be as large as wind waves at the site.
   -   Unplanned multiple use; this condition may or may not be anticipated, often is recreation oriented, and may lead to questions of safety.

FLOATING TIRE BREAKWATER

3.   The predominant type of floating breakwater that has been used on the Great Lakes and the East Coast of the U.S.A. has been the floating tire breakwater (FTB).   More specifically, a design developed in the early 1970's by the Goodyear Tire and Rubber Company and now known as the Goodyear FTB design, has been by far the most common floating breakwater.   In a 1982 survey of floating breakwater projects in the eastern United States (ref.2) 75 percent of all identified floating breakwater projects were Goodyear FTB's. FTB's dissipate wave energy primarily by transforming it into turbulence within and around tires; wave reflection is minor.

4.   The Goodyear design consists of modules, each containing 18 tires, interconnected to form a flexible mat as shown in Figure 1.   State-of-the-art construction guidelines are available (ref.3).   Many early Goodyear FTB installations failed for two main reasons: lack of sufficient reserve buoyancy and inappropriate choice of module binding and fastening materials (ref.2).

5.   Some early FTB's relied solely on air trapped in the tire crowns to provide flotation.   It is now recognised that supplemental flotation must be provided to ensure continued flotation.   The buoyancy of an FTB may decrease with time for the following reasons:-

–   Increase in weight due to marine growth on the tires

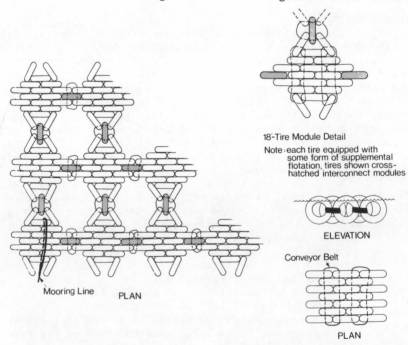

18'-Tire Module Detail

Note : each tire equipped with some form of supplemental flotation, tires shown cross-hatched interconnect modules

ELEVATION

Conveyor Belt

PLAN

Mooring Line   PLAN

Fig.1. Arrangement of tires in a Goodyear FTB (ref.3)

300

-   Increase in weight due to accumulation of sediment in the bottoms of tires.
-   Loss of trapped air and/or effectiveness of supplemental flotation decreases, e.g. polyurethane foam in the tire crowns (the most common supplement) breaks up and/or absorbs water.

6.   Field testing of module binding materials has led to recognition that conveyor belting is preferable in most situations (ref.4). The belting should be at least 12mm thick, 80mm wide, and should have 3 or more synthetic fabric plies. Overlapping conveyor belt ends can be fastened with a pattern of at least 2 bolts, nuts, and washers. Steel hardware is suitable for freshwater FTB's, while nylon hardware, dyed black, is more durable in saltwater.

7.   Table 1 provides summary information on several successful floating breakwaters, 11 of which are Goodyear FTB's. All but one of the FTB's used conveyor belting for module binding. Except where noted, all breakwaters listed were still in use in late 1987. This list is not a complete catalogue of successful floating breakwaters in the region, but represents those familiar to the authors. From Table 1, several points may be noted for these installations:

-   A-frame breakwaters have survived almost 20 years and are still in use.
-   Goodyear FTB's have survived 10 years and are still in use.
-   Maximum fetches are less than 20 km.
-   Car tires have been used more frequently than truck tires in Goodyear FTB's.
-   Three "old" Goodyear FTB's built without supplemental flotation are still being used successfully with regular maintenance to recharge the trapped air.
-   Supplemental flotation has been provided in a variety of ways.

8.   Most of the FTB's in the Great Lakes and East Coast regions have been installed to protect private marinas. Owners generally report satisfaction with them (refs.2,5). For FTB's built to state-of-the-art guidelines, the greatest maintenance requirements are removal of trapped debris and litter and removal and storage of the breakwater at locations where moving ice packs constitute a mooring problem. Occasional replacement of some belting and mooring lines is required.

BURLINGTON, ONTARIO FTB

9.   A large Goodyear FTB comprising 35,000 car tires (diameter D = 0.64m) was constructed in at Burlington, Ontario in 1981. A 64 module x 9 module (129m x 18.9m) FTB section was monitored in the field during 1981 and 1982 (Figure 2). Waves were measured with four bottom-mounted pressure transducers, two on each side of the FTB test section. Wave-induced loads on the steel chain mooring lines were measured with two electronic load cells and four mechanical "scratch" gauges. The field monitoring programme has been documented (ref.6) and

Table 1.   Summary information on some successful floating breakwaters in the Great Lakes and the East Coast U.S.A.

Location	Year Built	Max Fetch (km)	Max Depth (m)	Dimensions			Flotation	Comments
				Length (m)	Beam (m)	Draft (m)		
Thunder Bay, Ont.	1968	4.8	6.2	110	8.5	3.2	Pipes	Centreboard A-frame (ref.14)
Gananoque, Ont.	1969	5.5	4	61	8.5	3.2	Pipes	Centreboard A-frame (ref.14)
Newington, New Hampshire	1975	1.6	6+	46	6	0.5	Trapped air in tires	Goodyear - design car tire FTB, hauled onto beach each year for cleaning, replaced in 1985 by 2.4m beam concrete caisson (ref.2)
Westport, New York	1978	7.2	7.5	200	8	0.5	Trapped air	Goodyear FTB, car tires, air compressor used as required to recharge tire crowns, 9.5mm steel chain used to bind modules (ref.2)
Charlevoix, Michigan	1979	16	8	240	12,8,4	0.85 0.5	Sprayed polyurethane foam	Goodyear FTB, car and truck tires, constructed on ice, wood dock on top part of FTB
Catumet, Massachusetts	1980 1982	1.8 1.8	3.6+ 3.6+	17 122	6 6	0.5 0.5	Poured polyurethane foam	Goodyear FTB, car tires, prior installation in 1976, failed in 1978 due to flotation and binding material problems (ref.2)
South Portland, Maine	1980	13	9+	250	10,6	0.5	Trapped air	Goodyear FTB, air compressor used several times/year to recharge tire crowns (ref.2)
Burlington, Ont.	1981	4.4	12	490	18,10	0.5	Sprayed polyurethane	Goodyear FTB, 35,000 car tires (ref.7)
Lorain, Ont.	1981	*	3	183	24	0.5	Styrofoam blocks in polyethylene bags	Goodyear FTB, 20,000 car tires, to be removed in fall of 1987 due to completion of rubblemound breakwater
Cobourg, Ont.	1982	*	6	50	10	0.5	Sprayed polyurethane	Goodyear FTB, car tires
Erie, Pennsylvania	1982	3.2	3	37	12	0.85 0.5	Foamed pipes, poured polyurethane foam in some tires	Experimental pipe-tire floating breakwater
Marmora, New Jersey	1984	1.6	2.5+	120	10	0.5	Styrofoam wedges	Goodyear FTB, car tires, replaced floating concrete caisson that had failed
Harbor Springs, Michigan	1985	7.2	4.3	113	10	0.5	Preformed foam	Goodyear FTB, car tires,constructed on ice, wood dock on top of part of FTB
Belleville, Ont.	1985	2.2	4	120	12	0.5	Sprayed polyurethane	Goodyear FTB, car tires

+ Water depth at high tide
* FTB exposed to diffracted lake waves

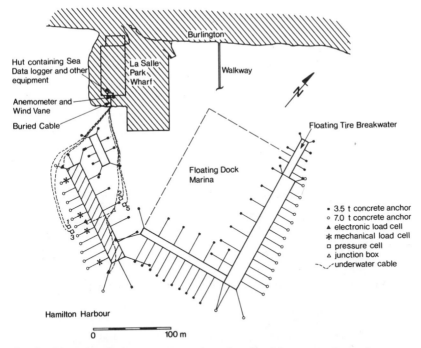

Fig.2. Layout of instrumentation for Burlington, Ontario,
       field monitoring study (ref.6).

breakwater construction details given (ref.7).

10.   Spectral analysis of the pressure records led to the
wave transmission design curve in Figure 3.   It is in good
agreement   with   earlier   two–dimensional   model   tests
(refs.8,9,10,11). The record of largest waves gave H_{m_0} = 0.65m
and T_p = 2.8s. For practical purposes, a breakwater is seldom
required unless wave attenuation of 50% or more is needed.
Figure 3 shows that to obtain a wave transmission coefficient
C_t of less than 0.5 the Goodyear FTB beam must be at least 1.2
times the design wavelength (L/B < 0.85). If the tire diameter
is more than one–third of the water depth (d), lower C_t values
may be obtained but reliable design data are not yet available.

11.   Wave–induced mooring loads on FTB's have been found to
increase with increasing wave height and length and decreasing
water depth.   From electronic measurements on the central
seaward mooring line, peak mooring loads ($F_{max}\cos\theta$) per unit
length (ℓ), where θ is the angle between the mooring line and a
perpendicular to the front face of the FTB and ℓ is the length
of breakwater frontage restrained by the mooring line, are
plotted versus the incident characteristic wave height in
Figure 4.   Also plotted are results from six–module–beam
two–dimensional prototype scale tests in 4m of water (ref.8),
for which the regular wave height has been substituted for H_{m_0}.

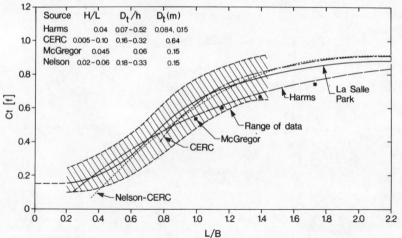

Fig.3. Wave transmission results from Burlington, Ontario,
compared with results from other investigators (ref.6).

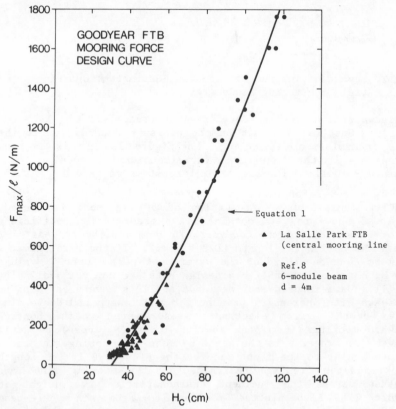

Fig.4. Peak mooring load data from Burlington, Ontario
(ref.6) and six-module beam data (ref.8).

Agreement is surprisingly good. A second order regression analysis of the 64 field test points and 39 model test points gives

$$\frac{F_{max} \cos\theta}{\ell} = -346 + 8.76\ H_{m_o} + 0.0789\ (H_{m_o})^2 \qquad (1)$$

where $F_{max} \cos\theta/\ell$ is in newtons per metre length and H_{m_o} is in centimetres. Equation 1 should only be used for values of H_{m_o} greater than 40cm and when D/d is less than 0.18.

12. Measured corner mooring line values of $F_{max} \cos\theta/\ell$ were found to be significantly larger than those at the central mooring line. This is thought to be due to three-dimensional effects at the corner such as oblique wave attack and diffraction. For design purposes, it is suggested that $F_{max} \cos\theta/\ell$ for a corner be estimated as twice the central mooring line load. Field experience with dragging anchors, especially at corners, substantiates this.

13. The cost of the breakwater, including materials, labour and profit, was $35 Canadian per square metre. The beam dimensions were 5 and 9 modules, giving an average cost per unit length of $500 Canadian per metre.

CONCRETE PONTOON BREAKWATERS

14. Concrete units, the most common type of floating breakwater on the (northern) West Coast, have typically been installed in saltwater where depths are greater, tidal ranges larger, and tidal currents stronger than at sites of tire structures on the East Coast. Another difference is that commercial fishing vessels as well as pleasure craft use the protection provided. Multiple use, both planned and unplanned, represents another difference where breakwaters are operated by local port authorities (ref.12).

15. In semi-protected waters of British Columbia and the States of Washington and Alaska, experience over the past 10 years shows that satisfactory wave attenuation performance is obtained with concrete pontoon units of width from 4.5-6.5m and maximum draft of less than 1.5m when exposed to incident waves of height up to 1.0m and periods up to 3-3.25 seconds; transmitted wave heights for these conditions are generally an acceptable 0.3-0.4m or less. Conditions typical of sites where floating breakwaters might be used involve significant wave heights between 0.6 and 1.2m, with periods from 2 to 4 seconds. Wave attenuation is mostly due to reflection from the rectangular pontoons, so the sea surface is rougher on the windward side of the breakwater than for a tire unit.

16. Concrete floor, deck, and sides typically are 0.10-0.15m thick; end walls may be thicker, especially if individual modules are post-tensioned together to form a longer, rigid unit. Welded wire fabric is common for reinforcement. Styrofoam blocks typically are used for interior forming and provide positive buoyancy as insurance

against flooding. Anchor line lockers, hawse pipes, and other hardware tied into the form are an integral part of each unit. Monolithic units weighing nearly 1,000 tons have been built.

17. The configuration of anchor lines crossed beneath the breakwater, as shown in Figure 5, is typical to provide keel clearance for vessel tie-up. The clump weights (usually concrete blocks) shown in the drawing are intended to produce a more even anchor line tension over the full tide range and thus reduce horizontal excursions of the breakwater, particularly at low tide levels. Line scopes of 1 vertical to 4-5 horizontal are common. Anchor line tensioning provides a mechanism for adjusting freeboard and alignment.

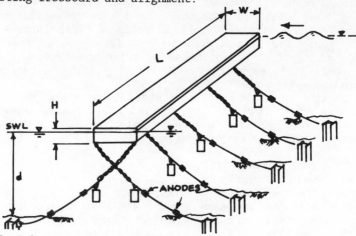

Fig.5. Schematic drawing of Friday Harbor breakwater

18. Connections between pontoons are a problem. Metal-to-metal connections have generally been unsatisfactory and often were under-designed. Connections incorporating large cylindrical rubber fenders appear more promising. More recent designs do not have connections between pontoons, which instead are fendered to protect against collision damage and depend upon anchor line restraint to hold them individually in position.

19. The U.S. Army Corps of Engineers, the lead government agency in the United States in the design and analysis of floating breakwaters, considers a 50 year design life in estimating costs of concrete units if reasonable quality controls are maintained. Galvanized steel cable anchor lines are anticipated to have a 50 year life if protected by anodes which will have to be replaced once, perhaps twice, over this life span. Without a corrosion protection system, the service life of 0.025m anchor chain is between 5 and 10 years in a temperate marine environment (ref.13); chains may have to be replaced 5 times during a 50 year structure life.

20. Field observations (refs.12,13) have led to some other conclusions, among them:

    —   Any large concrete float attached to shore will be

used for temporary moorage by large vessels if there is a shortage of dock space. This possibility should be considered in the design. Additional wind loads caused by the sail effect of larger vessels moored on the windward side can lead to increased float excursion.

- Boat waves may be more significant than wind waves, producing larger forces and perhaps overtopping of the float.

- Inadequate freeboard can lead to frequent overtopping, causing a slippery deck which can be dangerous if access to the breakwater is allowed. Breakwaters with shore access become popular fishing piers.

- Electrical services, if provided on the breakwater, should be properly designed for the marine environment.

21. Prior to the 1970's most floating breakwaters in the Pacific North West tended to be makeshift, with the exception of the A-frame unit at Lund, British Columbia (ref.14). Table 2, while not all-inclusive, shows the trend toward larger concrete units. The Lund A-frame unit was replaced in 1987 by three fendered but unconnected concrete pontoons each approximately 45m long by 7.6m wide by 2.7m high.

PORT OF FRIDAY HARBOR MARINA BREAKWATER

22. The 580 boat marina at Friday Harbor is located on the eastern shore of San Juan Island in the inland waters of northwestern Washington. The breakwater consists of 5 rectangular post-tensioned pontoons; three are 91.5m long by 6.4m wide by 1.8m high with a 1.4m draft, and the other two are 91.5m long by 4.9m wide by 1.5m high with a 1.1m draft. Respective design waves are shown in Table 2. Water depth at the site varies between 12 and 15m at MLLW. Tides are mixed with a duirnal range of 2.3m; maximum currents at the site are less than 0.5 m/s. Breakwater anchors are 52 steel H-piles embedded in firm silts, sands, and clays. Anchor lines (Figure 5) consist of 35mm diameter galvanized bridge rope with 9m lengths of 32mm link chain at the upper end. A 1-ton (submerged) clump weight is attached 15m from the upper end of each anchor line, initial tension is approxiomately 45kN, and 3 large aluminum anodes are attached to each anchor line to prevent corrosion (ref.13).

23. Part of one of the smaller pontoons is composed of two smaller units, each 22.9m long. These initially underwent prototype field tests for 18 months at an exposed site in Puget Sound, off Seattle. Water depth at the test site was about 15m at MLLW, the diurnal tide range was 3.5m, and maximum currents exceeded 60cm/sec. Anchor arrangements were basically those used later at Friday Harbor. Wave attenuation and anchor line forces were measured. Details of the monitoring programme and results obtained have been presented (ref.15), and some are given here. Three configurations were tested:

- Floats rigidly connected (effective single pontoon

Table 2. Summary information on some successful floating breakwaters on the West Coast U.S.A. and Canada

Location	Year Built	Depth MLLW (m)	Dimensions				Design Wave		Anchors		Comments
			L (m)	W (m)	Draft (m)	H (m)	Hs (m)	T (sec)	Wt (tons)	Line Size	
Lund, Brit. Columbia	1964	21	110	7.6	3.7	5.5	1.4	2.8	13	25mm chain	Centreboard A-frame. Replaced by concrete pontoons in 1987 (ref.14)
Tenakee, Alaska	1972	15	110	6.4	1.1	1.5	0.9	4.0	26	35mm chain	Catamaran; 18.3m post-tensioned modules, 0.9 x 1.5 x 4.6m indiv. pontoons (ref.12)
Sitka, Alaska	1973	14	293	6.4	1.1	1.5	0.9	4.0	Stake Pile	35mm chain	Catamaran; 18.3m post-tensioned modules, 0.9 x 1.5 x 4.6m indiv. pontoons (ref.12)
Port Orchard, Wash.	1974	14	473	3.7	0.5	0.9	0.6	2.5	Stake Pile	12mm Nylon rope	Rectangle; 19.3m post-tensioned modules (ref.12)
U.Wash, Friday Harbor, Wash.	1979	18	119	4.6	0.9	1.4	0.9	3.5	27	25mm chain	Rectangle; 3 modules (refs 12,13)
Ketchikan, Alaska	1980	26	293	7.0	1.4	1.8	1.0	3.5	60,18	chain	Catamaran; 4 post-tensioned modules (ref.12)
Semiahmoo, Wash.	1981	6	1067	4.6	0.9	1.4	0.6-0.9		Stake Pile	25mm Nylon rope	Rectangle; 18.3m modules, clump weights on anchor lines (ref.13)
Brownsville, Wash.	1981	9	110	5.5	1.2	1.5	1.0	3.4	Stake Pile	38mm chain	Rectangle; 24 4.8m units post-tensioned together into a single unit (ref.13)
Olympia, Wash.	1982	4	213	4.9	1.2	1.7	0.6	2.8	Timber anchor piles thru sleeves in floats		Rectangle; 7 modules (ref.13)
Friday Harbor, Wash.	1984	18 / 12	302 / 183	6.4 / 4.9	1.4 / 1.1	1.8 / 1.5	1.0 / 0.8	3.2 / 2.6	Stake Pile	12mm Nylon rope	Rectangle; 3 modules, 1-ton clump weights; Rectangle; 2 modules, 1-ton clump weights (ref.13)

length of 45.8m) with clump weights.
- Floats rigidly connected, without clump weights.
- Floats flexibly connected, without clump weights.

24. Data were treated by spectral analysis. A 0.08Hz low pass filter removed long period effects and a 1.0 Hz filter removed effects of short period phenomena such as ripples. Results of the spectral analysis were used to calculate significant wave heights, and transmission coefficients C_t were determined through comparison of incident and transmitted wave spectra. Waves were measured by resistance–wire wave staffs mounted on spar buoys having natural heave periods greater than 12 seconds. Anchor line forces were measured by load cells.

25. Wave transmission data are shown in Figure 6. The two–dimensional monochromatic wave laboratory model data (ref.16) were for a pontoon of equal width and draft. For the limited range of periods covered by the prototype data, agreement between prototype and model results was reasonable. The transmission coefficient centred on 0.4 but because of the limited range of wave periods no definite conclusions could be made about wave period, effects anticipated from model tests. Neither was there apparent influence of breakwater configuration.

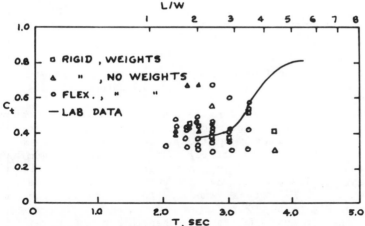

Fig.6. Wave transmission results from field monitoring study of concrete breakwater for Friday Harbor (ref.15) compared with model test results (ref.16).

26. Peak wave force (taken as statistical value of the highest one percent, after filtering) results are shown in Figure 7. There was no strong dependance upon wave height, nor for that matter (not shown here) on wave period. Forces shown are values in excess of the current drag and pretensioning, which was 22.4 kN in each of five lines with clump weights and 6.7 kN without. Force/unit length values were calculated assuming each anchor line carried one–fifth of the load and that incident wave crests were parallel to the breakwater face; the latter, due to breakwater orientation, was a reasonable

approximation and the breakwater was not long compared to wave crest lengths. Measured anchor line forces were much smaller than anticipated; pretensioning could have been the cause.

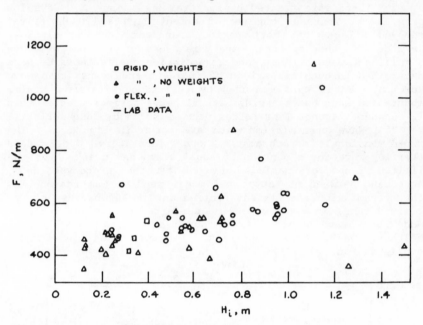

Fig.7. Peak anchor line forces from field monitoring study
of concrete breakwater for Friday Harbor (ref.15).

27. The test breakwater was subjected to limited boat wave tests. Wave transmission data are given in dimensionless form in Figure 8 for waves generated at three speeds V_s shown by a 15-ton U.S. Coast Guard utility boat (ref.17). Maximum height of the boat waves was 0.6m,; although not shown in Figure 8, incidence angles between wave crest and breakwater face varied between 5 and 35 degrees. Model data are for the same model shown in Figure 6. For longer wave lengths, C_t values are larger than in the model tests. When larger test waves (H_i to 0.9m maximum) from a 193 ton displacement marine tug acted on the breakwater there was significant overtopping and water on the breakwater deck.

28. The Friday Harbor breakwater was monitored closely in the period December, 1984 - June, 1986. Both winters were exceptionally calm, so the breakwater was not subjected to near-design wave conditions. The breakwater hosts large numbers of transient boats in the summer and is a popular fishing pier for the local population. Breakwater excursions were monitored. Maximum longitudinal motion was about 0.8m, most likely due to sail effects of transient boats moored to the floats during winds in excess of 18 m/s but which, because of orientation, did not generage large waves against the breakwater; maximum lateral motion was about 0.15m. All anchor

line components were in excellent condition; surface corrosion

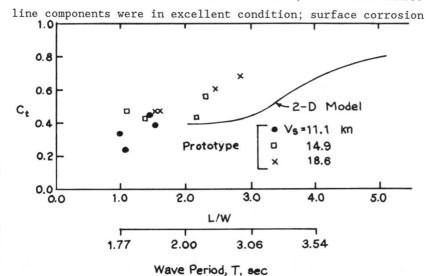

Wave Period, T, sec

Fig.8. Boat wave transmission data from field monitoring study of concrete breakwater for Friday Harbor (ref.17).

of the anodes had begun. After 2.5 years of operation some maintenance problems persisted. Stanchions on the breakwater to provide electrical service to transient boats are vulnerable to collision with bowsprits of docking boats. Electric junction boxes mounted flush with the deck are subject to water damage, and hardware providing support for electrical wiring has corroded. There was little or no wear or damage to the fenders separating the floats (ref.13).

29. The 1984 cost of the breakwater, when installed, was $4,020 U.S. per metre of length.

CONCLUSIONS

30. Field experience in North America with successful tire and concrete breakwaters has led to design and construction techniques which now exist to make these floating structures more economically and structurally viable and reasonably maintenance free, and they have become an attractive alternative in the design of harbours and marinas at limited-fetch locations. At some sites floating breakwaters are economically competitive with fixed structures although risks and costs associated with the still higher levels of maintenance must be recognised.

REFERENCES

1. NORTHWEST HYDRAULICS CONSULTANTS. Study to Determine Acceptable Wave Climate in Small Craft Harbours. Canadian Manuscript Report of Fisheries and Aquatic Sciences No. 1581, Department of Fisheries and Oceans, Ottawa, Ontario. 1980.
2. BAIRD A.V. and ROSS N.W. Field Experiences with Floating Breakwaters in the Eastern United States. U.S. Army, Coastal

Engineering Research Center, Misc. Report 82-4. 1982.
3. BISHOP C.T., DEYOUNG B., HARMS V.W. and ROSS N.W. Guidelines for the Effective Use of Floating Tire Breakwaters. Information Bull. 197, Cornell University Cooperative Extension, Ithaca, New York. 1983.
4. DAVIS A.P. Jr. Evaluation of Tying Materials for Floating Tire Breakwaters. Marine Technical Report 54, University of Rhode Island, Narraganssett, Rhode Island. 1977.
5. BISHOP C.T., BRODERICK L.L. and DAVIDSON D.D. (eds). Proceedings of the Floating Tire Breakwater Workshop 8-9 November, 1984. U.S. Army, Waterways Experiment Station, Technical Report CERC-85-9. 1985.
6. BISHOP C.T. Field Assessment of a Floating Tire Breakwater. Canadian Journal of Civil Engineering, 12(4), 1985: 782-795.
7. BISHOP C.T. and GALLANT B.A. Construction of a Goodyear Floating Tire Breakwater at La Salle Marina, Burlington, Ontario. Proc. Second Conference on Floating Breakwaters, University of Washington, Seattle, 190-207. 1981.
8. GILES M.L. and SORENSEN R.M. Prototype Scale Mooring Load and Transmission Tests for a Floating Tire Breakwater. U.S. Army, Coastal Engineering Research Center, Technical Paper 78-3. 1978.
9. McGREGOR R.C. The Design of Scrap-Tyre Floating Breakwaters with Special Reference to Fish Farms. Proc. Royal Society Edinburgh, 76B, 1978: 115-133.
10. HARMS V.W. Design Criteria for Floating Tire Breakwaters. Journal of the Waterway, Port, Coastal and Ocean Division, ASCE, Vol.106, No.WW2, 1979: 149-170.
11. NECE R.E. and NELSON W.L. Some Scale Effects in Model Tests of Floating Breakwaters. Proc. Second Conference on Floating Breakwaters, Univ. of Wash., Seattle, 87-98.1981.
12. RICHEY E.P. Floating Breakwater Field Experience, West Coast. U.S. Army, Coastal Engineering Research Center, Misc. Report 82-5. 1982.
13. NELSON E.E. and HEMSLEY J.M. Monitoring Completed Coastal Projects, Floating Breakwater Operational Assessment. U.S. Army, Waterways Experiment Station, Misc. Paper CERC-87--1987.
14. WESTERN CANADA HYDRAULIC LABORATORIES LIMITED. Development of Manual for the Design of Floating Breakwaters. Prepared for Department of Fisheries and Oceans, Small Craft Harbours Branch, Port Coquitlam, British Columbia. 1981.
15. NELSON E.E. and BRODERICK L.L. Floating Breakwater Prototype Test Program: Seattle, Washington. U.S. Army, Waterways Experiment Station, Misc. Paper CERC-86-3. 1986.
16. CARVER, R.D. Floating Breakwater Wave Attenuation Tests for the East Bay Marina, Olympia Harbor, Washington, Hydraulic Model Investigation. U.S. Army, Waterways Experiment Station, Technical Report HL-79-13. 1979.
17. NECE R.E. and SKJELBREIA N.K. Ship-Wave Attenuation Tests of a Prototype Floating Breakwater. Proc. Nineteenth International Conference on Coastal Engineering, ASCE, 1984: 2514-2529.

17. Risk insurance analysis

M. C. FARROW, Excess Insurance Group, London

SYNOPSIS   Why is there a need for insurance on breakwaters?
How do insurers consider them?   What are current results, and
how can we collectively improve them.

1. OPERATING ENVIRONMENT
(i)   The prime function of any underwriter is to ensure
that the premium and other income derived therefrom exceed
the total claims in any one year.   Therefore the objective
is the same as all companies to ultimately make a profit.   I
do not need to tell you that the basic premise of insurance
is that the many pay for the losses of the few.

(ii) Currently the worldwide market subject to the usual
nationalistic requirements means that although there are a
large number of insurers underwriting contractor all risk
policies, when talking about a specific category like the
insurance of breakwaters and associated structures, the actual
numbers involved reduce to a relative few.

(iii) Competitio n therefore is certainly not as wide as
most people anticipate and remember although a specific company may be seen to issue a policy in a territory, it is the
support of the co-insurers which tends to prevail.

2. COVERAGE AND UNDERWRITING CRITERIA
(i)   Every contractors all risks policy indemnifies the
insured in respect of physical loss or damage to the contract
works subject always to the specific exclusions and limitations
applicable to each risk.

Obviously when we are dealing with breakwaters there
are a number of circumstances which are inevitable, given the
natural characteristics of the work and construction programme.
Here I am thinking of the naturally occuring elemental perils
which provide a large known factor which may or may not be
capable of measurement.

(ii) Any underwriting is dependent upon loss experience
that a particular underwriter of a company has suffered in the
past. Some insurers, such as ourselves, who have underwritten
this class for over 20 years, have accumulated considerable

<u>SAN CIPRIAN PORT – SPAIN</u>

<u>REBUILDING AND REFURBISHMENT</u>

5 YEARS' DATA

MONTH	AVERAGE/MAXIMUM W/HEIGHT	HIGHEST	MY ESTIMATE OF NORMAL
JANUARY	8.40	10.10	8.0
FEBRUARY	7.77	10.0	7.20
MARCH	7.71	9.80	7.80
APRIL	6.45	8.20	5.50
MAY	5.83	8.0	5.50
JUNE	3.96	4.35	4.0
JULY	3.92	4.65	4.0
AUGUST	4.04	5.05	4.0
SEPTEMBER	6.10	7.65	5.50
OCTOBER	7.19	10.45	6.50
NOVEMBER	7.04	9.30	6.50
DECEMBER	9.45	14.86	9.0

RESTRICTIONS THAT WORK CAN ONLY BE DONE ON LAND BETWEEN NOVEMBER AND MARCH

ALSO COVERING THE BREAKWATER WHICH HAS BEEN REFURBISHED

data, and can recognise certain specialist contractor's,
designer's or engineer's particular experience.
It is this background knowledge which generates
the underlying underwriting approach.

3. SPECIFIC CLAUSES

It may be useful to explain in greater depth some of the more
specific clauses which can be encountered on an insurance
policy covering a major project.

I will repeat that it should be appreciated that since the
basic premise of a contractors all risk insurance is physical
loss or damage to the works from any cause whatsoever, it is
necessary to eliminate the inevitable, inherent and trade
risks. All of these should be either the responsibility of
one of the principals or are contrary to the basic laws of
insurance.

(i) A time clause of normally 72 hours which states that
if there is more than one storm within a 72 hour period, it
will be assessed as one storm and only one excess will be
payable by the insured.
The purpose of this is to clarify the situation so
that a client is clearly aware of the position and this can
eliminate a source of dispute at a later stage.
(ii) Exclusion of Loss or damage by normal action of the sea
The idea behind this exclusion is to exclude claims which are
caused by normal wave heights, given that time of the year and
the area involved that a claim actually occurred.

In my opinion this is an anomoly, as I have yet to meet anybody that can give a definition of "normal action" and whether
there is any legal definition to this phrase.

I prefer to agree in advance an estimate of "normal" prior
to the commencement of the contract as we have done on the rebuilding of San Ciprian (see Exhibit 1)
(iii) Exclusion of dredging or redredging costs
As we are all aware whenever dredging work is undertaken in
the open sea, tidal action will aways attempt to refill a void.
I appreciate that additional seabed movement does occur during
heavy weather period, but to my knowledge it is virtually impossible to measure the difference between the normal and the
exceptional. It is this lack of clearly defined measurement
that can cause a dispute when a claim occurs and as far as I
am concerned it is more satisfactory to make the position
clear at the outset so avoid potential problems.
(iv) Exclusion of Loss of fill in reclaimed areas unless
a breach in the permanent works.
This to some degree is a natural continuation of the line of
logic explained above. It is only when there is a clearly
defined breach in the permanent works whether it be a rock
bund, piled walls or others that the measurement of loss is
reasonably easy to ascertain.

(v) <u>Exclusion of loss or damage due to insufficient
compaction.</u>
This is what would have been called in past times a trade
risk whereby the construction method is inadequate for the
purpose.

I have to say at this stage judging from recent contracts I
have seen the work methods are sufficiently detailed and we
tend only to apply this type of clause when we have doubts
about the experience of the contractor.

(vi) <u>Warranty limited length of unarmoured run of
breakwater or sea defence</u>
It should be clearly understood that a warranty in insurance
terms is a serious imposition. It is a clear instruction
that something must or must not be done and a breach of such
condition makes the policy voidable.

Every underwriter needs to calculate accurately his major
exposure and it is quite obvious that on most wet contracts
the breakwater is likely to be the target risk.

The purpose of the clause which is only applied after studying the material details of the contract and generally in
conjunction with the design criteria and construction programme
is to restrict to acceptable proportions the unprotected work.

The actual figures will of course vary from contract to contract depending upon the specific facts applying.

4. RISK INFORMATION
<u>Material Damage</u> - First of all there is no general ideal
quality or quantity of information as this will depend upon
the relevant circumstances of each risk. Quite clearly,
there is a substantial difference between a small structure
in a predominately sheltered area and a major structure in an
area known for its volatility.

We have all become aware of the number of major incidents
which have occurred to breakwaters in the past ten years.

We realise that similar to many other high technology in-
dustries, time and material volume are the prime factors and
anybody that can show a saving in either is likely to be
awarded the contract.

It is no co-incidence that many structures built by Victorian
Engineers, which by modern parameters were over-designed, are
still standing.

Therefore considering the current state of the art most insurers have tended to become highly conservative.

Some idea of the level of information a competent underwriter
needs is as follows:-
(i)   Past experience and history of contractor/designer/
engineer to carry out the type of work being undertaken.
(ii) A clear breakdown of the contract price between major
elements to ensure that a fair price is given for the job.

(iii)   A full breakdown of contractors plant and equipment
   to be utilised in the construction.

(iv)   A clear bar chart/work programme to confirm thatthe
more hazardous elements are completed before the weather
exposes the work abnormally.

(v)   Full information relating to past weather and tides in
the area clearly showing the maximum and minimum past record
with probably extrapolations over the anticipated periods
(say 5 years).

(vi)   Adequate reserves in both materials, labour and time
programme, knowing full well that we are dealing generally in
a hostile environment.

(vii)   Method of construction together with length of un-
armoured/unprotected run. of breakwater.

(viii) Conditions of seabed with particular emphasis on the
area of the foundations.

(ix)   Any unusual characteristics i.e. cofferdams, or use
of floating plant.

(x)   Details of storm warning devices and precautions to be
taken to minimise damage in the event of severe conditions.
Here for example we would be considering the ability to withdraw plant to a safe position.

(xi)   Plans and elevation drawings showing sequence of construction.

(xii)   Situation of land based yards and quarry sites in relation to contract sites.

(xiii) Seismology reports.

(iv)   Contract conditions

(xv)   Any available independent feasibility/ technical
studies which may have taken place.

Third Party

(i)   Third party exposure in relation to existing facilities.

(ii)   Tracks and frequency of marine vessels using the area.

(iii)   Exact location of any underwater pipes or cables and
specific information of precautions to be taken to prevent
loss.

General

Obviously this is the ideal level of information which will
enable an underwriter to consider the risk implications
fully.

It should be stressed at this stage that we are predominately
not technical experts and on occasions will call for a report
from a suitably qualified professional for an independent
review.

5. CLAIMS CIRCUMSTANCES

It is quite clear that results in this particular category
of insurance have tended to worsen over the past 15-20 years
This to some degree could have been anticipated since as
the world becomes more industrialised , the demand for break

waters protecting harbours, marinas and land based industrial developments becomes greater.

The choices of sites is often dependent upon commercial considerations and appears not necessarily to be the most suitable. The inevitable result is that more hazardous locations are being used.

There are numerous reports of classic failures - Sines - Ras Lanuf - San Ciprian - Port D'Arzew - Ben Ghazi - Skikda and others, but we in the insurance industry have literally thousands of claims worldwide which add to the general loss record.

In some, such as Richards Bay, plant losses reached astronomical proportions probably due to an unsound system of work relating to the control of relatively inexperienced drivers and handlers.

Djen Djen in Algeria appears to be suffering similar problems.

Another point causing considerable concern to insurers is a relative inability of many forms of armouring, and these seem to grow larger in number every day, to adequately protect the main structure.

Dolos, Tetrapods, Roblocs amongst many other designs have been used to attempt to stave off the elements, with in all honesty very little effect judging from our claims statistics.

It is this seeming uncertainty on behalf of the Engineers that gives insurers a distinct lack of confidence, that we actually know, many cases, exactly how to protect the works and so keep losses to a minimum.

I am sincerely hoping that one of the real benefits from this Conference will be to detail positive steps which are being taken to clarify the position.

6. MARKET OUTLOOK
(i) Future Developments
We in the insurance market have become used to change and we are conscious that it is our duty to attempt to provide insurance whenever it is required.
We have always prided ourselves in trying to understand new work processes and doubtless in the foreseeable future we will be tested time and time again.
(a) Floating breakwaters being one of these developments that we see becoming more widely used.
(b) Tidal Barriers - breakwaters and power stations developing further.
(c) As technology continues to expand there is obviously a serious need for engineers, contractors, principals and insurers to work more closely and with the growth of modern communication systems this will surely become easier.

(ii)   Underwriting cycle
        It is an unfortunate fact of insurance that market
trends go in cycles and we very often forget about the previous
incidents we have endured.

This tends to give the impression to outsiders that we are
inconsistent in method and approach.

We will, hopefully, in the future, not expose ourselves again
to the over competition that has occurred in the past.

During the past three years we have emerged from the worst
trough in my insurance memory.

Peaks and troughs can never be avoided but with improved
statistical and data capture give us more experience and
therefore give the insured a more consistent and methodical
approach.

(iii)   Computerisation
The comments in (ii) above are further helped by computerisation, that is important to capture the correct data and
input in a way that will give the correct information.

As a company we are highly computerised and have extended
our systems to include "computer assisted underwriting".

We have these systems both on contractors all risk and
professional indemnity insurance.   For example our underwriting system was designed to follow the thought processes
that an underwriter goes through in evaluating the premium
and coverage terms we require.   In respect of Professional
Indemnity when an underwriter is offered the insurance for
a consulting engineer he immediately thinks of his basic
percentage rate on fees and thereafter adjusts this up to or
down based on certain factors.   Such judgment criteria
would include the obvious items such as limits of indemnity
and size of fees, but also includes adjustment because of
(a)   type of contract or operations
(b)   territorial limits
(c)   claims experience
(d)   quality of practice
(e)   coverage extent
(f)   self insured excess
(g)   professional discipline.

We built this into a computer program utilising tables of
adjustment from the mean.

Obviously this premium assessment process cannot entirely be
mechanised and therefore our rating formula requires underwriters' discretionary judgment in respect of factors like
claims and quality.   We have, however built in edits for
"increases and reductions" to maximise control and minimise
poor underwriting.

The underwriter cannot key in vastly reduced or inflated prices.

You might be surprised that we put a limit on increasing the premiums thus in simple terms if the underwriter wishes to lead the risk by 100% then we have to ask ourselves is this the sort of practice we want on our books.

We are the first in the market to advance to this stage but clearly the progress of computerisation will accelerate to alter and improve all aspects of our lives!

In closing, it is clear that insurers and their clients must grow closer to appreciate more clearly each others view points and needs. Our aim is to provide satisfactory protection at a reasonable cost but this relies upon your satisfactory performance.

Failures on your part will cause insurers to continue to reconsider this type of risk and perhaps some more will even withdraw from the market.

We must both get it right!

Discussion on Papers 15-17

PAPER 15

MR N. W. H. ALLSOP, Hydraulics Research, Wallingford
With reference to Paper 15, I was responsible for the
laboratory testing mentioned. Data from the test results
provide a most useful data set on the effects of random waves.
However, it should be noted that detailed differences between
this and other wave screens may give rise to marked
differences in performance. Advanced mathematical modelling
methods to describe the hydraulic performance of simple
screens are now being developed at Brunel University, under
the guidance and support of Hydraulics Research.

In other tests of these types of structure, we have noticed
effects of water level and wave angle. We have also noted
potential uplift problems due to wave interference effects
between front and rear screens below the capping slabs.

How did the Authors take account of the different water
levels in this project? Would they comment on the effect of
angled wave attack? How did they design for uplift on the
crest slabs?

MR G. CRAWLEY, Fairclough Howard Marine, Chatham
In 1983, we were asked by Camper Nicholson (C&N) to submit a
scheme for a new breakwater to protect their Marina at
Gosport. They had an existing pier, comprising a tubular
steel structure with a sheet piled wall on the outside, which
was nearing the end of its life, and they were looking to
extend their Marina.

Figure 1 shows Portsmouth Harbour with the new and the old
breakwaters. The client's requirements which had to be met in
our proposals were: low cost; narrow width to maximise the
space within the Marina with a fixed boundary position; an
effective wave attenuation; rapid construction so that the
Marina could be opened for the beginning of the 1984 season;
no siltation as the extension to the Marina had to be dredged;
and a 20 year life.

Having looked at the alternatives of a reinforced concrete
wall and a steel sheet pile wall, it was apparent that the way

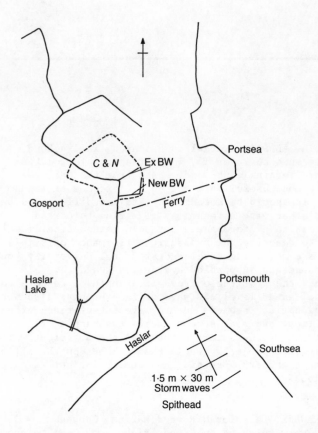

Fig. 1

to meet the client's budget was to use a partial permeable breakwater. Partial refers to the fact that the breakwater does not extend down to the bed of the Harbour and permeable refers to the holes in the wall. The concept was based on a similar Jarlan wall design used by Howards for concrete gravity structures in the North Sea such as the Ninian central platform.

The design data for the breakwater were: for a storm wave H = 0.5 m, L = 4 m, and combined with H = 0.2 m and L = 10 m; for an extreme storm wave H = 1.5 m, L = 30 m; and for a 1 in 50 year wave of H = 1.75 m, L = 55 m. The Harbour bed is a dense silty clayey soil with a C of 150 kN/m^2. The water depth at mean high water springs was 12.7 m at the head of the breakwater and 5.3 m at the start of the breakwater.

Having reviewed the available literature, we found that R. L. Wiegel had produced formulae for solid partial breakwaters and permeable full height breakwaters, the second case being for vertical slots. Our derivation of the total effect of a partial permeable breakwater is shown in Fig. 2, and the section through the design proposed to the client is shown in Fig. 3. This consists of a vertical tubular steel pile supporting the prestressed precast concrete dog bone planks which provide the slot, and a raking tubular pile at the back of the structure connected by universal beam bracing.

As the resulting design was based on interpretation of experimental data which was not strictly comparable, we recommended to the client that a model test be carried out to verify our predictions. The test was carried out for us by Queen Mary College, using the wave flume at City University, and a reasonable correlation was obtained between theoretical and test results, as is shown in Fig. 4. The marina waves for the scheme finally adopted were 0.3 m for the storm conditions and 0.6 m for the extreme storm. The longitudinal profile of the breakwater as constructed is shown in Fig. 5.

Since construction, the breakwater has performed as predicted. The majority of the motion in the Marina results from the bow waves of the shipping moving inside the Harbour, which enter the Marina beyond the extent of the breakwater.

MR C. DINARDO, Dinardo and Partners, Paisley
Did the Authors consider the possibility of introducing more porosity in the deck and upper section of the solid inner screen wall in order to reduce the build-up of (trapped) pressure?

MR P. LACEY, Ove Arup and Partners, London
I would be interested to know at what angle the waves are expected to attack the structure. There was little on the hydraulic test findings and I would appreciate a comment on what overtopping was found at high water and what transmission effects occurred at maximum overtopping.

Case 1.   Solid partial (P)

$$\frac{H^P_t}{H^P_i} = \sqrt{\left(\frac{E^P_t}{E^P_i}\right)} = \sqrt{\left(\frac{4\,\pi\,t/L + \sinh 4\,\pi\,t/L}{4\,\pi\,d/L + \sinh 4\,\pi\,d/L}\right)} = \sqrt{[F\,(L,\,t,\,d)]}$$

$$E^P_t = F\,(L,\,t,\,d)\,E^P_i$$

Case 2.   Permeable full height (for vertical gaps) (g)

$$\frac{H^g_t}{H^g_i} = \sqrt{\left(\frac{E^g_t}{E^g_i}\right)} = k\,\sqrt{(A_r)} \qquad (k = 1 \cdot 25)$$

$$E^g_t = k^2\,A_r\,E^g_i$$

Case 3.   Reflected energy permeates through pores

$$E^P_r = E^P_i - E^P_t = [I - F\,(L,\,t,\,d)]\,E^P_i \qquad \text{(Reflected energy)}$$

$$E^g_t = k^2 A_r E^P_r = k^2 A_r [I - F\,(L,\,t,\,d)]\,E^P_i \quad \text{(Transmitted energy)}$$

Total effect

$$ET_t = E^P_t + E^g_t$$

$$= F\,(L,\,t,\,d)\,E^P_i + k^2 A_r [I - F\,(L,\,t,\,d)]\,E^P_i$$

$$\frac{ET_t}{EP_i} = k^2 A_r + (I - k^2 A_r)\,F\,(L,\,t,\,d)$$

$$\frac{H_t}{H_i} = \sqrt{\left(k^2 A_r + (I - k^2 A_r)\,\frac{4\,\pi\,t/L + \sinh 4\,\pi\,t/L}{4\,\pi\,d/L + \sinh 4\,\pi\,d/L}\right)}$$

Fig. 2.   The design of a partial permeable breakwater based on formulae derived by R.L. Wiegel

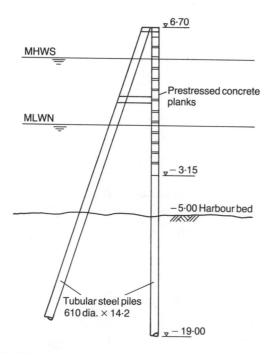

Fig. 3.   Gosport breakwater

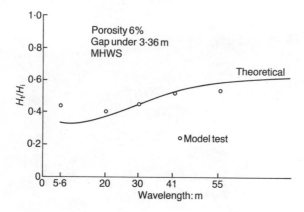

Fig. 4

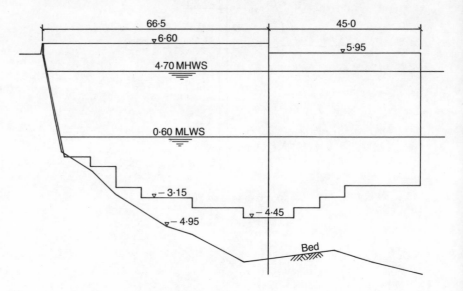

Fig. 5.   Long section of breakwater

MR J. LOVELESS, King's College London
One of the mechanisms which may give rise to scour at a
vertical breakwater is the presence of a potential flow path
under the structure. In the case of the Plymouth Marina
breakwater, this would seem to be a possibility that ought to
be guarded against.

Could the Authors give more details about the nature of the
design near the toe? The assumptions as to the value of the
wave transmission coefficient depend heavily on the porosity.
If there is scour at the toe, the porosity will be changed
considerably, and the assumptions made in stability
calculations will also be changed.

MR R. McKINNON, Sir William Halcrow and Partners, Swindon.
I was the resident engineer on the construction of Plymouth
Marine Events Base breakwater.

During the later stage of the construction, I lived on my
boat in the marina. During several storms from the south, the
boat movement within the marina resulting from the transmitted
wave was not uncomfortable. The orientation of boats into the
prevailing wind conditions no doubt helps. The most
uncomfortable waves for moored yachts came from passing
trawlers coming out of Sutton Harbour, which is in the lee of
the marina breakwater. This problem has now been resolved by
the Harbour Master's enforcing the existing speed limits.

MR GARDNER and MR TOWNEND (Paper 15)
In reply to Mr Allsop, tests were carried out at only one
water level. At low water, the breakwater was partially
protected by the Mountbatten breakwater; therefore, only the
high water case was critical.

Angled wave attack was not examined in the flume testing.
On the sections of the breakwater where angled wave attack was
experienced, the reflection and transmission of coefficients
for normal attack were used. However, the wave loading on the
structure was reduced in proportion to the cosine of the angle
between wave approach and normal to the breakwater.

Uplift was considered critical only in the region where
there were double screens. In this location, the precast deck
planks were anchored in position.

In reply to Mr Dinardo, as mentioned in the response to
Mr Allsop, the uplift pressure was resisted by anchor bolts.
Additional porosity was not desirable because pedestrians
would be walking on the deck. Where a single screen was used,
the width of the deck was small compared with the wave length,
so that trapped air pockets were unlikely and the wave crests
would be lower than at the double screen section.

In reply to Mr Lacey, wave attack on the central part of the
structure, where a double screen was used, was normal to the
screen. Although overtopping was not measured in the flume
tests, it was not observed to be significant. In addition,

because the width of the structure at the central section is approximately 10 m, this further reduces the potential for overtopping into the marina and any associated increase to the wave heights.

In reply to Mr Loveless, some of the flume tests examined the effect of a one metre gap beneath the wave screen, and it was found that the increase in wave transmission was unacceptable. Therefore, in the final design there is no gap beneath the screens. Scour was considered to be a risk in front of the central section and between the two screens. A scour prevention blanket, some 500 mm thick and constructed of rock, was therefore placed on a filter fabric in these areas.

PAPER 16

MR G. CRAWLEY, Fairclough Howard Marine, Chatham
The Authors refer to the low initial cost of floating breakwaters and have given a typical cost of a tyre breakwater as 500 $/m. Could they give a figure for the maintenance costs of floating tyre and concrete pontoon breakwaters?

MR P. LACEY, Ove Arup and Partners, London
Could the Authors please say what monitoring was carried out in the finished harbours? Do the floating breakwaters, especially in shallow water, excite siltation? What excursion distances did they allow in the design stage? For example, if moorings are too flexible the excursion distance becomes too great, if they are made too rigid then the anchor forces increase.

PROFESSOR NECE, MR NELSON and MR BISHOP (Paper 16)
Mr Crawley inquires about maintenance costs of floating breakwaters. Maintenance costs for tyre breakwaters are very site specific, depending to a large extent on the level of technology used in the breakwater construction. For earlier, smaller units, including some do-it-yourself projects, annual maintenance costs ranged between 2.2 and 18.2 % of the construction cost (reference 2 of Paper 16). The Burlington tyre breakwater, now in its eighth year, was constructed using state-of-the-art technology and is still in excellent condition. It is removed each winter and towed to a nearby sheltered harbour to protect it from winter ice damage. The cost of this operation is about Canadian $18 000. The Burlington breakwater was originally budgeted for an annual maintenance cost of 10% of construction cost. The concrete breakwater at Friday Harbor, which is a US federal government project, has average annual maintenance costs of US $18 000, which includes construction, engineering and design. This is approximately one percent of the construction cost.

Mr Lacey asks about monitoring carried out in the finished

harbours and whether floating breakwaters excite siltation in shallow water. Monitoring of projects has generally been restricted to visual observations and qualitative assessments. All US federal projects have an annual inspection and maintenance. The wave and anchor line force measurements at Burlington and off Seattle, for the tyre and concrete breakwaters respectively, were special prototype investigations which have not been repeated elsewhere. There is little evidence of floating breakwaters causing local sediment accumulation. Concrete breakwaters in the Pacific Northwest are generally located in water deep enough so that the pontoons have no impact on local currents. Tyre breakwaters are generally not used, because of environmental considerations, in water depths of less than 2.5 m.

Mr Lacey also raises pertinent questions concerning excursion distances and anchor line forces. Possible excursion distances are also site specific, depending on anchor line scope, initial tension, tide range and water depth. A typical anchor line scope is 5:1, providing a theoretical maximum excursion equal to nearly one water depth. Actual observations indicate that excursions for both tyre and concrete breakwaters are much smaller than this. In the prototype tests of the Friday Harbor pontoons the maximm observed excursion was less than 0.6 m, even when currents were running at 2 knots and anchor line tensions were reduced through removal of the clump weights. Design limits for the Friday Harbor installation were 1 m of lateral horizontal movement (to keep ramps between floats from being displaced) and 0.5 m differential vertical excursion between adjacent floats. As noted in the Paper the maximum lateral excursion noted to date is 0.15 m, but near-design wave conditions have not been experienced.

During the prototype tests of the Friday Harbor pontoons, conducted off Seattle, an anchor line stiffness test was performed, in which a marine tug pulled on the breakwater with varying loads, while surveying instruments measured horizontal displacements of the breakwater and load cells measured anchor line forces and tow loads. The anchor line stiffness varied from 4 400 N/m when the load in the anchor line was 8.9 kN above ambient, to 24 000 N/m at a load of 22 kN above ambient; the horizontal displacement was 1.6 m at the largest pull load (reference 15 of Paper 16). The rapidly increasing anchor line stiffness measured in this test explains why observed excursions of the breakwater were small even in fast currents and/or in high waves.

PAPER 17

MR P. LACEY, Ove Arup and Partners, London
How does the insurance industry view model test reports, and does it differentiate between physical and computational tests? I would have thought that the engineer behind the

tests and the quality of input data would have been paramount. Are they viewed as indicative, but still to be tempered with engineering judgement?

What I missed in the Paper was the clear lead as to what is not insurable. The reference to innovative blocks being suspect with respect to claims is worrying, as all of them have been supported by model tests usually extolling their virtues. Is it then that the engineers are trying to achieve too much economy at the risk of engineering judgement?

Has the Sea heard of 1 to 1.5 slopes?

MR J. E. CLIFFORD, Consulting Engineer, London
Does the Author feel that we are getting the benefits of competition amongst underwriters and insurers in obtaining quotations for premiums?

MR FARROW (Paper 17)
In reply to Mr Lacey, on any major wet works contract with its inevitable exposure to the elements, the results of model testing, either physical or computational, have a direct bearing on the terms of the insurance.

Equally, the results will reflect only the professional ability of the engineers behind the tests and the quality of the input data.

The real exposure to underwriters is during the stages of construction in relation to the altering weather patterns; and all too often the test results are limited in this area.

The question of an underwriter's differentiating between physical and computational testing is more difficult to answer, but bearing in mind the ever larger capabilities of computers, it seems likely that these will play a larger role in the future.

Apart from any clear moral hazard, the uninsurable elements will often depend on the inevitable or unquantifiable losses contained within the contract.

My reference to innovative blocks being suspect was purely the consequence of studying loss statistics where, it seems to me, the results are some way from those claimed. As a layman in regard to engineering, it does appear to me that there have to be sound physical reasons before a new system is selected.

It is a natural progression that, as we continue to develop our world, greater natural hazards have to be overcome. Surely we are constructing breakwaters and harbours in less suitable areas as a result of economic pressures and the lack of more suitable sites. Equally, it does seem that weather conditions are deteriorating and sea states continuing to reach higher levels year by year.

Mr Lacey's final comment of 1 to 1.5 slopes is obviously tongue-in-cheek, but perhaps it is worth considering that these do occur at the land/sea junction, but then only with

massive backup support; and still erosion attempts to reduce
the scale. I believe that engineers are always seeking to
achieve economy, since presumably weight must have a direct
bearing on cost and in the modern world it is the cost which
decides on the financial viability.

In reply to Mr Clifford, except in areas where one insurance
company has a monopoly, and it certainly does not apply in the
London Market, all major contracts are the subject of
competition. It is possibly because of the type of risk that
has been considered and the limited number of
underwriters/companies willing to insure that a different
impression has been given.

There will always be competition for good business, and if
breakwater/harbour construction losses were reduced then I am
sure there would be more underwriters and companies prepared
to offer terms.

P1. Application of computational model on dynamic stability

J. W. VAN DER MEER, Delft Hydraulics Laboratory, and M. J. KOSTER, Road and Hydraulic Engineering Division, Rijkswaterstaat, Delft

SYNOPSIS. The development of a computational model on dynamic stability was given by Van der Meer and Pilarczyk (Ref. 1). The model is able to predict profiles of slopes with an arbitrary shape under varying wave conditions. The boundary conditions which are required to use the model are described first. Then the model is used to design a berm breakwater in relatively deep water (18 m) and for severe wave conditions ($H_S = 7.6$ m). First the dimensions of the berm breakwater were optimized with respect to the amount of required armour stone. Then the influences of water depth, stone class and wave climate were investigated. Finally the stability after the first storms was analyzed in more detail.

DYNAMIC STABILITY

1. Most breakwaters and revetments are designed in such a way that only little damage is allowed for in the design criteria, damage being defined as the displacement of armour units. These criteria demand large and heavy rock or artificial concrete elements for armouring. A more economic solution can be a structure with smaller elements, profile development being allowed in order to reach a stable profile. Such structures are amongst others: berm or mass armoured breakwaters, S-shaped breakwaters, rock and gravel beaches.

2. The $H_S/\Delta D_{n50}$ parameter can be used to give the relationship between different structures, where: H_S = significant wave height, Δ = relative mass density and D_{n50} = nominal diameter of average stone mass. Small values of $H_S/\Delta D_{n50}$ give structures with large armour units. Large values imply gravel beaches and sand beaches. The following rough classification can be given:

- statically stable breakwaters: $H_S/\Delta D_{n50} = 1 - 4$
- berm breakwaters and S-shaped profiles: $H_S/\Delta D_{n50} = 3 - 6$
- dynamically stable rock slopes: $H_S/\Delta D_{n50} = 6 - 20$
- gravel beaches: $H_S/\Delta D_{n50} = 15 - 500$
- sand beaches: $H_S/\Delta D_{n50} > 500$

3. Van der Meer and Pilarczyk (Ref. 1) described a computational model for the profile development of rock slopes and

gravel beaches. More basic background is given recently by Van der Meer (ref. 3). The area given by $H_s/\Delta D_{n50}$ = 3-500 is covered by the computational model. The required model parameters will be described first and the model will then be applied to the design of a berm breakwater. This means that the application is focussed on the area given by $H_s/\Delta D_{n50}$ = 3-6.

GOVERNING VARIABLES

4. The size of armour units or gravel is referred to as the average mass of graded rubble or gravel, W_{50}, or the nominal diameter, D_{n50}, where:

$$D_{n50} = (W_{50}/\rho_a)^{1/3} \tag{1}$$

where: D_{n50} = nominal diameter (m)
   W_{50} = 50% value of the mass distribution curve (kg)
   ρ_a = mass density of stone (kg/m³)

The relative mass density of the stone in water can be expressed by:

$$\Delta = \rho_a/\rho - 1 \tag{2}$$

where: Δ = relative mass density (-)
   ρ = mass density of water (kg/m³)

The grading of the stone is expressed here by D_{85}/D_{15}, where the subscripts refer to the 85 and 15 percent value of the sieve curve, respectively. The shape of the stone can be angular, rounded or flat. The initial profile can vary from a uniform slope to a berm profile or a structure with a low crest.

5. The governing load variables are:
significant wave height H_s, average wave period T_z, storm duration given by the number of waves, N, the angle of wave attack, ϕ, and water level (tide).

6. Static stability is largely dependent on the initial slope, as is clearly expressed by the well known Hudson formula. Of course, for dynamically stable structures which are almost statically stable, the initial slope has also influence on the profile. It can be stated that, for $H_s/\Delta D_{n50}$ < 10 - 15 the initial slope has influence on the profile.

THE MODEL

7. The boundary conditions and possibilities of the model are:
- $H_s/\Delta D_{n50}$ = 3-500
- Arbitrary initial slope, given by x-y coordinates or by a previous computed profile.
- A crest above the water level.
- Computation of a sequence of storms (or tides) by using the previous computed profile as the initial profile.

8. For computation the following variables are required:
- The nominal diameter, D_{n50} (m)
- The grading of the stone class, D_{85}/D_{15}
- The relative mass density, Δ
- The significant wave height, H_s (m)
- The average wave period, T_z (s)
- The number of waves, N
- The water depth, d (m)
- The angle of wave attack, ϕ (degr.)
- The initial slope given by x-y-coordinates.

9. The computed profile for a berm breakwater with the boundary conditions used, is given in Fig. 1.

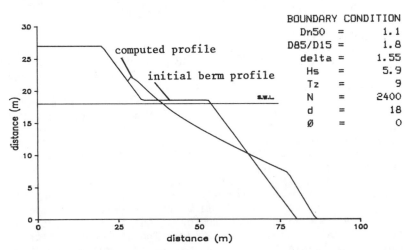

Fig. 1 Example of computed profile on berm breakwater

APPLICATION TO BERM BREAKWATER
10. The berm breakwater can be regarded as an unconventional design. Displacement of armour stones in the first stage of its life time is accepted. After this displacement (profile formation) the structure will be more or less statically stable. The cross-section of a berm breakwater consists of a lower slope 1:m, a horizontal berm with a length b just above high water, and an upper slope 1:n. The lower slope is often steep and close to the natural angle of repose of the armour. This means that m is roughly between 1 and 2.
11. During the design of a berm breakwater the following aspects should be considered:
- Optimum dimensions of the structure: m, n, b and crest height, obtained for chosen design conditions.
- Influence of water depth.
- Influence of stone class.
- Influence of wave climate.
- Stability after first storms.

12. The following boundary conditions are chosen for the design of a berm breakwater which were in fact taken from a project in the Spanish Mediterenian:
- water depth up to 18 m.
- no tidal range
- wave climate: 1/1 year : H_s = 4.7 m T_z = 8.2 s
                1/5 years : H_s = 5.9 m T_z = 9.0 s
                1/50 years: H_s = 7.6 m T_z = 10.0 s
- storm duration of 6 hours
- available stone classes: 0.5 - 9 t, D_{n50} = 1.01 m
                           1 - 9 t, D_{n50} = 1.11 m
                           3 - 9 t, D_{n50} = 1.19 m
- relative mass density: Δ = 1.55
- berm 0.5 m above the still water level (SWL).

Design of berm profile

13. The optimum values of m, n, b and crest height will be established for a water depth of 18 m and the 1/50 years wave conditions, H_s = 7.6 m and T_z = 10.0 s. This means that the structure is designed for $H_s/\Delta D_{n50}$ = 4.9. The optimum value of b can be established for various combinations of m and n and for the stone class with D_{n50} = 1.01 m. The criterion for the optimum value of b was the minimum value for which the upper point of the structure's crest was not a part of the erosion profile. In fact the upper point of the crest should lay on the initial slope, in order to prohibit erosion of the crest of the initial profile.

14. For each combination of m and n the minimum value of b was obtained iteratively, using the computational model. Figure 2 shows the minimum lengths of the berm as a function of the upper and lower slope angles. The berm length decreases almost linear with increasing lower slope, m. The same conclusion can be drawn for the upper slope, n.

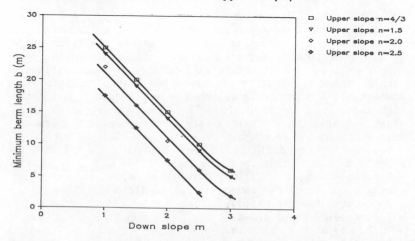

Figure 2 Minimum berm length as a function of down slope and upper slope

15. Figure 2 gives no information on optimum values for m and n. In fact Figure 2 gives various structures with more or less the same stability (no erosion on the upper slope). Therefore an other criterion is introduced. The amount of stones required for construction can be minimized, giving the cheapest structure. The height of the upper point of the crest amounted from 7 to 9 m. The crest height of the initial profile was chosen at a fixed level of 9.5 m above SWL which is about 1.25 times the significant wave height. The area of the cross-section from the crest to the toe of the structure gives a measure of the amount of stones required. This amount, B, was plotted versus the lower slope and for various upper slopes in Figure 3. The berm lengths are the same as in Figure 2.

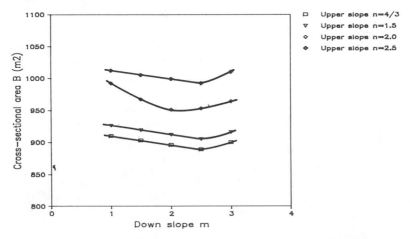

Figure 3   Cross-sectional area as a function of down slope and upper slope

16. From Figure 3 it can be concluded that a steeper upper slope reduces the amount of stones required. The difference is small for the steepest slopes of n = 4/3 and 1.5. The lower slope has less influence on the amount of stones required. Based on Figure 3, the lower and upper slopes were chosen for m = n = 1.5. The berm length becomes b = 19 m (Fig. 2). This choice of berm breakwater dimensions and the profile after design conditions is shown in Figure 4.

Influence of water depth
17. With the lower and upper slopes fixed at 1:1.5 the berm length becomes 19 m for a water depth of 18 m. The berm length can be reduced in shallower water using the same design conditions. Figure 5 shows this reduction of b for shallower water.

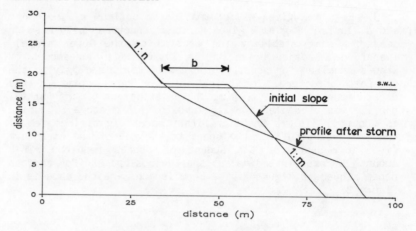

Figure 4   Berm breakwater with m = 1.5, n = 1.5, b = 19 m and
            profile after 1/50 years storm

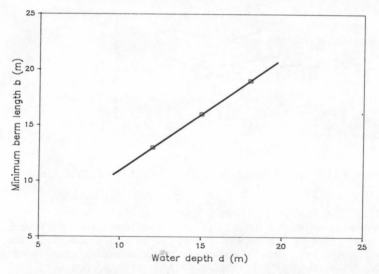

Figure 5   Influence of water depth on minimum berm length

Influence of stone class

18. Upto now the wide gradation of 0.5-9 t with D_{n50} =
1.01 m was used. Heavier stone will show less displacement of
material. Therefore the profiles under design conditions were
computed for the stone classes 1-9 t and 3-9 t according to
available output curves of a quar. Figure 6 shows all three
profiles. As the differences in D_{n50} are small, the differences in profile are small too. It can be concluded that the
wide (and cheaper) class of 0.5-9 t is satisfactory.

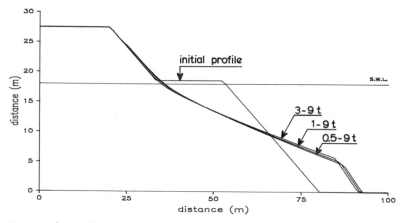

Figure 6   Influence of stone class

Influence of wave climate

19. The berm breakwater was designed for H_s = 7.6 m, T_z = 10.0 s and a storm duration of 6 hours, being the 1/50 years condition. The structure, however, will show profile changes for much lower wave heights. Therefore, the profile was calculated for a wave height of H_s = 4.7 m, being the 1/1 year wave height. It is furthermore interesting to know the influence of a higher wave height than the design wave height. Another profile was calculated for H_s = 9.2 m, being the 1/400 years wave height. The profiles are shown in Figure 7. The highest wave height shows some erosion of the upper slope, but under these circumstances some erosion can be allowed. The armour layer should be thick enough, however.

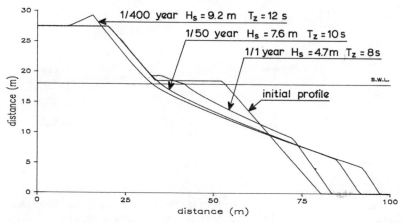

Figure 7   Influence of wave conditions

20. The wave period has influence on the profile and is often an uncertain factor in the design. The profiles for a lower wave period (T_z = 8 s) and a higher wave period (T_z =

339

12 s) were calculated and shown in Figure 8, together with the period of 10 s. The longer period gives again some erosion of the upper slope.

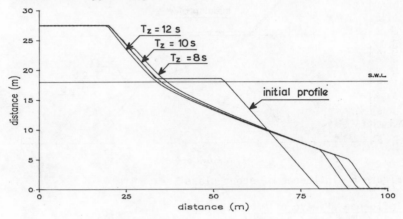

Figure 8   Influence of wave period

21. Finally the influence of the storm duration can be investigated. Profiles for a storm duration of 6, 12 and 24 hours are shown in Figure 9. The influence is very small. Erosion increases a little and the material is transported downwards.

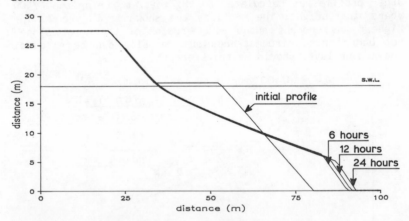

Figure 9   Influence of storm duration

Static stability
22.   The computational model is valid for structures under dynamic stability. This means that if the wave height is too small, or the diameter too large, a profile can not be calculated with this model. The structure must then be considered as statically stable. New stability formulae (Van der Meer (Ref.2)) can be applied in that case. The transition from dynamically to statically stable structures depends on the

$H_S/\Delta D_{n50}$ value, but also on the equivalent slope angle. A gentler slope is more stable than a steep slope.

23. First profile changes to the berm breakwater will occur for relatively low wave heights. More severe storms will change the profile again. But how stable is the berm breakwater after its first profile changes? Consider the profile after the 1/1 year storm with H_S = 4.7 m. The profile is shown in Figure 7. The equivalent slope of the profile around the still water level is about 1:3.6. Choosing this equivalent slope, a damage curve can be drawn for static stability (Van der Meer (Ref. 2)). This curve is shown in Figure 10. Start of damage occurs for S = 2-3 and about one layer of stones is removed for S = 8-12. From Figure 10 follows that after the 1/1 years storm the structure will act as statically stable upto H_S = 6 m. In that case about one layer of armour stones will be displaced. For higher wave heights the profile will change more and will become dynamically stable again.

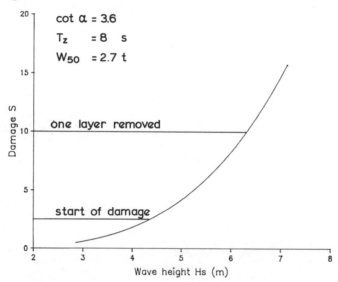

Figure 10 Damage curve for a statically stable homogeneous structure with an equivalent slope 1:3.6

OTHER APPLICATIONS

24. The computational model can be used to describe the behaviour of rock and gravel beaches, including the influence of storm surges and tides. It can also be used to design a two-layer S-shaped breakwater. The length and slope of the gentle part of the S-shaped breakwater can be estimated and also the steeper upper and lower slopes. Another application is the prediction of the behaviour of core and filter layers under construction when a storm hits the incomplete part of a breakwater.

CONCLUSIONS

25. The boundary conditions and possibilities of a computational model on dynamic stability have been summarized. This model was used to design a berm breakwater. The optimum dimensions of the breakwater were calculated with respect to the minimum amount of stone required for construction. The 1/50 years design conditions were used for this procedure. Influences of water depth, stone class, and wave climate were investigated. Finally the transition from dynamic to static stability was studied in more detail.

REFERENCES

1. VAN DER MEER, J.W. AND PILARCZYK, K.W. Dynamic stability of rock slopes and gravel beaches. Proc. 20th ICCE, Taipei, 1986. Also Delft Hydraulics Communication No. 379-1987.

2. VAN DER MEER, J.W. Stability of breakwater armour layers - Design formulae. Coastal Eng., 11, 1987, pp. 219-239.

3. VAN DER MEER, J.W. Rock slopes and gravel beaches under wave attack. PhD. thesis, Delft University of Technology, 1988. Also Delft Hydraulics Publication No. 396.

P2. Developments in the analysis of armour layer profile data

Dr J. P. LATHAM, Dr A. B. POOLE and M. B. MANNION, Coastal Engineering Research Group, Queen Mary College, London

SYNOPSIS. Breakwater armour profile data from models and prototypes are classified according to the type and quality of the height information. High resolution profiles from models and 'break-of-slope' profiles from prototypes are illustrated. Results of potential hydrodynamic significance are analysed and compared. It is suggested that by referring to a high resolution profile or a precisely defined 'break-of-slope' profile, height and thus thickness corrections for less precise survey methods can be made. These corrections are related to the roughness of the surface which is shown to depend on the type of armour.

INTRODUCTION

1.   The construction and monitoring of both prototype and model structures under wave attack require reproducible measurements of armour layer profiles.

2.   On real structures, the quality control during construction, and the monitoring of storm damage and subsidence may have to be performed under unfavourable survey conditions. Divers, suspended depth probes and sonar or echosounder (ref.1) systems have all been used.

3.   In the laboratory flume, profile data is used to compute erosion damage areas (refs 2,3) and layer thicknesses and to quantify the readjustment of dynamic profiles (ref. 4).

4.   A more detailed consideration of what might be represented by profile data is given below. There is often considerable disagreement as to precisely how the 'true' profile of the armour layer relates to the surveyed profile and to the design profile. Armour units or blocks introduce both random and periodic components to the profile.

5.   It has also been suggested that the profile data need not be restricted simply to determining the extent or position of the armour layer, but that the detailed character of the profile roughness resulting from different armour shapes, construction methods and storm damage merit further investigation (ref. 5).

6.   This study has provided a large number of new results at high resolution but the problems of applying the same methods to full scale structures has proved difficult to solve.

DESIGN OF BREAKWATERS

This paper seeks to present some of the more interesting
visual and numerical reference material, linking it with
some suggestions as to possible applications that might be
developed from it.

7.   Since many of the techniques used are new, it is
necessary to define the terms used and consider the implications of methods of data acquisition before proceeding to
consider the data and its analysis.

PROFILING NOMENCLATURE AND DEFINITIONS

8.   The reference frame used in this work, shown in Fig. 1,
is for linear sections of a structure. The height z_{ij} at the
point (x_i, y_j) is shown at the intersection of two
profiles - an AC or downslope profile and a BC or alongslope profile. The xy grid refers to the horizontal plane
and relates to the most convenient origin.

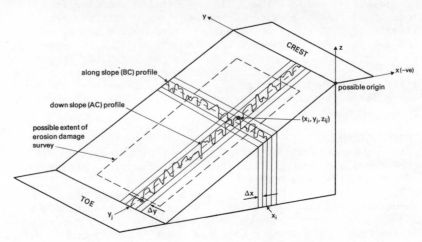

Fig. 1. Reference system for breakwater trunk
profile surveys

9.   This reference frame allows standardised definitions
of layer thicknesses, erosion damage areas or volumes and
roughness descriptors (ref. 5). However, any of these
measured parameters will also be dependent on (i) the sampling
intervals, i.e. Δx and Δy, (ii) the width and slope of the
probe foot, (iii) the measurement accuracy. The parameters
with dimensions of length are most conveniently expressed
in terms of the nominal diameter D_{n50} (ref. 3) of the armour
blocks or units under survey.

Probe Width

10.   The basic effects of a finite probe width are those
summarized in Fig. 2.

344

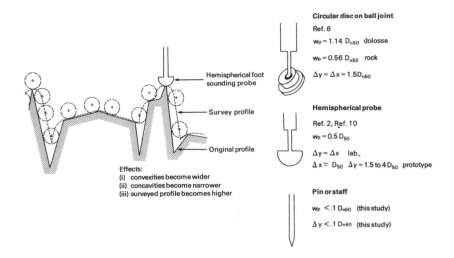

Circular disc on ball joint
Ref. 6
$w_p = 1.14 D_{n50}$ dolosse
$w_p = 0.56 D_{n50}$ rock
$\Delta y = \Delta x = 1.5 D_{n50}$

Hemispherical foot
sounding probe

Survey profile

Original profile

Effects:
(i) convexities become wider
(ii) concavities become narrower
(iii) surveyed profile becomes higher

Hemispherical probe
Ref. 2, Ref. 10
$w_p = 0.5 D_{50}$
$\Delta y = \Delta x$ lab.,
$\Delta x \simeq D_{50}$ $\Delta y = 1.5$ to $4 D_{50}$ prototype

Pin or staff
$w_p < .1 D_{n50}$ (this study)
$\Delta y < .1 D_{n50}$ (this study)

Fig. 2. Distortion of profile by probe foot

Fig. 3. Examples of probe foot and sampling interval combinations

11. The most commonly used laboratory sounding probe is the hemispherical foot, although Zwamborn (ref. 6) found a ball-and-socket mounted disc to be preferable (see Fig. 3). The aim of using these devices is to provide a systematic means of obtaining a general picture of the armour layer surface. The detailed effect of probe width, w_p, may not be widely appreciated. Even the experimental sounding method of Hudson (ref. 7) upon which the armour packing density and layer thickness formulae of the Shore Protection Manual were based, is not generally known.

12. Depending on the armour type, a finer probe, $w_p < 0.1 D_{n50}$, could lead to average profile height measurements of up to $0.5 D_{n50}$ lower than by using standard profiling procedures for models and for prototypes (see Fig. 4).

13. The three-dimensional shape of the probe foot inhibits valley (i.e. gap) penetration even more than in two-dimensional representation.

Sampling interval

14. When using a disc probe of $w_p \simeq D_{n50}$, a doubling of the sampling intervals Δx and Δy from 1.5 to $3 D_{n50}$ had negligible effect on the calculated thicknesses of dolosse armour layers (ref. 6). For a constant w_p, this can be theoretically predicted. If the average height is all that is required, sampling interval is not critical provided that the extreme effects of too infrequent sampling are avoided.

15. For more detailed insight into the character of the

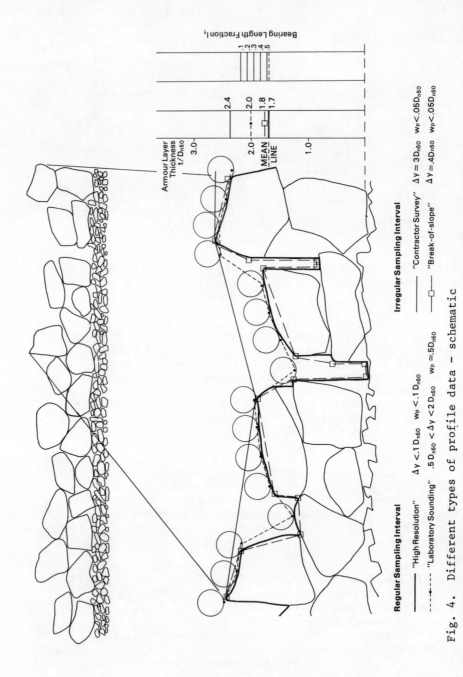

Fig. 4. Different types of profile data - schematic

Regular Sampling Interval

"High Resolution" $\Delta y < .1 D_{n50}$ $w_p < .1 D_{n50}$

"Laboratory Sounding" $.5 D_{n50} < \Delta y < 2 D_{n50}$ $w_p \approx 5 D_{n50}$

Irregular Sampling Interval

"Contractor Survey" $\Delta y \approx 3 D_{n50}$ $w_p < .05 D_{n50}$

"Break-of-slope" $\Delta y \approx 4 D_{n50}$ $w_p < .05 D_{n50}$

armour surface based on high resolution profile data, the theoretical work of Longuet-Higgins (ref. 8) on ocean surface statistics is most relevant. Theoretical and experimental developments by Sayles and Thomas (ref. 9), who used rough metal surfaces, have demonstrated the strong dependence on sampling interval of roughness parameters such as peak and zero crossing density as well as average slopes, curvatures and peak or valley heights. Unfortunately their relationships are not easily applied to give sampling interval corrections to armour profile statistics.

16. However, sampling with a fine probe at a regular interval of less than 0.3 D_{n50} would comply with the suggested restriction (ref. 8) of one third of the least wavelength of interest. Taken over five sample lengths of at least 25 D_{n50} appears to give sufficient accuracy for a meaningful comparison of roughness parameters of the static laboratory models.

Types of profile survey data

17. The types of survey relevant to both models and prototypes have been schematically represented in Fig. 4, where the influence of the survey method on the average height is illustrated.

18. For a regular sampling interval, two types of survey can be defined:

(1) 'High resolution' profiles - currently impractical for prototypes, give effectively all the peak and valley information, and can be used for the calculation of a range of roughness parameters and void coefficients.

(2) 'Laboratory sounding' profiles - will introduce the effects shown in Fig. 2, generally cutting out valley rather than peak details. The wider probe leads to higher average thickness determinations than in (1), see Fig. 4. This survey type was also applied to the proto-type rip-rap study (ref. 10) so that methods of erosion damage measurement would be more compatible with the laboratory study of ref. 2.

19. With prototypes, the remote techniques of sonar and echo-sounding make below water level surveys more practical though perhaps less accurate than above water surveys. Generally, above water surveys taken for the contractor are sampled at irregular intervals down the slope. Two further types of data acquisition are therefore considered.

(3) 'Contractor survey' profiles - are taken to be those where only peak heights of the most conveniently located armour blocks falling within a sighted profile line are measured. The resulting profile is a very loosely defined enveloping profile which may even be below the design profile.

(4) 'Break-of-slope' profiles - only applicable above water, introduced here as a possible practical alternative to (3). It may be defined to yield a standard form of profile data that contains sufficient detail to give

verifiable values for average profile height, roughness and a two-dimensional void coefficient (Table 1 and Fig. 5). The definition of the 'Break-of-slope Profile suggested here is as follows:

TABLE 1 : DEFINITION OF ROUGHNESS PARAMETERS

SYMBOL	NAME	DEFINITION	COMMENT
σ	Root-mean-square Roughness	$\sigma = \left(\frac{1}{L}\int^L z^2\,dx\right)^{1/2}$	Equals the standard deviation of the profile height distribution $p(z)$
R_a	Centre-line-average Roughness	$R_a = \frac{1}{L}\int^L |z|\,dx$	
L	Sampling Length		
ML	Centre-line-average Mean Line	The straight line dividing the profile with equal area of solid above to void below	Prior to roughness parameter evaluation the analysis fits a mean line and refers z to a zero at this line
Y_{pm}	Mean Peak Height	Base on 3 consecutive points	Sensitive to sampling interval
Y_{vm}	Mean Valley Height	Base on 3 consecutive points	Sensitive to sampling interval
C_{pm}	Mean Peak Curvature	Base on 3 consecutive points	Sensitive to sampling interval
D_p	Peak Density	Peaks(or valleys)per unit length	Sensitive to sampling interval
X_o	Zero Crossing Density	Zero crossings per unit length	Sensitive to sampling interval
$\overline{\theta}$	Average Absolute Slope Angle	Slope defined between two consecutive points	
λ	Average Wavelength	$\lambda = 2\pi R_a / \tan\overline{\theta}$	Referred to in Ref.8
l_f	Bearing Length Fraction at distance from ML	Fraction of solid traversed by line drawn parrallel and a fixed distance from the ML	See Figure
l_{fm}	Bearing Length Fraction at ML		
v	Void Coefficient	$v = \dfrac{\frac{1}{2}R_a}{l_{fm}}$	Proposed here and based on fluid volume below a line divided by the bearing area at that line (in this case the ML,see figure)

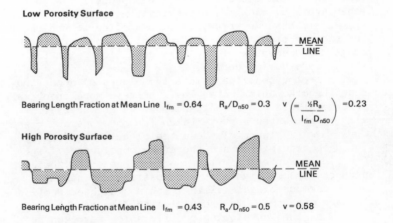

Low Porosity Surface

MEAN LINE

Bearing Length Fraction at Mean Line $l_{fm} = 0.64$ $R_a/D_{n50} = 0.3$ $v\left(= \dfrac{\frac{1}{2}R_a}{l_{fm}\,D_{n50}}\right) = 0.23$

High Porosity Surface

MEAN LINE

Bearing Length Fraction at Mean Line $l_{fm} = 0.43$ $R_a/D_{n50} = 0.5$ $v = 0.58$

Fig. 5. Illustration of the two-dimensional void coefficient.

'Within the vertical plane of the line being profiled,
the vertical height of each consecutive block traversed
should be represented by between two and four points.
These points will include the extreme ends of each block
and positions of break-of-slope such as block edges
which best define the block outline. (Note: for multi-
legged armour, up to six points may be required per block.)
20. Thus valley gaps, perhaps going down to the sublayer,
will be sampled along the profile.

Roughness parameters from profile analysis
21. The concept of quantitative measures of roughness
and some of the parameters used were introduced in an
earlier paper (ref. 5) and are given in Table 1. The
two-dimensional void coefficient is introduced in Fig. 5.
22. For high-resolution or linearly interpolated break-
of-slope profiles, the roughness parameters obtained from
analysis of the profile height data will be of greater
significance than for other profiling methods. All profile
data are referred to a horizontal straight mean line prior
to analysis.
23. Downslope (AC) profiles when referred to the sloping
mean line, give roughness parameter results that are quite
similar but not exactly equivalent to the parameters obtained
from an equivalent surface laid horizontally and then surveyed.
24. Alongslope (BC) profiles will usually give a mean line
of negligible slope. However, the parameter results will be
affected by the armour layer slope. As with layer thickness,
it is necessary to multiply by the cosine of the slope
angle to obtain the roughness parameters normal to the layer.
This further correction is not necessary when comparing
identically sloping armour layers.
25. Filtering at D_{n50} using a high pass filter (ref. 5)
has the effect of removing the longer wavelength roughness
components from the profile so that the analysis results, for
example, do not reflect the influence of the S-shape often
developed in an AC profile but only represent roughness
features resulting directly from the block shapes themselves.

ROUGHNESS PARAMETER RESULTS
Static models
26. A 500 pin profiler was built to sample 500 mm long
model profile sections with w_p = 0.8 mm and Δy = 1.0 mm. The
vertical pins are released to fall onto the model's
surface and are then locked in position. The pin heights are
measured by a computer/video camera system which is adjusted
so that it just fails to resolve the gaps separating the pins.
27. The models were built on a 1:2 slope with a sublayer
of D_{n50} = 10 mm. For most models, the sublayer, first and
second layer were surveyed using 5 profile lines separated
by Δx = 30 mm. Except where stated, placement was loose,
random and designed to reproduce construction conditions
above water level.

28. Difficulty was experienced building the glass spheres model to a representative double layer because of instability on a 1.2 slope and the low D_{n50} ratio of armour to sublayer.

29. Results for the roughness parameters are given in Table 2 in normalized form. The direct visual comparison of

TABLE 2 : STATIC MODEL DATA - ROUGHNESS PARAMETERS

SYMBOL	GLASS SPHERES	CUBES		DOLOSSE		FRESH Tight Pack		ROUND Tight Pack		FRESH Loose	ROUND Loose
	n = 2	n = 1	n = 2	n = 1	n = 2	n = 1	n = 2	n = 1	n = 2	n = 2	n = 2
σ	0.583	0.477	0.357	0.539	0.621	0.370	0.385	0.446	0.374	0.362	0.325
R_a	0.471	0.367	0.252	0.417	0.492	0.286	0.287	0.339	0.287	0.265	0.248
D_p	1.61	2.18	2.11	2.25	1.97	2.42	2.52	2.22	2.48	2.47	2.58
Y_{pm}	0.156	0.137	0.115	0.161	0.225	0.135	0.133	0.147	0.095	0.102	0.085
Y_{vm}	-0.296	-0.185	-0.139	-0.241	-0.273	-0.155	-0.140	-0.182	-0.171	-0.116	-0.100
C_{pm}	21.0	39.6	27.6	41.3	39.6	48.6	44.8	33.2	28.0	37.5	29.1
$\bar{\theta}$	54.7	54.8	45.6	59.8	62.6	54.5	54.7	56.3	52.4	46.7	42.6
L_{fm}	0.581	0.660	0.610	0.525	0.581	0.548	0.579	0.575	0.545	0.581	0.586
v	0.405	0.278	0.208	0.397	0.423	0.261	0.248	0.295	0.263	0.228	0.211
λ	2.10	1.62	1.57	1.52	1.60	1.28	1.28	1.42	1.39	1.57	1.69

Note : Normalised raw data

the raw data profiles in Fig. 6 is instructive when seen together with Table 2. Some features of interest are:

(i)   the roughness of the spherical model is exaggerated by the sublayer roughness;

(ii)   the roughness of double layers tends to be less than for single layers for rocks and cubes, but not for dolosse;

(iii)   dolosse, as expected, gives a much higher roughness and void coefficient than cubes and rocks;

(iv)   mean valley heights are always greater than mean peak heights;

(v)   the lower curvature for rounded compared with fresh rock is easily recognized by the mean peak curvature, but note the effect of the flat faces on the cubes.

29. The variability of the average roughness parameters is indicated in Table 3 from two models where 20 and 10 separate profiles were examined and means and standard deviations obtained.

TABLE 3 : ROUGHNESS PARAMETERS FOR FRESH AND ROUND ROCK

SYMBOL	FRESH (20 PROFILES) Loose Pack		ROUND (10 PROFILES) Loose Pack	
	MEAN	STD. DEV'N.	MEAN	STD. DEV'N.
σ	0.362	0.037	0.325	0.049
R_a	0.265	0.024	0.248	0.029
D_p	2.47	0.28	2.58	0.63
Y_{pm}	0.102	0.040	0.085	0.042
Y_{vm}	-0.116	-0.034	-0.100	-0.041
C_{pm}	37.5	9.8	29.1	7.7
L_{fm}	0.581	0.046	0.586	0.050

Note : Normalised raw data and standard deviations

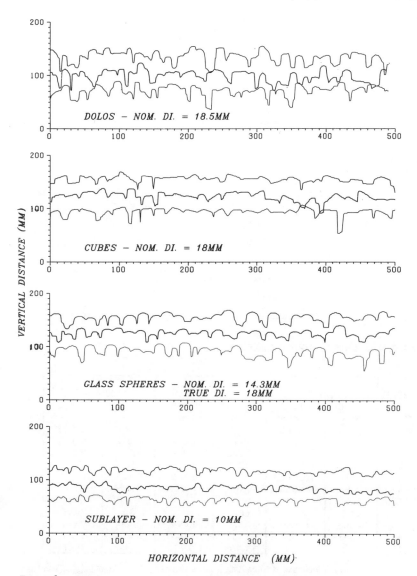

Fig. 6.   High resolution BC profiles from static models
(continued on next page)

Flume tested models

30.   Three high resolution AC profiles, two of cubes and
one of quarrystone, from Delft Hydraulics Laboratories were
analysed. Results are presented in Table 4 and profiles
shown in Fig. 7. To examine roughness resulting from block
effects rather than erosion damage, a High Pass filter of
44.3 mm was applied to the cube profiles. The 20% reduction
in roughness shown in the analysis suggests that after 3000
waves, the profile is smoother and may be nearer to

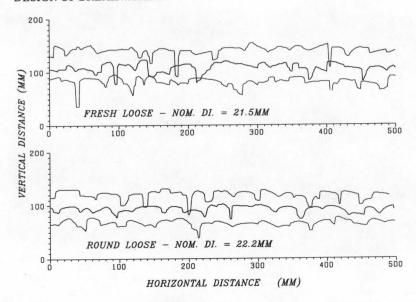

Fig. 6(ctd). High resolution BC profiles from static models

TABLE 4 : DELFT FLUME PROFILE DATA - ROUGHNESS PARAMETERS

SYMBOL	CUBES				ROCK	
	0 WAVES		3000 WAVES		3000 WAVES	
	PROFILE VDMP2		PROFILE VDMP1		PROFILE VDMP3	
	Raw	Filtered	Raw	Filtered	Raw	Filtered
Slope	-0.548	-0.548	-0.534	-0.534	-0.291	-0.291
σ	0.476	0.350	0.426	0.289	0.822	0.322
R_a	0.358	0.260	0.333	0.207	0.686	0.247
D_p	4.26	4.43	4.26	4.65	1.9	1.86
Y_{pm}	0.160	0.148	0.116	0.130	0.123	0.184
Y_{vm}	-0.052	-0.058	-0.089	-0.056	-0.324	-0.291
C_{pm}	96.6	107.0	107.0	99.5	37.1	39.5
$\bar{\theta}$	61.3	61.5	60.5	60.0	60	60.6
L_{fm}	0.603	0.573	0.537	0.494	0.494	0.552
v	0.297	0.227	0.302	0.193	0.694	0.223
λ	1.23	0.88	1.19	0.74	2.49	0.87

Note : Van der Meer`s normalised flume data ; Cubes -wp=$0.045D_{n50}$, Δx=$0.045D_{n50}$,D_{n50}=44.3mm
Rock -wp=$0.023D_{n50}$, Δx=$0.093D_{n50}$,D_{n50} =214mm

hydrodynamic equilibrium than when constructed.
31. Fig. 7(b) shows a profile taken from the Delft
Delta Flume after 3000 waves have caused high erosion
damage to the quarrystone armour. The raw and filtered
heights, referred to the mean line, are given in Fig. 7(c)
where the elimination of the S-shaped profile by filtering
is demonstrated. Comparing columns 4 and 6 of Table 4
suggests that rock gives a considerably greater roughness
than cubes.

352

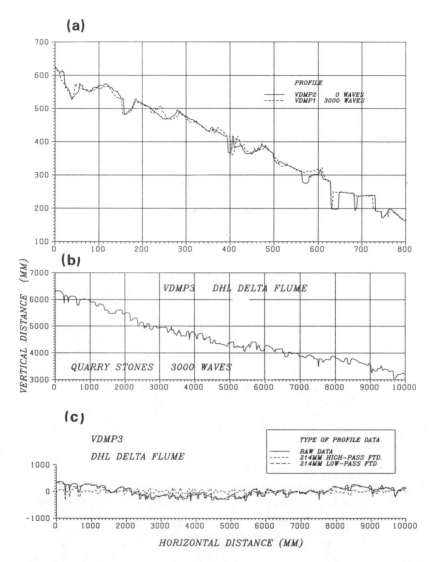

Fig. 7. High resolution AC profiles from Delft Hydraulics
         flumes.
   (a)   Cubes before and after 3000 waves
   (b)   Quarry stones after 3000 waves
   (c)   Effect of filtering at D_{n50} on middle profile

32. High resolution profiles of the Wallingford tests
(ref. 13) on shape effects of quarried rock were not
available and therefore data showing any influence of wave
attack on roughness parameters remains limited.

353

DESIGN OF BREAKWATERS

Prototype structures

33. Alongslope (BC) profiles were surveyed at Poole Marina, Dorset and at Dawlish, Devon, and a downslope (AC) profile was surveyed at Herne Bay, Kent. With D_{n50} less than one metre in each case, the survey could be performed using a Zeiss Electronic Tacheometer and manually placed two-metre staff with reflector.

34. Break-of-slope sampling was used as regular interval sampling was considered impractical. With low wind conditions, an 80 point survey line took two hours to profile.

35. Profiles are shown in Fig. 8. The Herne Bay AC profile (which appears to have an S-shape component) is shown in Fig. 8(c). Table 5 indicates that Poole shows a 12% greater roughness (R_a) than Dawlish Warren Upper, which is a further 14% greater than the Dawlish Warren Lower profile.

LAYER THICKNESS AND PACKING DENSITY

36. The widely accepted formulae for calculating layer thickness t and the packing density or number of blocks per unit area of slope N, (ref. 11) can be written (using D_{n50}^3 = unit volume, V) as

$$t = nk_\Delta \, D_{n50} \qquad (1)$$

$$N = nk_\Delta \, (1 - P_f/100) \, \frac{1}{D_{n50}^2} \qquad (2)$$

where k_Δ is the layer coefficient, P_f the fictitious porosity (i.e. porosity as laid) and n is the number of armour layers.

37. Suggested P_f and k_Δ values (ref. 11 and Table 6) give poor agreement between predicted and surveyed armour thickness if:

(i) the survey technique is not based on probe widths of
$$w_P \simeq \tfrac{1}{2} \, D_{n50} \quad \text{(assumed to apply in ref. 11)}.$$

(ii) an extreme packing density, resulting from either a very tight placement method or a very loose packing arrangement, was used in construction.

38. To help resolve (i), it may be useful to consider a new concept which might be termed 'bearing length fraction' in greater detail (Fig. 9). Assume that different profiling methods give identical sublayer heights and that the source of discrepancy is in the armour layer survey.

39. It is reasonable to suggest that vertical heights will be practically the same for 'break-of-slope' as for 'high resolution' profiling surveys, or only up to 0.1 D_{n50} higher (see Fig. 4). The heights above mean line for the 'contractor survey' and 'laboratory sounding' profiles are shown schematically in Fig. 4 as about 0.7 D_{n50} and 0.3 D_{n50} respectively, for the profile drawn.

40. The meaning of bearing length fraction may be illustrated by considering bearing lengths of l_f = 0.1,0.2,0.3 etc. For example, at l_f = 0.1, a slice parallel to the mean

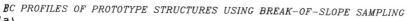

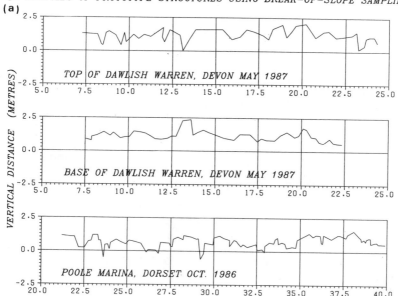

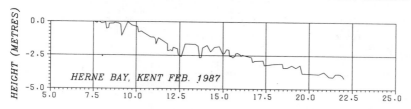

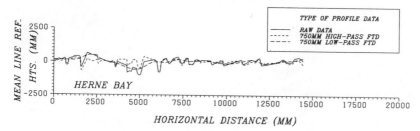

Fig. 8. Break-of-slope prototype profiles

line would cut through 10% solid armour, $l_f = 0.2$ corresponds
to 20% and so on.

DESIGN OF BREAKWATERS

TABLE 5 : PROTOTYPE PROFILE DATA - ROUGHNESS PARAMETERS

	BC PROFILES						AC PROFILES	

SYMBOL	DAWLISH WARREN				POOLE MARINA		HERNE BAY	
	UPPER		LOWER					
	Raw	Filtered	Raw	Filtered	Raw	Filtered	Raw	Filtered
Slope	0.009	0.009	-0.020	-0.020	0.022	0.022	-0.289	-0.289
σ	0.434	0.287	0.391	0.196	0.459	0.249	0.375	0.231
R_a	0.336	0.225	0.290	0.184	0.383	0.184	0.269	0.167
D_p	1.05	0.90	0.69	1.26	1.20	1.50	1.19	1.45
Y_{pm}	0.355	0.310	0.252	0.153	0.101	0.178	0.175	0.157
Y_{vm}	-0.221	-0.378	-0.295	-0.109	-0.414	-0.254	-0.323	-0.237
C_{pm}	73.9	89.7	39.0	35.2	38.4	58.3	46.9	52.8
$\bar{\theta}$	51.4	50.7	37.1	33.9	52.2	52.6	49.9	49.1
L_{fm}	0.518	0.548	0.453	0.481	0.522	0.568	0.564	0.574
v	0.324	0.262	0.320	0.204	0.367	0.219	0.332	0.201
λ	2.18	1.48	3.25	1.83	2.24	1.20	1.98	1.26

Note : Normalised data, D_{n50} (Dawlish)=850mm, D_{n50} (Poole)=800mm, D_{n50} (Herne Bay)=750mm

TABLE 6 : STATIC MODEL DATA - GEOMETRIC PARAMETERS

NAME	SYMBOL	UNIT	SPHERES	CUBES	DOLOSSE	FRESH		ROUNDED	
						Tight	Loose	Tight	Loose
GRADING	W_{85}/W_{15}					1.26	1.26	1.22	1.22
FOURIER SHAPE FACTOR	P_c					1.87	1.87	1.62	1.62
FOURIER ASPERITY ROUGHNESS	P_R					0.0128	0.0128	0.0039	0.0039
NOMINAL DIAMETER	D_{n50}	mm	14.30	18.00	18.50	21.50	21.50	22.20	22.20
LAYER THICKNESS n=1	t_1	mm		16.06	14.61	22.95		25.83	
LAYER THICKNESS n=2	t_2	mm	27.80	36.62	29.68	47.53	34.70	52.45	37.43
NORMALISED THICKNESS n=2	t_2/D_{n50}		1.94	2.03	1.60	2.21	1.61	2.36	1.69
UNIT VOLUME	V	cm3	2.901	5.832	6.318	9.962	9.962	10.963	10.963
BULK DENSITY	ρ		0.586	0.601	0.430	0.536	0.536	0.621	0.621
AS LAID BULK DENSITY	ρ_L		0.623	0.643	0.507	0.599	0.600	0.661	0.665
PACKING DENSITY n=2	N	cm	0.60	0.40	0.24	0.29	0.21	0.32	0.23
FICTITIOUS POROSITY	P_f	%	37.7	35.7	49.3	40.1	40.0	34.0	33.5
ROUGHNESS n=1	σ_1	mm		7.68	8.90	7.13		8.86	
ROUGHNESS n=2	σ_2	mm	7.45	5.75	10.28	7.41	6.96	7.43	6.46
NORMALISED ROUGHNESS n=2	σ_2/D_{n50}		0.521	0.319	0.556	0.344	0.324	0.335	0.291
LAYER COEFFICIENT	K_Δ		0.97	1.02	0.80	1.10	0.81	1.18	0.84
FICTITIOUS POROSITY(Ref.11)	P_f			47	56		37		38
LAYER COEFFICIENT (Ref.11)	K_Δ			1.10	0.94		1.00		1.02

Notes : See Ref.6 for dolosse volume, Ref.12 for fourier shape parameters
All thicknesses and roughnesses are normal to slope

TABLE 7 : CHANGE OF BEARING LENGTH FRACTION

	Static model tests							Prototype	

BEARING LENGTH FRACTION L_f	Normalised height above mean line (h/D_{n50}) Unfiltered								
	SPHERES	CUBES	DOLOSSE	FRESH Tight	FRESH Loose	ROUND Tight	ROUND Loose	POOLE	DAWLISH WARREN Upper
0.5	0.13	0.06	0.11	0.06	0.05	0.08	0.08	0.05	0.02
0.4	0.24	0.12	0.25	0.14	0.12	0.15	0.14	0.17	0.12
0.3	0.35	0.18	0.38	0.22	0.17	0.21	0.18	0.30	0.22
0.2	0.48	0.26	0.50	0.29	0.32	0.28	0.25	0.43	0.41
0.1	0.66	0.34	0.68	0.39	0.44	0.36	0.34	0.57	0.46
ROUGHNESS σ/D_{n50}	0.47	0.25	0.49	0.29	0.29	0.27	0.25	0.38	0.34
ROUGHNESS R_a/D_{n50}	0.58	0.36	0.62	0.39	0.37	0.36	0.33	0.46	0.43

Note : For prototypes, slope= 1:1.5, for models, slope= 1:2

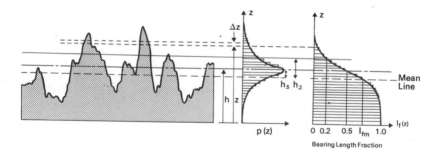

Fig. 9.   Bearing length fraction of a profile

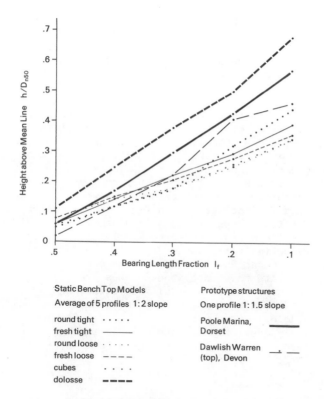

Fig. 10.   Relative height above mean line as a
function of bearing length fraction

41.   Table 7 and Fig. 10 give the relationship between
l_f and the height above mean line. For the static model
tests, this is described by the following equation:

357

$$\frac{h}{D_{n50}} = 2.5 \frac{R_a}{D_{n50}} (0.6 - l_f) \tag{3}$$

where $l_{fm} = 0.6$ and, for linearity, $l_f > 0.1$ have been assumed. An equation such as (3) could provide a standardized correction procedure for relating heights determined from various survey techniques with a 'true' (i.e. high resolution) profile height.

42. Comparing observed and previously suggested k_Δ values in Table 6 with bearing length fraction values in Table 7, for cubes, as an example, indicates that k_Δ recommended values are probably based on survey heights equivalent to $l_f \simeq 0.4$ of the high resolution profile.

43. If new sets of recommended k_Δ and P_f values are to be worked out, it is suggested that they should refer to either the high resolution height (giving compatibility with Table 6), or to the 20% bearing length fraction height ($l_f = 0.2$) for which the distance h above the high resolution height equals the roughness R_a. Such a value of l_f could correspond, for instance, to the mean height given by sonar scanning.

44. Considering the influence of packing and block shape on the geometric parameters in Table 6, it is interesting to note the very large differences in the layer thickness (and therefore k_Δ) due to loose and tight placement of the fresh and round rock. The tight placement, placing the longest axes downslope and the intermediate axes normal to the slope, produced the same fictitious porosity P_f as did the loose placement, which simulate the 'orange peel' grab method often used in practice and in flume model construction.

45. The fictitious porosity and bulk density values differ considerably for fresh and rounded rock. This is due to the extreme difference in stone roughness, expressed by the fourier asperity roughness P_R. The fourier shape factors of $P_C \simeq 1.75$ indicate that both stone shapes are less equant than the stone shapes typically specified and so there is a greater range of possible packing densities in these static models.

46. The standard deviation in the average thickness of the static models was found to be insignificant, typically less than $0.1 D_{n50}$.

DISCUSSION

47. Table 8 shows the average roughness parameters for static models with roughly similar shaped stones but with three different nominal sizes. The results are subject to sampling interval effects as the same 500 pin profiler was used in each case. The sampling intervals required for parameters such as mean peak curvature (which could indicate rates of wear) are impractical outside the laboratory. But such parameters as σ, R_a, l_{fm} and v can be more easily related for different survey methods on models and prototypes.

TABLE 8 : STATIC MODEL DATA - ROUGHNESS PARAMETERS FOR FRESH ROCK

SYMBOL	D_{n50}= 10mm n > 2	D_{n50}= 21.5mm n = 2	D_{n50}= 49mm n = 1
σ	0.478	0.362	0.298
R_a	0.391	0.265	0.22
D_p	1.24	2.47	3.89
Y_{pm}	0.193	0.102	0.03
Y_{vm}	-0.181	-0.116	-0.119
C_{pm}	18.1	37.5	101.7
$\overline{\theta}$	42.4	46.7	48.7
L_{fm}	0.526	0.581	0.558
v	0.372	0.228	0.201
λ	2.69	1.57	1.21

Note : Normalised raw data

48.   Table 9 compares high resolution model data and break-of-slope prototype data from various rock sources. BC and AC profiles have had a high pass filter of D_{n50} applied to remove the S-shape and corrected for slope prior to analysis. The roughness R_a is then found to be comparable over two orders of magnitude in size.

49. R_a is considered the most important profile parameter because of its simple physical interpretation in terms of areas above and below the mean line and its sampling interval independence.

50.   Observations from this study support the suggestion (ref. 14) that lower sublayer roughness resulting from

TABLE 9 : ROUGHNESS FOR VARIOUS FRESH ROCK SOURCES

ROCK	D_{n50} (mm)	CENTRE LINE AVERAGE ROUGHNESS R_a
SUBLAYER (BC)	10	0.169
FRESH (BC)	21.5	0.165
WALLINGFORD FRESH (BC)	49	0.148
VDMP3 (AC)	214	0.247
HERNE BAY (AC)	750	0.167
POOLE MARINA (BC)	800	0.153

Note : Normalised filtered data, R_a normal to slope

$W_{armour}/W_{sublayer}$ of at least 10 will help give 'smoother' armour layer surfaces. Also noted was that dolosse, which achieves its high stability through good interlocking and high void ratio, also shows high roughness. So high roughness values alone cannot be used to indicate the lower stability due to exposure of individual blocks to excessive hydraulic forces. Instead the profile height at a low bearing length fraction, e.g. 0.05 or 0.1, ought also to be considered.

51. Data from model tetrapod profiles of Port d'Arzew-El-Djedid breakwater were discussed in ref. 15 where, for the equivalent prototype, $D_{n50}/W_p = 9$, $\Delta x = 1$ metre, and $D_{n50} = 2.7$ metres. This showed the average surface to be about two metres below the design surface (leaving a more exposed crownwall). The design surface thickness was probably calculated using eq. (1) with $k_\Delta = 1.04$ and $n = 2$. Either the k_Δ value should be nearer to 0.7 or the designer should use the appropriate thickness correction relating 'true' average surface to design surface.

52. This example further illustrates the need to investigate k_Δ, P_f, unit shape and placement with reference to survey methods as suggested in this paper. Also prototype and model packing need to be compared for scale effect. It is suggested that the two-dimensional void coefficient could be as important as P_f with regard to hydraulic efficiency.

53. The balance between the cost in time of performing a given survey method during construction and the benefit yielded by the quality of that survey information has yet to be established.

CONCLUSIONS

54. Provided careful consideration is given to the method of sampling a profile, detailed information on roughness, thickness, void characteristics, packing and possibly grading of the armour layer can be used to study their influence on stability, run-up and reflection coefficients.

55. Precise and reproducible means of specifying layer thickness and rock quantities can be developed given the essential practical guidance of experienced contractors.

ACKNOWLEDGEMENTS
56. We are grateful to J. W. van der Meer for providing high resolution profile data from the Delft Hydraulics Laboratories and for his interest and comments on their analysis. Thanks are also due to staff at Hydraulics Research Ltd for their advice and comments. The analysis subroutines were written by R. S. Sayles of Imperial College. The research is funded by the S.E.R.C. Reference GR/D/00832.

REFERENCES
1. READ, J. The control of rubble-mound construction, with particular reference to Heguvik Breakwater in Iceland. Proc. Conf. Breakwaters 1988. Inst. Civ. Engrs, 1988.
2. THOMPSON, D. M. and SHUTTLER, R. M. Riprap design for wind wave attack. A laboratory study in random waves. Wallingford, EX707, 1975.
3. VAN DER MEER, J. W. and PILARCZYK, K. W. Stability of rubble mound slopes under random wave attack. Delft Hydraulics Laboratory, Publ. no. 332, 1984.
4. VAN DER MEER, J. W. and PILARCZYK, K. W. Dynamic stability of rock slopes and gravel beaches. Proc. 20th ICCE, Taipei, 1986.
5. LATHAM, P-P. and POOLE, A. B. The quantification of breakwater armour profiles for design purposes. Coastal Eng., 1986, 10 pp.253-273.
6. Z WAMBORN, J. A. Measurement techniques, dolos packing density and effect of relative block density. CSIR Research Report 378, Stellenbosch, South Africa, March,1980.
7. HUDSON, R. Y. Design of quarry-stone cover layers for rubble-mound breakwaters. Research Report 2.2, US Army Corps of Engineers, July 1958.
8. THOMAS, T. R. (Editor). Rough Surfaces. Longman Inc., New York NY, pp.261, 1982
9. SAYLES, R. S. and THOMAS, T. R. Measurement of the statistical microgeometry of engineering surfaces. Journal of Lubrication Technology, 1979, 101, pp.409-417.
10. YOUNG, R. M., PITT, J. D., ACKERS, P. and THOMPSON, D. M. Riprap design for wind wave attack: long term observations on the offshore bank in the Wash. CIRIA Tech. Note 101, 1980.
11. Shore Protection Manual, US Army Corps of Engineers, Coastal Engineering Research Centre, 1984.
12. LATHAM, J-P. and POOLE, A. B. The application of shape descriptor analysis to the study of aggregate wear. Quarterly Journal of Engineering Geology, London, 1987, 20, pp. 297-310.
13. BRADBURY, A. P., ALLSOP, N. W. H., LATHAM, J-P., MANNION, M. and POOLE, A. B. Rock armour for rubble mound breakwaters, sea walls and revetments: recent progress. Hydraulics Research Report SR150, March 1988.
14. SIMM, J. D. and HEDGES, T. S. Pore pressure response and stability of rubble-mound breakwaters. Proc. Conf. Breakwaters 1988. Instn. Civ. Engrs, 1988.
15. JENSEN, O. J. A monograph on rubble mound breakwaters. Danish Hydraulic Institute, Nov. 1984.

P3. Performance of single layer hollow block armour units

Dr S. S. L. HETTIARACHCHI, University of Moratuwa, Sri Lanka, and Professor P. HOLMES, Imperial College of Science and Technology, London

SYNOPSIS. This paper refers to investigations performed on breakwaters constructed with a single layer of hollow block armour units as the primary armour, of which the cubic (Cobs and Sheds) and hexagonal (Seabee) shapes are two examples. A characteristic feature of such units is their placement in a pre–determined manner which together with their fixed geometry allows close control of the voids matrix of the primary armour layer. The results of this study are derived from extensive model investigations performed on typical breakwater sections with layered fill and armoured with hollow block armour units.

INTRODUCTION

1. The emphasis in most research work on rubble mound breakwaters has been on the hydraulic stability of armour units. Dynamic forces under wave attack, the material properties of the units, the influence of the voids matrix of the primary armour layer, and the hydraulics of wave motion within the porous structure have received less attention. Breakwater cross–sections constructed with model armour units are subjected to design wave conditions in the laboratory to confirm their suitability for use in the prototype. A major design consideration is the stability of individual model armour units with reference to their displacements from the original position. This enables the identification of different levels of damage and the definition of stability coefficients (K_D) for different types of armour units (ref. 1).

Types of armour units

2. Breakwater designers have developed various shapes of artificial armour units in order to obtain high hydraulic stability at a relatively small armour block weight. The different types of artificial armour units used in practice can be broadly classified into three types

(1) Bulky
(2) Slender, interlocking
(3) Hollow block

3. Bulky armour units rely mainly on their weight for stability. Slender, interlocking units have the advantages of greater hydraulic stability due to interlocking effects and a relatively reduced weight. Armour units belonging to these two types are usually placed at random.

It is important to note that for these units the voids which contribute to the dissipation of wave energy are established between the armour units in a random manner.

4. Hollow block armour units are of more recent origin and are somewhat different to the other two types in that the voids are built into the individual units in the required form. Armour units belonging to this type are usually placed as a single layer to a predetermined form. Thus the resulting voids matrix of the primary armour is geometrically well-defined in contrast to that of the other two types. A typical cross-section of a breakwater constructed with hollow block armour units is illustrated in Fig. 1.

Characteristics of the hollow block armour unit

5. The important aspect of the hollow block concept is the systematic analysis of the voids matrix of the primary armour layer. This allows absolute control of the geometry of the voids within the confined boundaries of an individual armour unit or a group of units to produce a cost effective primary armour layer which is very efficient with respect to wave energy dissipation. It is evident that the stability of a breakwater consisting of hollow block armour units does not depend on the degree of interlocking between the units and as a result the weight of the individual armour units can be reduced considerably. These units have been produced in various external shapes of which the cubic form has been more popular (ref. 2).

6. One of the main difficulties in using hollow block armour units is establishing a design criterion. Unlike other types of armour units they have proved to be extremely stable during hydraulic model tests and as such the definition of a stability coefficient on the basis of Hudson's approach is not applicable. Values of K_D greater than 80 have been observed for these units which have found to be more stable on steeper slopes, rather than less stable (ref. 3).

7. When attempts were made to identify the failure mechanism it was observed that excessive overtopping of a breakwater having a relatively mild slope and with an unsupported crest dislodged several units. However, this problem was overcome by using appropriate restraining measures (refs. 3, 4). A comparatively loose laying pattern also resulted in the movement of a few units mainly by rocking or lifting at high incident wave amplitudes. Once a unit is extracted from the armour assembly there exists an opportunity for other units to fall over or to be lifted from their positions. The resulting instability will be characterised by lifting, rocking and rolling of armour units. This state corresponds to one of the possible failure mechanisms for hollow block armour units provided that the crest and the toe wall of the breakwater remain stable.

8. An increased number of breakwater failures recently has made it necessary to consider in detail the structural integrity of armour units together with other design factors. Until this stage designs were based only on hydraulic stability tests for which the strength of armour units was not scaled. This led to the development and use of comparatively large interlocking type of armour units. Recent failures indicate that the limits of applicability of these units have been exceeded mainly because due consideration was not given to other aspects, particularly dynamic forces and the capability of armour units to withstand such loads.

OBJECTIVES OF THE STUDY

9. This paper refers to investigations performed on breakwaters constructed with a single layer of hollow block armour units as the primary armour. The study was broadly classified into two sections.

10. The first section relates to the measurement of wave induced forces, perpendicular and parallel to the slope, acting on a single hollow block armour unit of cubic shape. These two forces were identified as "lift" and "along–slope" force. The second section relates to the measurement of reflection, run–up and run–down coefficients on different types of hollow block armour units.

Importance of lift and along–slope force

11. When randomly packed armour units of different shapes are placed on a slope they are free to move in almost any direction around any of their axes. However, the same is not applicable in the case of hollow block units of which most are cubical with rectangular vertical faces and are placed in a pre–determined compact form. Under such circumstances the study of lift and along–slope force is adequate to understand the forces acting on such a unit. The assessment of the upward lift component and the impact components of both lift and along–slope force is of particular relevance to the long term durability of materials and the understanding of possible failure modes for the breakwater.

12. The importance of the upward lift force in relation to the stability of hollow block armour units has already been introduced in paragraph 7.

Dynamic hydraulic loads on an armour unit

13. The importance of impact forces on armour units can best be understood by considering the characteristic features of dynamic hydraulic loads acting on them. These loads are essentially of two types. The first are oscillatory forces which are gradually varying or quasi–static loads due to wave action on the slope. These oscillatory forces are usually exerted during uprush and downrush of waves. The second type of dynamic hydraulic loads are forces due to direct wave impact. The presence of these forces and their magnitude will be dependent to a great extent on the type of wave profile at the point of impact.

Influence of dynamic impact loads

14. Impact forces due to direct wave action influence the armour unit in two ways. Firstly, they impose hydraulic impact loads of high magnitude acting over a very short time interval. This is of particular relevance to armour units placed in the vicinity of the still water level. Secondly, they cause the movement of a given armour unit which in turn will strike neighbouring armour units, thus imparting structural impact loads. Rocking, rolling and collisions between armour units and parts of one broken unit striking another unit are some of the main effects of this type of load.

15. In the case of hollow block armour units which are placed to a predetermined layout, each unit is in contact with the neighbouring units such that the contact surfaces are well defined and controlled. Hence the influences of rocking, rolling and collisions are reduced to a great extent. In addition, the slope corresponding to the upper surface of the

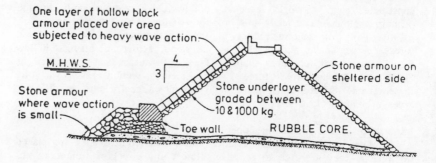

Fig. 1. Typical cross-section of a breakwater in Jersey using 'Cobs'

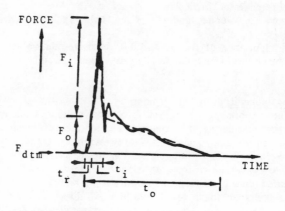

F_{dtm} = force corresponding to still water depth (datum force)

F_o = gradually varying or quasi-static force

F_i = impact force (in excess of F_o)

t_r = rising time of F_o

t_i = impact duration

t_o = duration of gradually varying force

Fig. 2. Typical schematization of time-dependent wave impact forces by periodic water waves

armour units is aligned throughout the breakwater and it is very unlikely that parts of units will be removed and displaced over a considerable distance. Hence for this type of unit forces due to direct wave impact play a vital role. They impose impact components in the directions parallel and perpendicular to the slope.

16. If hollow block units are not placed with care it is possible that long narrow spaces may be generated between the units. This will encourage the movement of armour units and also create regions in which wave pressure can concentrate. These conditions contribute to the potential occurrence of structural impact loads, particularly in the upward direction, and abrasive forces. It should be noted that it is quite possible that narrow crevices, however small, are formed within acceptable standards of construction. These may also develop due to the settlement of the underlayer. Attention should also be focused on armour units placed on the roundhead of the breakwater. Due to the external curvature of the structure it is quite difficult to achieve perfect alignment between the units in this region, thus allowing movement under wave attack.

Wave reflection, run-up and run-down

17. Wave reflection, run-up and run-down are three important parameters used to assess the interaction between waves and coastal structures. For this part of the investigation different types of hollow block armour units were used for the primary armour. The objective was to obtain information of the respective coefficients over a wide range of incident wave conditions for different geometrical configurations of the primary armour layer. Results from previous investigations (ref. 3) were also re-analysed to provide empirical relationships for prediction purposes. In particular, attention was focused on wave reflection because observations of wave-induced scour at the toe of coastal structures have revealed a strong correlation between the scouring process and the reflected energy. Reflection coefficients also form a good indicator of the energy dissipated by the structure.

EXPERIMENTAL WORK

18. Experimental measurements were made on a typical breakwater section consisting of a single layer of hollow block armour, a secondary layer and a central core. Two dimensional tests were performed for different wave periods and wave heights using regular waves.

Force measurements

19. Force measurements were made on model Shed and Cob armour units by adopting a specially designed strain-gauged transducer. It should be noted that hollow block units placed at the boundaries of a breakwater such as at the crest do not represent typical units for force measurements. Such units demand closer examination particularly with regard to adequate support and overtopping. Hence the instrumented unit was located centrally and positioned such that it was not in contact with the neighbouring units or with the underlayer (ref. 5). This unit which was located near the still water level was subjected to wave impact forces.

20. Series of tests were also performed for varying still water levels in order to study the influence of the relative position of the

instrumented armour unit on the recorded forces. Another aspect which was investigated in this study was the influence of the permeability of the underlayer. For this purpose force measurements were also made with the primary armour resting on an impermeable underlayer. Thus data available relates to two different cases based on the characteristics of the underlayer.

Reflection, run–up and run–down measurements

21. Wave reflection was measured by using a moving wave probe and tracing the wave envelope in front of the structure. The accepted loop–node technique was used for the analysis of data (ref. 6). Run–up and run–down were measured by the use of a stationary probe.

22. In collecting data the datum for both force and wave height measurements was maintained at still–water–level free from any disturbances due to waves. The readings obtained thus represent variations above and below the values corresponding to the initial state. The measurements were recorded digitally and on a UV recorder. The latter was included to ensure that all peak forces were recorded correctly.

METHOD OF ANALYSIS

23. For the analysis of force traces a typical schematization of the time–dependent force record of either the lift or along–slope force is presented in Fig. 2. This is based on a similar illustration by Stive (ref. 7) for wave impact pressures at approximately prototype scale on a uniform slope.

24. At time $t = 0$ the force corresponds to that of the initial state. The time dependent force trace is divided into two parts. The first corresponds to a gradually varying part of duration t_0, of the order of the wave period and having a maximum value of F_0. The second corresponds to the impact component of duration of t_i and peak value of F_i in excess of F_0. The total peak force recorded will thus be the sum of F_0 and F_i. The time t_r in Fig. 2 relates to the rise time of force F_0.

25. It should be noted that the typical schematization of the force record presented in Fig. 2 refers to a general case. For a given armour unit the characteristic features identified will depend mainly on its relative position in relation to the still water level, the overall geometry of the structure and the incident wave climate. The latter together with the slope of the breakwater and the sea bed topography will determine the type of breaker which will be dominant for a given wave climate.

26. In analysing the force traces the important aspect was the determination of the peak force whenever it was found to be present. When waves impact on the armour unit, the force transducer records a peak value followed by damping of the recorded signal. Vibrations of the transducer are superimposed on the force record. Because of the presence of these vibrations it is quite likely that the recorded peak impact is slightly greater than the true value. However if a smoothing procedure is adopted it is quite likely the peak value of the smoothed record will be lower than the true value. A moving point smoothing procedure was adopted for the analysis of force records. It was expected that the smoothed record obtained would be representative of the original record and contain reduced peaks. Both measured and

smoothed records were analysed.

27.   To analyse the relative magnitude of the respective force components they were represented in dimensionless form with respect to the corresponding components of the submerged weights of the armour units.   These ratios thus represented the magnitude of wave induced dynamic forces in comparison with static forces which would otherwise be present.   Reflection, run–up and run–down were similarly expressed as coefficients defined with respect to the incident wave height.

28.   For the analysis of wave height measurements, values above and below the still water depth were considered positive and negative respectively.   For force measurements the upward along–slope force and the upward lift force were considered positive (Fig. 3).   Based on this sign convention, forces at the instant of wave impact on the armour unit were characterised by a positive along–slope force and a negative lift force, the latter acting in the direction of the core.

29.   When interpreting the results from hydraulic model studies of force measurements due consideration should be given to the method of instrumentation used for the experimental work and the effects of air entrainment.

30.   The absolute value of the wave impact component and to greater extent the force trace thereafter will be influenced by the method of instrumentation.   The dynamic characteristics of the force transducer should be such that it is able to record impact forces acting over a very short time interval.   For example, if the natural period of the transducer is much greater than the impact duration, the recorded peak could be lower than the true peak.   The effect of vibrations of the transducer in this respect has been already discussed in paragraph 26.   It should be noted that when an analog–digital covertor is used for data collection a high sampling frequency should be adopted.   If not, it is possible that the peak value of the impact component will not be recorded, thus underestimating its value.

31.   An important parameter which affects the magnitude of the impact component is the degree of air entrainment.   It has been observed that surface tension in a model is too large resulting in lower air absorption during wave breaking.   This contributes to increased impact forces in the model giving conservative estimates.

DISCUSSION OF RESULTS

32.   An overall analysis of the results indicated that the smoothed values were more stable than the measured values.   The smoothing procedure reduces the magnitude of peak signals corresponding to impact forces.   An analysis of the results indicated that for more than 75% of the experimental data the reduction in the magnitude of the peak signal was less than 35%.

33.   For all the tests performed repeatable records were obtained for both lift and along–slope forces.   These records were also consistent with the typical schematization of a force record presented in Fig. 2.   The characteristic features of the force records were heavily dependent on the properties of the incident wave and the position of the instrumented armour unit relative to the still water depth.

34.   For the experimental conditions used for this study it was observed that the upward normal force which tends to lift the armour unit out of the primary armour assembly occurred during the run–down

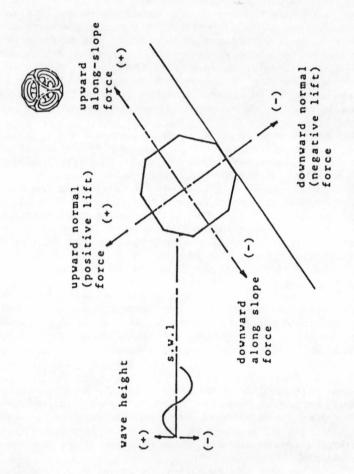

Fig. 3. Sign convention for wave height and force measurements

phase when the water level on the breakwater slope was between the maximum value of run–up and the still water depth.

35.   The study revealed that the critical aspects of the force traces were the upward normal component of the lift force and the impact components of both lift and along–slope force.

36.   Results from tests using regular waves indicated that for a given armour unit – depending on its relative position and incident wave conditions – impact loads were superimposed on gradually varying or quasi–static loads.   For armour units located in the immediate vicinity of the still water level impact loads were observed in both along–slope and normal components corresponding to the point of impact, with impact increasing with increasing wave steepness.   The positive along–slope force was found to be the dominant loading force.

37.   As the wave steepness decreased the force traces gradually became free from sharp peaks corresponding to impact loads and were characterised only by the gradually varying type of dynamic loads.

38.   The intensity of the impact force was reduced as the degree of submergence increased.   This was mainly due to the fact that at greater levels of submergence most of the impact energy was absorbed by the row of armour units positioned above that of the row containing the instrumented armour unit.   Hence under these conditions the instrumented armour unit was not fully exposed to direct wave impact forces.

39.   One of the main observations is that the introduction of an impermeable underlayer did not cause a significant change in either the lift force or the along–slope force.   Characteristic features of the force records remain the same for both structural configurations.

40.   This test series also highlighted several design aspects regarding the placement of such armour units on geotextiles.   It is important that materials used for this purpose be firmly attached to reduce the effects of movement due to uplift pressure forces acting on them.   This phenomenon is identified as breathing of the underlayer and its consequences can be particularly important when placing comparatively lightweight porous armour blocks on geotextiles.   It is important to ensure that the component of the submerged weight of the armour units perpendicular to the slope be large enough to resist possible movement of the underlayer.   These observations also focus attention on the importance of the permeability of geotextiles materials used for this type of purpose.   By adopting porous materials the pressure build–up can be reduced to a great extent.

41.   Figs. 4 to 6 represent plots of the measured variables for wave periods 1.0, 1.5 and 2.0 secs respectively.   The plots clearly illustrate the variation with time of the different variables and reference is made in particular to the characteristic features of the impact loads for waves of short period and purely oscillatory loads for waves of longer period. Fig. 7 represents a plot of the smoothed variables corresponding to Fig. 4.   In comparison with measured variables in the same test the smoothed force records are free of vibrations from the transducer arrangement and the magnitude of the peak impact load is lower.

42.   Fig. 8 illustrates the variation of the along–slope force for different incident wave conditions at a constant still water depth.   It illustrates that the magnitude of the impact load reduces as the wave height decreases at constant period and when the wave period increases

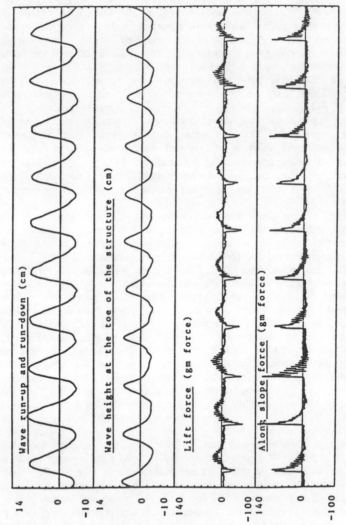

Fig. 4. Wave height and force measurements (measured values)

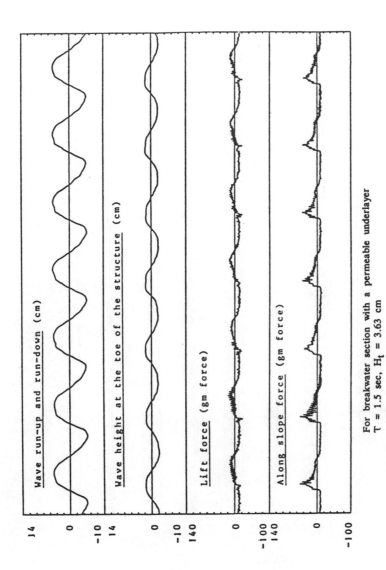

Fig. 5. Wave height and force measurements (measured values)

373

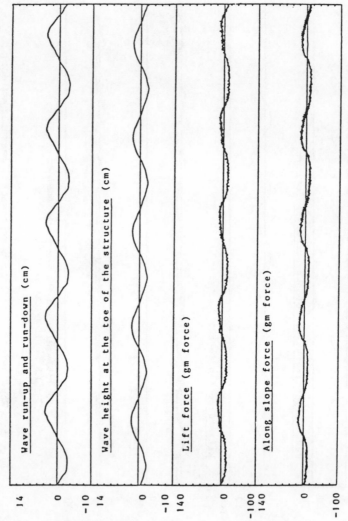

Fig. 6. Wave height and force measurements (measured values)

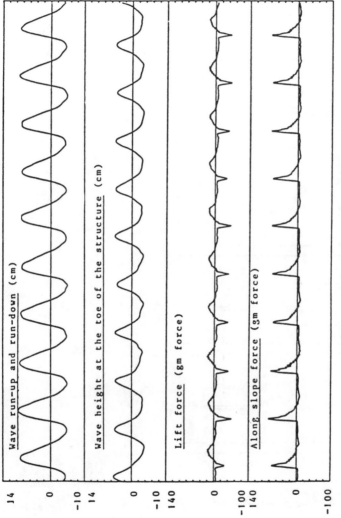

Fig. 7. Wave height and force measurements (smoothed values)

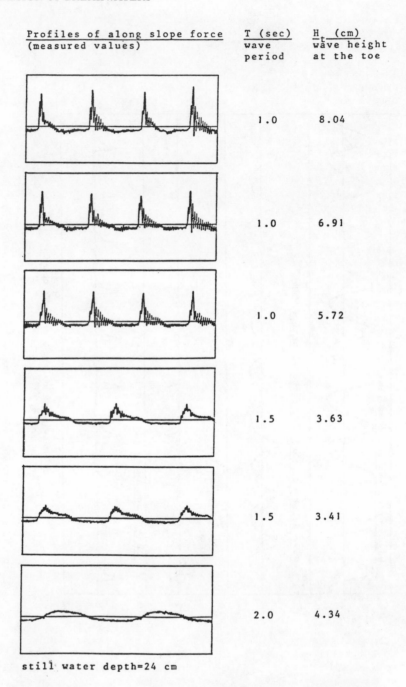

Fig. 8. Variation of along slope force (measured values) for different incident wave conditions

for similar wave heights. Significant changes in the force records are observed as the wave period varies from 1.0 to 2.0 secs.

43. An overall analysis of the results of both the complete breakwater section and that with the impermeable underlayer permitted an assessment of the respective force components. The relative magnitude of the along–slope impact load recorded a maximum of just over 5 for the breakwater section. The maximum value of the downward normal impact load was of the order of 2.5 for the structure having an impermeable underlayer. The maximum value of the positive lift force was of the order of 1.5 for the same structure. On most occasions this component was very much less than unity. Downward along–slope force recorded a maximum of the order of 2.0 for the structure having an impermeable underlayer. It should be noted that impact loads were not present in the traces of the positive lift force and the downward along–slope force, they consisted of purely oscillatory forces.

44. The values of the relative magnitude given in the previous paragraph are those based on measured data and as such may overestimate the maximum values due to the vibrations of the transducer being superimposed on the recorded signal. The corresponding values based on smoothed data represent a lower bound and may even be underestimates. However, the results from both measured and smoothed data provide an assessment of the order of magnitude of the respective force components acting on a typical hollow block armour unit under different incident wave conditions. Although forces corresponding to critical states of instability would not be achieved due to limited wave heights under the available laboratory conditions, the results would correspond to service loads encountered by the armour unit. Detailed results are presented in ref. 5.

45. It is important to note that the relative magnitude of the porous lift force was found to be within acceptable limits, for the experimental conditions used for this study, and the hydrodynamic forces were not high enough to extract the unit from the armour slope. In reality this force will be resisted by the weight of the unit plus frictional forces acting between the contact surface of the units.

46. The importance of the downward along–slope force acting on a hollow block armour unit can be assessed in relation to the design of the toe beam of the breakwater. For a breakwater constructed with hollow block armour units, the toe should be able to withstand the downward component of the static weight of the armour units and additional wave–induced forces. In comparison with the static weight corresponding to a typical breakwater section having 8 to 10 rows of armour units, the wave–induced forces in that direction are small.

47. Reflection, run–up and run–down tests clearly indicated that the energy dissipation characteristics and the overall performance of a hollow block armour slope were dependent to a high degree on the external and internal structure of the individual hollow block armour unit. An armour slope consisting of units having lateral porosity and an interconnected voids matrix with continuous flow paths was found to be more effective in dissipating wave energy. The results of extensive tests performed on different armour units indicated that by the proper selection of governing parameters it is possible to design cost–effective hollow block armour units having increased porosity while optimising reflection, run–up,

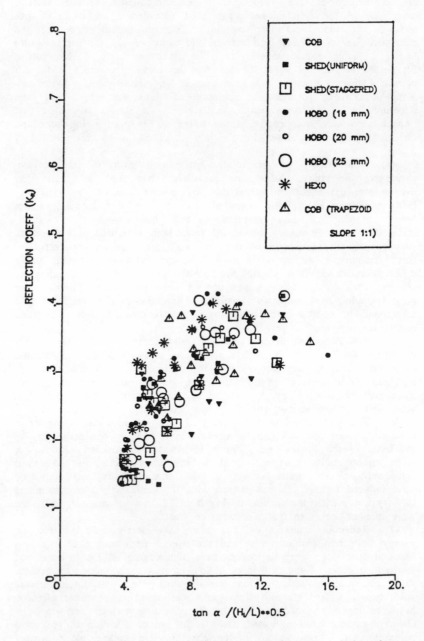

K_R vs tan α /(H$_i$/L)**0.5

FOR SLOPES (1:1 1/3) OF VARYING ARMOUR

Fig. 9. Reflection coefficient against surf similarity parameter

run–down and transmission through the structure.

48. It should be noted that although a certain amount of scatter was present in the experimental data, the results identified a certain domain in which the respective coefficients lie. In particular the scatter in the reflection coefficients was greater for increased wave heights. Under these extreme conditions of overtopping waves the wave records obtained by using the loop–node technique were not well defined in comparison with those obtained under mild incident wave conditions. Hence it is inevitable that a greater degree of scatter will be present.

49. Reflection coefficients decreased with increasing wave steepness and decreasing armour slope. Run–up and run–down coefficients exhibit more scatter and an overall comparison indicates that, in general both coefficients decrease with increasing wave steepness. Empirical equations were determined for the prediction of reflection, run–up and run–down coefficients and a comparison was made with previous studies on hollow block (ref. 3) and interlocking type of armour units (ref. 8). It was found that the following two relationships represented the best correlation between the variables.

$$Y = a.(Ir)^2/((Ir)^2 + b) \tag{1}$$
$$Y = a (1 - \exp(-Ir.b)) \tag{2}$$

where

Y = reflection, run–up or run–down coefficient
Ir = $\tan \alpha/\sqrt{H_i/L_0}$ = Iribarren Number
α = armour slope
H_i = incident wave height
L_0 = deep water wave length
a,b = constants to be evaluated from experimental data

50. Figs. 9 and 10 illustrate the variation of the reflection coefficient vs the Iribarren Number for selected hollow block units. Fig. 9 contains the results for different types of armour units. Fig. 10 refers to the performance of Cob units for varying armour slope, wave period and wave height. These results refer to tests performed with regular waves. Results from tests performed with random waves are given in ref. 9.

51. For design purpose an initial estimate of the reflection coefficients can be made by these methods and the recommended values of the empirical constants are given in references 5 and 9. Prediction equations should only be used within the flow range for which they have been developed and care should be exercised when extrapolating in either direction. For the final design detailed tests should be performed for the selected structural configuration and design wave conditions.

52. The results from this study and those from similar studies performed at different scales illustrate the importance of obtaining experimental data over a wider range so that more generalised formulae can be obtained. Some of the differences observed were attributed to scale effects. The results, to a certain extent illustrated the variation in the empirical coefficients in eq. 1 when determined from different sets of experimental data using various types of armour units.

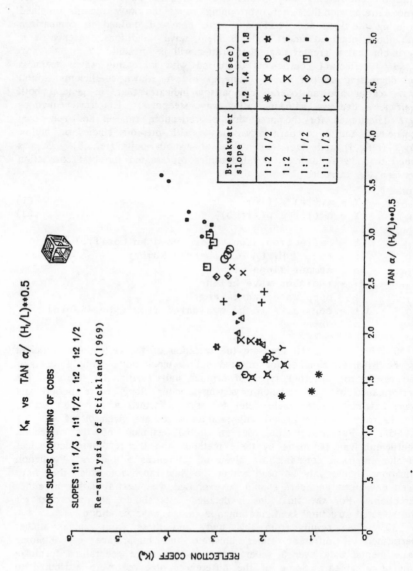

Fig. 10. Reflection coefficient against surf similarity parameter

REFERENCES
1. HUDSON, R.Y. Laboratory investigation of rubble mound breakwaters. Journal of the Waterways and Harbours Division. ASCE, 1969. Vol. 85, WW3, pp. 93–119.
2. WILKINSON, A.R. and ALLSOP, N.W.H. Hollow block armour units. Proc. ASCE conference Coastal Structures '83. Virginia, U.S.A. 1983, pp. 208–221.
3. STICKLAND, I.W. Cob units – Report on hydraulic model research. Wimpey laboratory. Ref. No. H/334, 1969.
4. HYDRAULICS RESEARCH STATION. The Shed breakwater armour unit, model tests on random waves. Report EX 1124, 1983.
5. HETTIARACHCHI, S.S.L. The influence of geometry on the performance of breakwater armour units. PhD Thesis, Imperial College, University of London, 1987.
6. MADSEN, O.S. and WHITE, S.M. Reflection and transmission characteristics of porous rubble mound breakwaters. Ralph. M. Parsons laboratory, Department of Civil Engineering, Massachusetts Institute of Technology, U.S.A. Report No. 207, 1975.
7. STIVE, R.J.H. Wave impact on uniform slopes at approximately prototype scale. Proc. symposium on scale effects in modelling hydraulic structures. Esslingen, Germany.
8. HYDRAULICS RESEARCH STATION. High Island Water Scheme in Hong Kong. Model tests of a Cob block wave protection cover for the inner face of the main dam. Report EX 632, 1973.
9. ALLSOP, N.W.H. and HETTIARACHCHI, S.S.L. Wave reflections in harbours. The design, construction and performance of wave absorbing structures. Hydraulics Research Ltd., Report No. OD 89, 1987.

P4.   New investigations into rubble-mound breakwaters

Professor H. W. PARTENSCKY, Franzius-Institute, University of
Hannover

SYNOPSIS.   The Paper summarizes the results of investigations
on rubble mound breakwaters obtained by a group of researchers
from the Franzius-Institute,University of Hannover,F.R.G., in
a smaller wave flume(11om/2.2 m,with H ⩽0.5 m) as well as in
the Large Wave Channel of the Universities of Hannover and
Braunschweig (32om/5m, with H⩽2.5 m). Since 1983, the investigations have concentrated on the stability of the cover layer
composed of different armor units, the definition of a new damage criterion as well as on the influence of the wave period
and the unit weight of the armor layer elements.

RESULTS OF THE INVESTIGATIONS

1.New damage criterion

In order to determine the damage in the cover layer, an overlay method with positive-negative photos taken between tests
runs was used. All movements of armor units were evaluated and
tabulated in one of six classes (Table 1), each class corresponding to an ever greater combination of rotation or displacement, the latter being given as a portion of the elements significant height H or length. In addition, each class is associated with a weight factor W_j,which is proportional to the
severity of damage likely to result from the movement of the
armor unit.

CLASS	ROTATION	DISPLACEMENT	WEIGHT FACTOR
1	<< 5°	<<	1
2	5° - 15°	<< - H/6	4.
3	15° - 30°	H/6 - H/3	9
4	30° - 45°	H/3 - H/2	16
5	45° - 90°	H/2 - < H	25
6	> 90°	ROLL	36

Table 1: Definition of damage classes

The resulting damage to the cover layer can be expressed by a
damage index J,which is determined by the sum of the partial
damages of all classes:

$$J = \sum_{j=1}^{j=6} U_j \times W_j \qquad (in \ \%) \qquad (1)$$

where U_j = dislocation in % of all elements in class j

The tests were carried out with three different types of armor units(tetrapods,dolos and cubes). The results showed that the damage distributions of all three types of armor units were very similar(Fig.1). A maximum allowable percentage of damage was then defined for each of the six classes which reflects the susceptibility to breakage of each element in question(Fig.2).

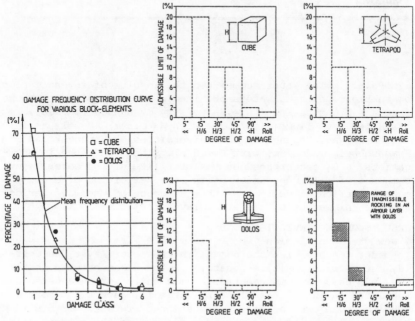

Fig.1: Normalized damage   Fig.2: Admissible limits of damage
distribution   for different armor units

With these limits defined per class, the permissible damage index J becomes shape-dependent, i.e. for dolos $J \leq 155$, for tetrapods $J \leq 243$ and for cubes $J \leq 436$ (ref.1).

2.Influence of the wave period.

The results of the tests showed that greater damage in the cover layer was most often associated with a wave period of intermediate length(ref.2).

3.Influence of the unit weight of the armor elements.

Test comparisons between armor elements of different unit weight with the same total block weight came out clearly in favor of the larger elements of lesser density.

REFERENCES
(1) PARTENSCKY,H.W. New investigations on vertical and rubble mound breakwaters. Proc.of 2nd Cinese-German Symposium,vol.II, paper C1,University of Hannover,1987
(2) PARTENSCKY,H.W. et al. On the behavior of armor units in the cover layer. Proc.of XXth Conf.of Coastal Eng.,paper 87,1986.

P5. Hydraulic effects of breakwater crown walls

A. P. BRADBURY and N. W. H. ALLSOP, Hydraulics Research, Wallingford

SYNOPSIS. This paper summarises recent advances in the hydraulic design of breakwater crown walls to limit wave overtopping. An extensive series of hydraulic model tests have been carried out under random waves (Ref 1). The tests examine the effects on overtopping discharge, of a number of crown wall and armour crest configurations, for a wide range of sea states. The results have been compared with prediction methods from previous studies. Design formulae for the prediction of overtopping discharges over breakwaters with crown walls have been advanced.

INTRODUCTION
1.   A crown wall can increase the overall effectiveness of a breakwater in limiting and directing wave overtopping, and can provide access for maintenance and services. The total cost to achieve a given level of hydraulic performance may be reduced significantly by the use of a carefully designed crown wall. Current design methods are however unreliable in their prediction of the effectiveness of different crown wall and armour crest configurations. Similarly very little information is available to support the estimation of wave forces on the front face of a crown wall.
2.   The design must ensure that the overtopping discharge, or wave action excited by overtopping, is kept below a tolerable level. Overtopping flows must also be directed away from the rear face armour, or other vulnerable parts of the structure. The crown wall is often also expected to provide access for maintenance, carry services and pipelines and in some instances provide public access. In carrying out these functions, the crown wall must also resist forces due to wave action.
3.   Little information is presently available to the engineer, on either estimation of overtopping discharges or forces on breakwater crown walls. Present design methods do not give confidence in calculations of wave forces or overtopping for even the most simple armour/crown wall configurations.

DESIGN OF BREAKWATERS

4. The stochastic nature of storm waves implies that a crown wall would have to be uneconomically large to prevent all overtopping. The probability of some overtopping occurring should therefore be allowed for in design. The level of overtopping permitted will vary according to a range of factors, including the crest and rear slope protection, frequency of the use of berths to the lee of the breakwater and the construction of other structures close behind the breakwater.

Previous work

5. The prediction of overtopping discharge under random waves has been addressed by relatively few researchers and the general application of those results available, to crown walls, is somewhat uncertain. Owen (Ref 2) has developed an empirical method for calculation of overtopping discharges for simple sea walls under random waves, although the effect of a crown wall element in relation to the simple slope is not discussed. More recent work by Steele & Owen (Ref 3) has described the performance of complex sea walls in terms of an efficiency factor, relating the hypothetical performance of a smooth plain slope to a slope with a crest wall of complex form. Ahrens & Heimbaugh (Ref 4) discuss results of random wave tests for sea walls of various geometry in shallow water, and derive an expression that appears similar to Owen's for plain slopes. None of these studies were however explicitly designed to examine the effects of geometric changes to the armour/crown wall detail on overtopping discharge.

HYDRAULIC MODEL TESTS

6. A model test programme was designed to measure the effects on overtopping discharges and wave forces of geometric variation of the crest detail of a breakwater, both of armouring and of the shape and height of the wall. The effects of the following parameters were examined in the model.

a) Incident wave climate: the influence of wave height, wave period and wave steepness were studied by using nine different wave conditions.

b) Water level: tests used two different water levels. A constant foreshore gradient of 1:52 was used throughout the study. Effects of uncertainties in the wave transformation near to the structure were minimized by measurement of wave conditions at the site of the structure.

Model test sections

7. The structure geometry and construction type of different breakwaters and their crown walls may vary considerably. Factors such as seaward slope of the breakwater; porosity, permeability, and roughness of the

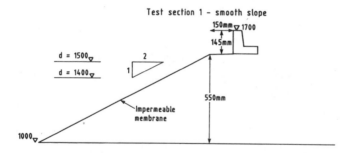

Test section 1 - smooth slope

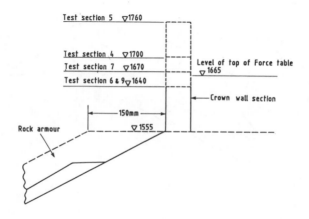

Comparison of test sections 4 - 7
Variations of crown wall crest level

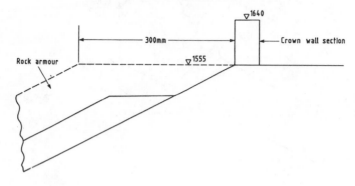

Test section 11 wide rock berm

Figure 1. Example of model test sections

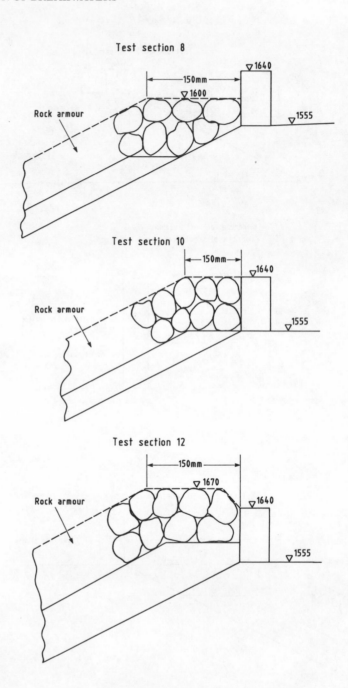

Figure 2. Comparison of armour configuration at crest

front slope armouring; positioning of the armour relative
to the crown wall; elevation and geometry of the wall; all
have significant effects on the hydraulic performance.

8.   The test sections were designed to incorporate as
many of these parameters as practicable. The effects of
slope roughness were examined by comparing rock armoured
structures with smooth slopes. In all cases the core was
impermeable, representing a worst case for wave run-up. A
single seaward slope of 1:2 was used throughout the study.
Whilst the slope affects the form of the wave striking the
crown wall, it was felt that a 1:2 slope was reasonably
typical, and also generally represents the most severe
case for wave run-up.

9.   Geometric changes to the structure were
concentrated around the crest area, at the interface
between the crown wall and the armouring, and at the crest
of the crown wall. The effects on hydraulic performance
of the following geometric parameters were included:
   a) Freeboard;
   b) Height of crown wall;
   c) Level of armouring and berm width of armouring
relative to crown wall;
   d) Profile of crown wall.
Examples of some of the crest configurations tested are
given in Figures 1 and 2. Details of the construction of
all test sections are given in Table 1.

Test procedure
   10. Overtopping discharges were collected in calibrated
tanks behind the test sections. The test procedure
adopted was based on that developed by Owen for seawalls
(Ref 2). A total of 500 T_m waves, collected in 5 batches
each of $100T_m$, were sampled for each wave condition.
This allowed calculation of a mean discharge and standard
deviation.

11. Horizontal forces on the crown wall were also
measured on a total of 8 different test sections, using
the same wave and water level conditions as the
overtopping tests. The forces were recorded using a
simple force table. The results of this work are
discussed in Reference 1.

ANALYSIS OF OVERTOPPING MEASUREMENTS
   12. The discharge data from the model tests was
analysed using a variety of empirical methods suggested by
other authors, including Owen (Ref 2) for smooth sea
walls, and Ahrens & Heimbaugh for structures in shallow
water.

13. The dimensionless framework suggested by Owen for
simple sea walls defines dimensionless freeboard R* and
dimensionless discharge Q*. Owen suggests a relationship
of the form:

TABLE 1 **Model test section construction**

TEST SECTION	SLOPE TYPE (Cot α=2)	SLOPE CREST LEVEL	WALL CREST LEVEL	R_c (m)	F_c (m)	A_c (m)	G_c (m)
1	smooth	0.555	0.70	0.20	0.145	0.055	0.15
2	smooth	0.700	0.70	0.20	0.000	0.200	0.00
3	smooth	0.555	0.70	0.20	0.145	0.055	0.00
4	armoured	0.555	0.70	0.20	0.145	0.055	0.15
5	armoured	0.555	0.76	0.26	0.205	0.055	0.15
6	armoured	0.555	0.64	0.14	0.085	0.055	0.15
7	armoured	0.555	0.67	0.17	0.115	0.055	0.15
8	armoured	0.555	0.64	0.14	0.040	0.100	0.15
9	armoured	0.555	0.64	0.14	0.085	0.055	0.15
10	armoured	0.555	0.64	0.14	0.000	0.140	0.15
11	armoured	0.555	0.64	0.14	0.085	0.055	0.30
12	armoured	0.555	0.68	0.18	0.000	0.180	0.15
13	armoured	0.555	0.64	0.14	0.085	0.055	0.15

All levels are relative to the toe of the test section (m)

TABLE 2 **Summary of empirical coefficients**

Test Section	A	B	Correlation Coefficient R^2
1	5.0×10^{-7}	-3.098	0.93
2	3.4×10^{-6}	-2.033	0.81
3	1.4×10^{-5}	-1.848	0.70
4	6.7×10^{-9}	-3.457	0.81
5	3.6×10^{-9}	-4.368	0.93
6	5.3×10^{-9}	-3.514	0.84
7	1.8×10^{-9}	-3.600	0.96
8	1.6×10^{-9}	-3.182	0.84
9	1.3×10^{-8}	-2.585	0.67
10	3.7×10^{-10}	-2.920	0.73
11	1.0×10^{-9}	-2.823	0.61
12	1.3×10^{-9}	-3.817	0.80
13	5.9×10^{-10}	-3.154	0.71

$$Q^* = A \exp(-BR^*/r) \tag{1}$$

$$Q^* = \bar{Q}/T_m \, g \, H_s \tag{2}$$

$$R^* = (R_c/H_s)(2\pi)^{\frac{1}{2}} \tag{3}$$

A and B are coefficients for different slope angles cot α, and r is a relative run up or roughness coefficient. An example of results from this study analysed using the above techniques are illustrated in Figure 3. The coefficient of regression achieved for the fit of the data to this relationship was not particularly good, even for structures of the simplest geometry.

14. Ahrens & Heimbaugh present overtopping data for a number of structures in shallow water by an equation of slightly different form:

$$Q = Q_o \exp(C, F') \tag{4}$$

Where Q_o is a coefficient with the same units as O (volume/unit time per metre run of wall), C_1 is a dimensionless coefficient, and Ahrens' dimensionless freeboard is defined in terms of the local wave length of the peak period, L_{ps}:

$$F' = \frac{R_c}{(H_s^2 L_{ps})^{1/3}} \tag{5}$$

Comparison of the results of this study with a simplified version of the above equations (using the shallow water wave wavelength $L_p = T_m \sqrt{gh}$), suggest a better relationship than that discussed by Owen.

15. On careful examination of graphs of R^* against Q^*, for values measured in this study, it was noted that there was a stronger dependence on dimensionless freeboard R_c/H_s than on wave steepness. A dimensionless relationship incorporating this function was derived with a new definition of dimensionless freeboard:

$$F^* = R^* \left(\frac{R_c}{H_s}\right) = \left(\frac{R_c}{H_s}\right)^2 \left(\frac{s}{2\pi}\right)^{\frac{1}{2}} \tag{6}$$

$$Q^* = A \, F^{*B} \tag{7}$$

It appears that equations of this form may give a slightly better description of the relationship of Q^* to F^* than an equation of form of (1) above. Coefficients A and B have been calculated for each model section tested. These are given in Table 2. The effect of increasing the weighting of the function R_c/H_s draws the data closer to

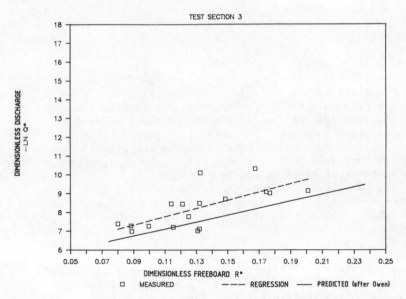

Figure 3. Dimensionless freeboard (R*) against dimensionless discharge (Q*), test section 3

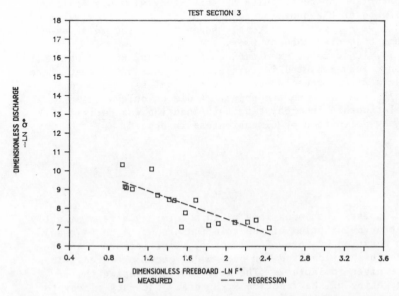

Figure 4. Dimensionless freeboard (F*) against dimensionless discharge (Q*), test section 3

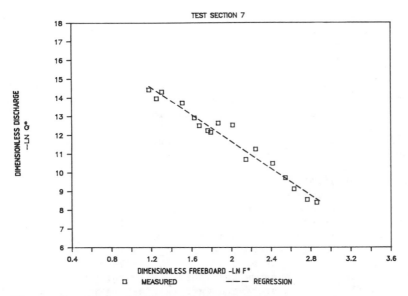

Figure 5.   Dimensionless freeboard (F*) against dimensionless discharge (Q*), test section 7

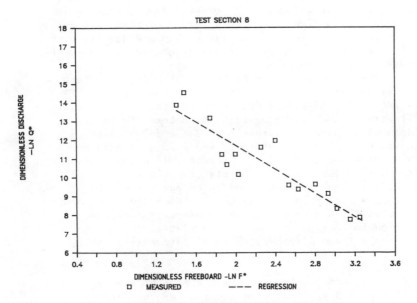

Figure 6.   Dimensionless freeboard (F*) against dimensionless discharge (Q*), test section 8

the regression line, improving the correlation coefficient significantly, particularly for higher discharge events.

16. Examples of the relationship between ln Q* and ln F* are demonstrated in Figures 4-6. It should be noted that the results of this study were derived in relatively deep water conditions. Ahrens' prediction method, which was based largely on results in shallow water, may be more appropriate at lower water levels, because of the local wave length effects considered by the inclusion of L_{ps} in the equation for F'.

17. None of the analysis methods described above explicitly take account of the structure geometry. The use of empirically derived coefficients involves a significant simplification of the description of the overtopping processes.

18. The effects of berm width, armour crest location and crown wall freeboard are most significant for a limited range of wave and water level conditions. When the crown wall is inundated, relatively small geometric variations at the crest have little noticeable effect on the discharge. It appears that the relationship $Q_* = AF*^B$ holds well for the most severe events. Results from this study indicate that this relationship is valid at least for the limit $Q_* > 2 \times 10^{-5}$. Whilst the threshold is not clearly defined, the relationship becomes weaker below this level. This may be due to the following factors:

a) Low mean discharges may be subject to significant variation by occasional large waves. The confidence in the use of low discharge events for prediction purposes is therefore much reduced. The large standard deviation of the five samples measured for low discharge events in this study, supports this.

b) The crown wall geometry has a significant effect on the discharge, and a more complicated function than F* is required to describe the freeboard parameter. Figure 7 illustrates the main geometric parameters that might be included in such an expression. The model studies indicate that the ratios F_c/A_c and F_c/G_c are parameters which might be included in a more detailed description of freeboard. The relative effects of various crown wall geometries have been examined by comparison of the performance of each test section. Space does not permit detailed discussion of the performance of each test section in this paper, but further details are given in Reference 1 and a summary of the findings given below.

CONCLUSIONS

20. The overtopping performance of a breakwater crown wall can be described by an equation of the form

$$Q_* = AF_*^B$$

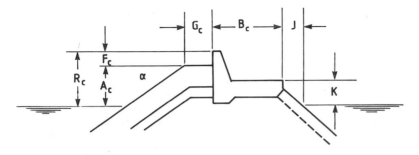

Figure 7. Crown wall geometry

Coefficients for A and B for the crown wall configuration
tested are given in Table 2. This method of prediction
provides a better description of overtopping than
equations of the form suggested by Owen for simple slopes
(Ref 2) and Ahrens' & Heimbaugh (Ref 4) for a revetment
and wave wall. The model tests confirm that the following
geometric parameters have a significant effect on
overtopping discharge.
    a) Increasing the freeboard of the vertical wall (F_c)
results in a reduction in discharge.
    b) Increasing the rock armour berm width (G_c) results
in a reduction in discharge.
    c) Concave seaward faces of the crownwall give a better
performance than vertical crown walls of the same height.
    d) Increasing the freeboard of the rock armour (A_c) and
reducing the ratio F_c/A_c results in a reduction of
overtopping discharge for most conditions. The same
geometric variation may however result in slightly worse
performance in very extreme events.

ACKNOWLEDGEMENTS
    This paper describes the results of work funded jointly
by the Department of the Environment and the Ministry of
Agriculture Fisheries and Food. The work was carried out
in the Maritime Engineering Department of Hydraulics
Research, Wallingford. The authors would like to thank
their colleagues for their assistance in the preparation
of this paper.

DESIGN OF BREAKWATERS

NOTATION

A_c	Elevation of armour crest relative to static water level
B_c	Structure width, in direction normal to face
$C, C_1,$	Empirical or shape coefficients
$F*$	Dimensionless freeboard parameter
G_c	Width of horizontal armour crest berm
H_s	Significant wave height, average of highest one-third of wave heights
h	Water depth
J	Geometric parameter, rear face
K	Geometric parameter, rear face
L_p	Deep water wave length of peak wave period
L_{ps}	Wave length of peak period in water depth in front of structure
Q	Overtopping discharge, per unit length of sea wall
$Q*$	Dimensionless overtopping discharge
R_c	Crown wall freeboard, relative to static water level
$R*$	Dimensionless freeboard
r	Roughness value
s	Wave steepness, H/L_o
s_m	Steepness of mean period $2\pi H_s/g T_m^2$
T_m	Mean wave period
T_p	Spectral peak period, inverse of peak frequency
α	Structure front slope angle

REFERENCES
1. Bradbury A P, Allsop N W H & Stephens R V. "Hydraulic performance of breakwater crown walls". Report SR 146, Hydraulics Research, Wallingford, 1988.
2. Owen M W. "Design of seawalls allowing for wave overtopping". Report EX 924. Hydraulics Research, Wallingford, June 1980.
3. Steele A A J & Owen M W. "Prestatyn coast defence: prediction of discharge over a proposed seawall". Report EX 1335, Hydraulics Research, Wallingford, 1985.
4. Ahrens J P & Heimbaugh M S. "Irregular wave overtopping of seawalls". Proc Conf Oceans '86, IEEE, Washington, September 1986.

P6. Scour of the sea bed in front of vertical breakwaters

J. H. LOVELESS, King's College London

SYNOPSIS. The phenomenon of scour at vertical breakwaters near to the toe of the structure is an important design consideration. The paper aims to throw some new light on the subject by a comparison of theory, model studies and prototype experience. Possible design improvements to mitigate the problem are also discussed.

BACKGROUND

1. Nothing so enrages the sea as a vertical impermeable wall. Thus, the many vertical breakwaters that have been built are a sure testament to the courage of the design engineers who conceived them. Perhaps therefore, it is not surprising that coastal engineers have frequently chosen rubble mound designs to fulfil the increasingly demanding requirements of port and harbour authorities around the world.

2. However, recent failures with rubble mound designs show that engineers have not fully comprehended the vulnerability of these designs to certain modes of failure. Ackers (ref.1), for example, has shown that large Dolos units are vulnerable to fracture and this is now widely appreciated. However, there are also other potential problems with rubble mound designs which models may not reveal because of scale effects and which theory cannot yet predict. Smith (ref.2) postulates one such failure mechanism as a possible cause of the collapse of the Sines breakwater whereby the fluidisation and consequent failure of the toe zone of the breakwater could have occurred.

3. Vertical breakwaters do not suffer from uncertainties of this type. That is to say the strength of the structure, at least, may be assessed with a high degree of certainty. The main uncertainty, which remains today, is the extent and severity of any sea bed scour which may occur over the life of the structure. In practice almost all vertical breakwater designs nowadays include bed protection. Even so, scour problems may still be encountered as, for example,

in the case of the Brighton Marina described by Ganly
(ref.3).

WAVE REFLECTION

4.   The oustanding feature of a vertical wall is that it
can reflect 100% of the approaching wave energy.   In this
event, for normal approach of a regular sinusoidal wave a
pure standing wave is created.   Assuming linear wave theory
applies, the height of the standing wave will then be twice
that of the original progressive wave; the wave energy in
front of the wall will be doubled and the amplitude of the
sea bed velocity will also be twice as great.   When the
approaching wave is quite steep the wave heights at the
antinodes are greatly enlarged by the phenomenon of
clapotis, whereby vertical water spouts are created as the
oncoming and reflected waves collide.   These changes
obviously have a dramatic effect on the potential for scour
in front of the wall.

5.   Real waves are neither sinusoidal nor regular and they
may not approach the wall normally.   However, except for the
cases of clapotis and wave breaking near the wall, it may
still be assumed that nearly all of the wave energy is being
reflected.   For non- normal approach the reflection creates
the short-crested sea state and this is a familiar sight in
the vicinity of any vertical breakwater.   Of course
non-vertical structures also reflect wave energy as, for
example, has been shown by Wilkinson and Allsop (ref.4), who
found that even a permeable slope of 1:1.5 generally
reflected 35% of the wave energy of a JONSWAP spectra.

WAVE ABSORBTION

6.   Naturally therefore engineers have sought to find ways
of reducing the amount of reflection created by the vertical
structure.   To this end it has frequently been proposed to
absorb some of the wave energy using chambers backed by
either permeable or solid walls and fronted by permeable
walls.   Terrett et. al. (ref.5) for instance, reported a
model study of the design originally proposed for the
Brighton Marina.   It was found that, although the structure
would reduce wave reflection considerably for prototype
waves with a period of 7 seconds it had less effect for wave
periods under 5 seconds and over 10 seconds.   Furthermore,
wave forces, which it was hoped would also be reduced, were,
in the case of longer period waves, found to be the same
with both solid and permeable walls.

AN ALTERNATIVE APPROACH : TRAINING THE VERTICAL VELOCITIES

7.   All the kinetic energy of a normal wave incident upon
a vertical impermeable wall must be converted into vertical
velocities.   Therefore, an alternative approach, might be to
consider how these vertical velocities could, with safety,
be destroyed.   Little attention, it seems, has been given to

this possibility. Nagai and Kakuno, however, have reported (ref.6) the design of a slitted box-type wave absorber which has been constructed at Osaka in Japan. This includes a perforated horizontal wall with a porosity of 14% which must of course modify very greatly the vertical velocities in the vicinity of the solid vertical wall.

8.   Three approaches to the above objective seem possible. These are:
a. Destroy the vertical velocities at the wall by introducing very large roughness elements.
b. Direct the vertical velocities so that they separate from the wall in the form of jets which are then dissipated safely either by dispersion in air or by turbulence.
c. Induce the formation of controlled vortices having horizontal axes of rotation so that energy is dissipated in turbulence.

Fig.1 shows examples of each of the structures which might be designed to fulfil these three conditions. The latter two have been tested in the present research and are described in more detail subsequently.

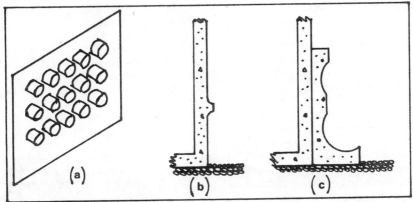

Fig.1   Breakwater profiles to control vertical velocities at the wall

REPORT OF SCOUR PROBLEMS WITH CAISSON DESIGNS
9.   One rather dramatic caisson failure which arose as a result of scour was reported (ref.7) by Ichikawa at Tagonoura Port, Suruga Bay, Japan. Here an 18.0m diameter caisson was subjected to an estimated 6.5m of scour at its seaward toe causing it to sink 2.5m and lean forward at an angle of 30°. The typhoon waves which gave rise to this depth of scour were noted as being particularly steep waves having heights of 8m and periods of 14 seconds. The original water depth at the caisson was about 11m and the bed material was sand. The inshore bed slope was 1 on 7 and the offshore bed slope was much steeper at about 1 on 4.

10. Another type of scour problem with vertical caisson breakwaters was reported by Donnelly and Boivin (ref.8). Here, a breakwater constructed from rectangular caissons and founded directly on a sandy bottom experienced settlements of up to 1.5m as a result of scour in the vicinity of the open joints between adjacent caissons. However, this problem was created by the flows set up in the joints by differential water levels on either side of the structure and thus it was easily solved by sealing the joints.

11. An example of a third type of scour problem which may be encountered is that described by Ganly (ref.3). In this type a narrow scoured trench is created immediately adjacent to the toe. This scour has, in this case, cut through over 600mm of solid concrete.

12. These three types of scour, which may be encountered (with vertical caisson breakwaters) are generated by three different mechanisms. The first occurs when waves break at or near the toe of the structure. The second occurs when there exists a sufficiently permeable flow path through the caisson near the foundation area. Finally, the third occurs when a clapotis at the vertical wall returns vertically downwards as a jet with sufficient energy to penetrate to the bed. Having identified the mechanisms which create the scour, appropriate remedies can then be devised.

DETAILS OF MODEL STUDIES

13. The experiments reported now were devised with the specific objective of investigating the nature of the two direct scouring mechanisms described above and of finding appropriate and feasible remedies. The tests were of an exploratory nature and therefore require to be followed up by more comprehensive studies.

14. All the tests were carried out in a 1.5m wide 1.0m deep 16m long wave flume. A test section 0.8m wide, placed centrally, was employed to reduce reflections. Both random and regular wave conditions were used. Tests were conducted:
(a) with a plain vertical wall
(b) with a section as in Fig.1(b)
(c) with a section as in Fig.1(c)
(d) with a model of part of the Brighton Marina as shown in Figure 2.

15. Wave heights were meausred with two twin wire probes and, for the random wave tests, the output from these was fed directly to a data logger. The significant wave height and power spectral density was then calculated using specially tailored computer programs. Wave velocities were measured with a two component, 5.5cm diameter electromagnetic current meter having a peak frequency response of 2.0Hz.

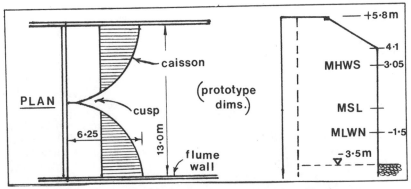

Fig.2. Model of part of the Brighton Marina

16. Two sorts of bed material were used. The first, coal (with a D_{50} of 6.5mm) was intended to represent shingle and the second, sand (with a D_{50} of 1.2mm) was intended to represent coarse sand. Yalin (ref.9) proposed a method for scaling beach shingle whereby the permeability of the prototype material and the drag forces on individual particles could be reproduced in the model. To meet these requirements Yalin proposed that it was necessary for the following functions to be the same in the model and prototype:

$$\frac{v^2}{gL} \text{ - Froude Number}$$

$$\frac{U^2}{g(S_s-1)D} \text{ - Drag Number} \qquad (1)$$

$$f(Re)\frac{V^2}{gD} \text{ - Percolation Number}$$

where v and L are any velocity and length related to the wave motion, U is the mean orbital velocity at any point and time, D, is a typical particle size, such as D_{10} or D_{50}, V is the velocity of percolation and f(Re) is a function of Reynolds number depending upon the void ratio of the bed material. It was found that the coal particles having a specific gravity of 1.48 and a D_{50} of 6.5mm approximately satisfied the above conditions for a scale of 1:17, the scale of the Brighton Marina model.

THE THRESHOLD VELOCITY

17. Scour occurs as a result of fluid velocity; no scour can occur until the near-bottom velocity, during some part of the wave cycle, is sufficient to exceed the velocity required to dislodge a sediment particle. Once bed movement has commenced either accretion or erosion can occur at any

401

point and each will continue until some sort of equilibrium is established. Essentially this condition, for a given sediment, occurs when a drag number (like that given in Equation 1) exceeds a certain value. In these experiments this condition was found to be

$$\frac{U_\infty^2}{g(S_S-1)D} \geq 3.3 \tag{2}$$

where U_∞ is the amplitude of the horizontal velocity component just outside the boundary layer (in this case 3.0cms above the bed). This condition implies a threshold velocity of 23cm/sec for the sand and 16cm/sec for the coal sediment. No significant difference was detected in the tests between this value for waves with periods of between 1.0 and 2.5 seconds.

18. Many other experimenters have sought to determine the threshold condition in unsteady flow. Sleath has compared 18 of these formulae and has shown (Figure 6.1 of ref.10) that there is no good agreement between the various formulae. For example, for 0.8mm sand subjectd to a 3s wave, the predicted threshold velocity varies between 10 and 110cms/sec. In conclusion therefore it appears that the Shields condition still gives the best estimate and for the above case, according to Sleath, the Shields condition predicts a threshold velocity of 24cms/sec.

Table 1. Observed and predicted near-bed threshold velocities. (sand, $D_{50}=1.2$mm)

wave period (secs)	wave height (cms)	wave length (cms)	water depth (cms)	observed* threshold velocity (cms/sec)	velocity predicted from Eq.3 (cms/sec)
1	8.5	104	20	24	18
1.25	8.5	166	20	22	26
1.66	9.5	230	20	23	31
2.50	9.5	366	20	23	34

* measured 3.0cms above a plane horizontal bed at a mode point

19. In front of a vertical breakwater, for a pure standing wave, assuming linear wave theory, the maximum horizontal velocities occur at the node points and are given by:

$$(U_\infty)_{max} = \frac{\pi H}{T} \cdot \frac{\cosh 2\pi(z-h)/L}{\sinh 2\pi h/L} \tag{3}$$

Although the waves used in the threshold tests were not pure standing waves, nevertheless equation 3 gave theoretical near bed velocities which agreed fairly well with the observed values as shown in Table 1. Also, for shallow water, the form of equation 3 suggests that U_∞ is proportional to $H/h^{1/2}$ and this was in fair agreement with the results.

SCOUR DUE TO STANDING WAVES

20. In cases where the water depth at the breakwater is deep and maximum waves are small no wave breaking should occur. Design is then based on bed velocity considerations. Because a vertical breakwater creates standing waves it is likely that the threshold velocity of the sea bed sediment in front of the structure will be exceeded much more frequently after construction. For example, an offshore wave 1.0m high could produce a 2.0m standing wave and in water 5.0m deep would result in near-bed velocities of 150cm/sec. At this velocity, equation 2 suggests that all particles finer than D = 42mm would be in motion.

21. Fortunately scour resulting from standing wave conditions is rarely harmful. The coarse fraction of the bed is moved to the node points, the fine part to the anti-nodes and no significant scour usually occurs at the wall. The studies by De Best et.al. (ref.11) and Irie and Nadaoka (ref.12) cover this aspect of scour very thoroughly so it is not considered further here.

MAXIMUM SCOUR DEPTHS AT THE TOE

22. The most serious condition for scour at the toe of a vertical wall occurs when the diving jet of a breaking wave impinges at or near the toe. This condition appears to occur approximately when

$$d = H_b(1.3-3S) \tag{4}$$

where d is the still water depth at the wall and S is the offshore beach slope.

23. Ichikawa's results (ref.7) indicated that scour depths equal to the offshore wave height could occur where the offshore beach slope was 1:5 and that scour up to 80% of H_0 could be expected where S equals 1:10. Sato, Tanaka and Irie (ref.13) found that maximum scour was nearly equal to the offshore wave height. For a beach slope of 1:10

24. The results of the present study have indicated that:
(a) For the coal sediment and a 1:7 beach slope, maximum scour at the toe was 120% of the offshore wave height.
(b) For the sand sediment and a 1:7 beach slope, maximum scour at the toe was 80% of the offshore wave height.

Figure 3 shows the resulting profiles of these two extreme events. The duration of the test was 15 minutes in each case and regular waves were employed.

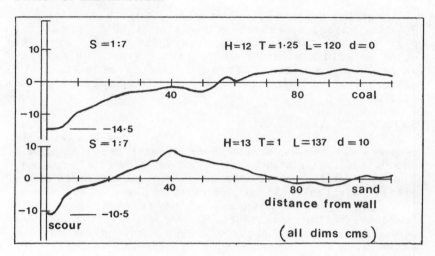

Fig.3.   Eroded profiles of the maximum scour conditions

PROTECTING THE TOE AREA FROM WAVE BREAKING

25. It is apparent from Fig.3 that only the area close to the toe is vulnerable to scour from the wave breaking mechanism. By altering the shape of the wall in the vicinity of the toe, for example as in Fig.1(c), this scour can be significantly reduced. The worst case encountered in tests with this shape gave scour at the new toe equivalent to 27% of the offshore wave height.

SCOUR BY CLAPOTIS

26. The scour problems at the Brighton Marina are created not from wave breaking but, as a result of a severe clapotis, especially in the cusps. This scour was reproduced in the model when the bed material was sand but, it could not be detected with the coal. The model results (with sand) appear to agree with the evidence available from the prototype.

27. Various solutions were examined in the model. The two most effective solutions were:

(a) placing a screen across the cusp

and (b) installing a deflector (e.g. as in Fig.1(b)).

The latter of these would be easier and cheaper to implement and would probably be more durable. Fig.4 shows the typical scour and accretion occurring along the centreline of a cusp in the model with and without the deflector. Further information regarding the details of the Brighton Marina design have been reported by Terrett et.al. (ref.14).

SCALE EFFECTS

28. Any model study of scour phenomena will be subject to scale effects. Since the worst scouring is associated with the plunging jet of a breaking wave it is suggested that

model tests of scour need to be conducted at some, as yet
undefined, minimum Weber number. It is well known that the
scouring efficiency of a jet varies considerably with scale
because of the different amounts of air-entrainment induced.

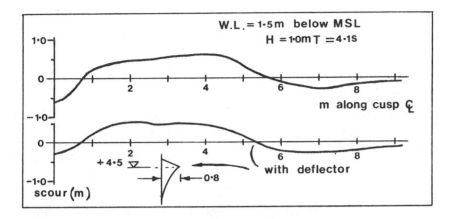

Fig.4.   Brighton Marina model bed profiles

CONCLUSIONS
   29. Scour of a sandy sea bed in front of a vertical
breakwater was found in these studies to reach a maximum of
80% of the approaching wave height for regular waves.
   30. It is proposed that concrete L-shaped blocks keyed
onto the face of a vertical caisson but, allowed to settle
independently, may be an economic way to mitigate the
effects of scour at the toe of the wall.
   31. Small deflectors are a simple way to reduce the scour
arising from the effects of clapotis.

ACKNOWLEDGEMENTS
   32. The author is grateful to Mr. M. Owen and
Dr. K. Powell of Hydraulics Research Ltd., for their
material assistance. Mr. Terrett, Mr. Ganly and Mr. Ross
provided much useful information about the Brighton Marina
and Mr. Gary Forde carried out many of the tests.

REFERENCES
1.   ACKERS P.   Hydraulics research: communication and
     consequences. Proceedings of the Institution of Civil
     Engineers, Part 1, 1984, vol.76, Nov. 1053-1068.
2.   SMITH A.W.S.   Large breakwater toe failures. Proceedings
     ASCE, 1983, vol.109, WW2, May.
3.   GANLY P.   Discussion. Proceedings of Conf. Breakwaters,
     design and construction. ICE, 1983, May.

4. WILKINSON A.R. & ALLSOP N.W.H. Hollow block armour units. Proceedings Conf. on the Design and Performance of Coastal Structures, ASCE, 1983, March.

5. TERRETT F.L. et.al. Model studies of a perforated breakwater. Proceedings 11th Coastal Engg. Conf. ASCE, 1968, Part 2, 1104-1120.

6. NAGAI S. & KAKUNO S. Sea walls in deep seas. Proceedings 17th Coastal Engg. Conf. ASCE, 1980, vol.11, Chapter 123.

7. ICHIKAWA T. Scouring damages at vertical wall breakwaters of Tagonoura Port. Costal Engg. in Japan, vol.10, 1967.

8. DONNELLY P. & BOVIN R. Pattern of wave-induced erosion under caisson-type breakwater. Proceedings 11th Coastal Engg. Conf. ASCE, 1968, Part 1, 599-604.

9. YALIN S. A model shingle beach with permeability and drag forces reproduced. Proceedings 10th IAHR Congress, 1963, London, 169-175.

10. SLEATH J.F.A. Sea bed mechanics, p.258, J Wiley & Sons, New York, 1984.

11. de BEST A. et.al. Scouring of a sand bed in front of a vertical breakwater. Proceedings of Conf. Port and Ocean Engg. under Arctic Conditions, Norway, 1971.

12. IRIE I. & NADAOKA K. Laboratory reproduction of seabed scour in front of breakwaters. Proceedings 19th Coastal Engg. Conf., 1984, Chapter 116.

13. SATO S. et.al. Study on scouring at the foot of coastal structures. Proceedings 11th Coastal Engg. Conf. ASCE, 1968, Chapter 37.

14. TERRETT et.al. Harbour works at Brighton Marina: investigations and design. Proceedings of the Institution of Civil Engineers, Part 1, 1979, vol.66, May, 191-208.